INTERFERENCE MITIGATION

IEEE PRESS
445 Hoes Lane, P.O. Box 1331
Piscataway, NJ 08855-1331

INTERFERENCE MITIGATION

Theory and Application

Rabindra N. Ghose

Technology Research International, Inc.

IEEE Electromagnetic Compatibility Society, *Sponsor*

The Institute of Electrical and Electronics Engineers, Inc., New York

This book may be purchased at a discount from the publisher when ordered in bulk quantities. For more information, contact:

IEEE PRESS Marketing
Attn: Special Sales
P.O. Box 1331
445 Hoes Lane
Piscataway, NJ 08855-1331
Fax: (908) 981-9334

Printed in the United States of America
10 9 8 7 6 5 4 3 2 1

ISBN 0-7803-0410-1
IEEE Order Number: PC3061

Library of Congress Cataloging-in-Publication Data

Ghose, Rabindra N.
Interference mitigation : theory and application / Rabindra N. Ghose.
p. cm.
Includes bibliographical references and index.
ISBN 0-7803-0410-1 (alk. paper)
1. Radio—Interference. 2. Radar—Interference.
3. Electromagnetic interference. 4. Shielding (Electricity).
5. Signal processing. 6. Adaptive filters. I. Title.
TK6553.G4875 1996
621.384 ' 11—dc20 95-46317
CIP

To My Parents,
The Late Dr. Phatick C. and Bimala B. Ghose

CONTENTS

PREFACE

As the signal density of the available spectrum proliferates, the desired signal of one becomes the undesired signal or interference of another. Often, the degree and nature of such interferences, including their frequency spectra or waveforms, are not *a priori* known. An interference could also be intentional and hostile, affecting the communication or radar operation of the victim. These interferences are not white noise, and special means and technologies are now available to eliminate them from receivers.

This book addresses principally the concept, theories, and technologies of interference cancellation involving selective subtraction of the undesired signal from an ensemble of desired and undesired signals in a receive line, leaving the desired signal substantially unaffected. This process differs markedly from conventional frequency domain filtering of the undesired signal, where the filter output yields a spectrum which is the product of the filter transfer function and the combined signal and interference spectra. Thus, in a conventional filter, the desired signal must inevitably be affected by the filter transfer function, although the degree of influence of the filter on the desired signal will depend on the characteristic difference between the desired and undesired signals and the characteristic of the filter. To further illustrate the difference between the conventional filter and an interference canceller, one may note that a conventional filter is ineffective when the desired and undesired signals occupy the same frequency band, while an interference canceller can eliminate the interference for such a case, as long as the desired signal and interference are not coherent or correlated.

To understand the effectiveness of interference cancellation and to determine when it will be preferred to a conventional filter or where it will be the only remedy to maintain a satisfactory communication or radar operation, a comparison of the characteristics and performance of conventional filters and interference cancellers is desirable. Such a comparison is addressed in Chapter 2 of this book.

As noted earlier, the exact characteristics of the interference, such as its magnitude, frequency spectrum, waveform, or its direction of arrival, may not always be *a priori* known. In addition, these parameters of the interference may change with time, either gradually or abruptly. To be effective, the interference cancellation for such cases has to be adaptive to accommodate the not-so-well-defined interference. The characteristics of adaptive interference cancellation are dealt with in Chapter 3.

Interference cancellations from collocated and remote interference sources are discussed in Chapters 4 and 5. Collocated interferences are encountered when a receiver, intended to receive a distant weak signal, has to be collocated with a high-power transmitter. Such problems are also common in vehicular communications involving simultaneous use of multiple radios. For such cases, the receiver can be overwhelmed by the transmitter signal coupled indirectly to the receive antenna line, and the receiver can be in jeopardy even when the frequencies of the transmitted and received signals differ widely. Often, conventional frequency-domain filters are of little value to revive a satisfactory receiver operation, as the intensity of the transmitted signal saturates the receiver front-end, making reception difficult, if not impossible. A receiver could also be severely affected by an interference of remote origin. Such interferences could be in the form of cochannel interferences where a common frequency band is shared by the interference and the signal of interest. They could also be adjacent channel interferences affecting the receiver mostly because of nonideal receiver selectivity to discriminate against the out-of-band undesired signals or interferences while receiving the signal of interest. Again, the interferences of remote origin could be hostile, in the form of jamming signals, or they could be inadvertent or unavoidable, although not hostile for the chosen scenario and mode of operation of the receiver. In any case, the characteristics of the interference will be unknown or they will be time-varying in most cases. The cancellation process, then, has to be adaptive to accommodate such interferences. Almost always, this process involves obtaining a sample of the interference and synthesizing a counter or cancelling interference signal from the sample such that the synthesized counter interference, when subtracted from the receive-line ensemble of the desired signal and interference, leaves only the desired signal as the residue. Means of obtaining the sample interference are usually different for collocated and remote-origin interferences. The basic concepts and requirements for cancellation of interferences of both collocated and remote origins, including the factors that affect the degree of cancellation in each case, are discussed in Chapters 4 and 5.

Often, in real life, a receiver encounters more than one interference. These multiple interferences could be similar in magnitudes and frequency spectrum, or they could be very diverse from one another, or a combination of both. If adaptive processes are to be used for their cancellations at the receive line, means must exist or be created to avoid interactions of these simultaneous processes which could create confusion and hence unsatisfactory cancellations. Similar problems also arise when equivalent multiple interferences evolve due to propagation multipaths of one or more interferences. Discrimination from one another, as is necessary for the cancellation of such interferences, is difficult since they have the same frequency spectra. Adaptive cancellation technologies involved in the cancellation of multipath and multiple interferences including the use of antenna arrays are discussed in Chapters 6 and 7.

Sometimes, like filters, a receiving antenna array can be used to mitigate one or more interferences from the receive line. Nulls can be created, for example, at the radiation pattern of such a receiving array along the directions of interference sources by adjusting the weights or phases associated with the array elements. When the directions of interference sources are not known or when they change with time, the weights need to change in an adaptive manner to effect the interference mitigation. The usefulness and limitations of adaptive arrays for the mitigation of arbitrary and harmful interferences are discussed in Chapters 5 and 6.

Finally, adaptive concepts and technologies which adapt to special types of not-so-well-defined interferences, including adaptive harmonic suppressors, spatial filters, adaptive direction finding, and the like are discussed in Chapter 8.

This book is intended for systems and design engineers, and for the engineering and operation management staff responsible for maintaining satisfactory radio communication, navigation, or radar operation, notwithstanding a friendly yet unavoidable adverse electromagnetic interference environment or in the presence of one or more hostile jamming signals. Often, the electromagnetic compatibility is an important, as well as a difficult problem area, and new and innovative thinking on the part of the responsible engineers may be needed. From a variety of interference problems and remedies as discussed in this book, it is hoped that suggestions of ideas on possible approaches may evolve so as to be the starting points for solving the complex compatibility problems of concern. An elementary knowledge of mathematics, including matrix algebra and simple expectations in random processes, and basic electrical engineering and communication practices is assumed on the part of the reader.

It is believed that a book of this type is needed because of the continuously growing requirements for adaptive interference cancellation and adaptive filtering, along with an increasing interest in the associated closed-loop control technologies of the engineering community to perform many functions and operations, with seemingly no book available in this field at present.

The author acknowledges, with sincere thanks, the advice of Mr. Walter A. Sauter, a former colleague, and Dr. Geoffrey Ghose, and the help and assistance of Mrs. Patricia Van Ballegooijen and Mrs. Jackie Frazier in the preparation of the manuscript.

Rabindra N. Ghose
Technology Research International, Inc.
Calabasas, California

1

INTRODUCTION

1.1 OVERVIEW AND HISTORICAL BACKGROUND

The electromagnetic spectrum suited for radio and radar, as wide as it is, is not unlimited, most limitations being imposed by engineering and cost constraints and, sometimes, by physical laws. As the need to use the spectrum and the number of users proliferate, causing a high density of signals in a given segment of the frequency spectrum, desired signals of many users become the undesired or even intolerable signals for others. Administrative controls, such as a strict and restricted frequency allocation or the control of radiation pattern of the transmitting antenna, are often not adequate. And even if they are adequate today, it is unlikely that, in some regions, they will remain effective in the future, considering the enormous growth of both uses and users. Also, the frequency spectrum conservation that allocates different frequencies for different users is often not enough to avoid undesirable signals because of the unintentional radiation of spurious signals or signals at the harmonics of the carrier frequency from the emitter source or because of the type of modulation. In pulse or spread spectrum modulation, for example, the modulation sidebands may often extend beyond the allocated frequency band. In addition, unwanted and intolerable signals may also result from hostile jamming that attempts to deny an effective radio or radar operation. These types of undesired signals are different in characteristic from white noise and special means are available to rid them from the receiver, thereby restoring the radio or radar operation again. In this text, we will refer to these undesired signals as interferences, regardless of their intentional or unintentional purposes.

Traditionally, a frequency-domain filter with a nonuniform frequency response characteristic in the frequency range of concern has been used to reject one or more interferences that appear at the receiver or receive line. Thus, if a number of interferences is likely to exist at frequencies near but higher than the frequency band of interest, a low-pass filter can be used to eradicate such interferences.

Similarly, if the interferences are likely to be encountered at frequencies near but lower than the frequency band of interest, the use of a high-pass filter would avoid such interferences. A bandpass or a band-reject filter can also be used to reject one or more interferences, depending on the characteristics of the interferences and the desired signal or the signal of interest. The use of a filter, however, is not always an effective remedy for interference rejection since a filter, while rejecting an interference, also affects the desired signal, depending on the characteristics of the filter and the desired signal. Obviously, if a filter's modification or distortion of the desired signal is not acceptable, the filter cannot be used. An example of such a situation is encountered when the desired and undesired signals occupy an overlapping frequency band.

Another means available to suppress an interference or an additive noise, while leaving the desired signal relatively unchanged, is a matched filter. The concept of matched filters evolved from an optimal filtering theory that originated with Wiener's pioneering work [1] and later was extended and enhanced by Kalman, et al., and Kailath [2–4]. In a matched filter, the filter transfer function is related to the spectral characteristics of the desired signal or interference. Such filters can be fixed or adaptive. The design of a fixed filter requires *a priori* knowledge of the desired and undesired signals. Adaptive filters, on the other hand, have the ability to adjust the filter parameters automatically. Hence, the adaptive filter design requires little or no *a priori* knowledge of the characteristics of the desired signal nor of the interference.

Unlike filters, an interference canceller suppresses the interference with a near-identical counterinterference, thus cancelling the interference without affecting the signal of interest. This counterinterference is usually derived from a sample of the interference to be cancelled by synthesizing and modifying the sample, as needed, for cancellation. If the synthesis is not correct, a residue of interference remains after cancellation. In an adaptive interference canceller, this residue is used as an "error" signal for a closed-loop control that operates to drive the "error" signal to zero by adjusting the synthesis process for the counterinterference.

Means for interference suppression in a receive line is not unique. The design of an ideal means may depend on: the characteristics of the desired signal and interference, the required degree of interference suppression, permissible distortion of the desired signal, and cost.

The concept of interference cancellation, a principal subject of this book, is simple, although its implementation may not always be simple. This is apparent from the brief history of its development. Although it is difficult to identify the first development and use of interference cancellation, and adaptive interference cancellation in particular, perhaps the earliest investigation on adaptive interference cancellation was undertaken at the General Electric Company between 1957 and 1960 [5, 6]. A system for antenna sidelobe cancelling was designed and built that used a reference input from an auxiliary antenna and a simple two-weight adaptive filter. In 1959, the least-mean-square (LMS) algorithm, which has been used subsequently for control in some adaptive interference cancellers, was devised [7]. Other activities relating to interference cancellation, although addressed in different contexts, have been pursued at the Cornell Aeronautical Laboratory [8], at the Institute of Automatics and Telecommunications in Moscow, U.S.S.R. by Aizermann and colleagues, and in Great Britain [9]. The work at Bell Laboratories [10] on adaptive filtering has also led to many commercial applications.

In the basic concept of interference cancellation [7], a sum of the desired signal S_D and an undesired signal S_U, uncorrelated with the desired signal S_D, constitutes the input of the canceller. A second signal S_R, also uncorrelated with the desired signal, but correlated with the first undesired signal S_U in some way, is used as the reference signal for the canceller. The reference signal S_R is "filtered" to produce an output y that is as exact a replica of S_U as possible. This output is then subtracted from the primary input $S_D + S_U$ to produce the system output $Z = S_D + S_U - y$. In an adaptive canceller, Z is referred to as an error signal, and adjustments are made in the "filter" to minimize the expectation of Z^2. A more in-depth treatment about this LMS algorithm is presented later.

In the late 1960s Ghose and Sauter [11, 12] investigated and devised various means to implement adaptive interference cancellers to obtain a high degree of interference cancellation in problems of radio communications and radars. For this class of cancellers, the undesired signal S_U at the canceller input differs from the reference signal S_R by only an amplitude factor K and a phase $\emptyset$, so that the "filter" referred to above becomes a signal controller capable of changing the amplitude and phase of the reference signal when controlled by the error signal. The interferences encountered in a vast majority of radio and radar receivers belong to this class. The reference signal, obtained by a sensor placed close to the receiving antenna or through a direct coupling from a collocated interference source, differs from the interference to be cancelled in the receive line, in most cases, by merely an amplitude factor and a phase, particularly when no multipath propagation is involved. Various signal controllers and closed-loop controls are now feasible for this class of interference.

A key element in the control of interference cancellers introduced by Ghose et al. is a synchronous detector with an input, an output, and reference ports. The detector yields a direct current (dc) signal as long as the input to the detector has a signal of the same waveform as that at the reference port. Thus, when a sample of the interference to be cancelled is placed at the reference port of the synchronous detector, one obtains a dc signal output of the detector as long as there is a residue of that interference in the error signals; that is, the signals that follow the subtraction of the counterinterference from the receive line. The presence of a desired signal in the error signals does not usually affect the operation of the synchronous detector. The dc signal output of the detector, in turn, is used to adjust the values of the amplitude and phase factors introduced by the signal controllers. A mathematical model of the interference canceller that addresses the factors that influence the degree of cancellation is provided by Ghose [12] and later in this text.

As electromagnetic waves of the interference of remote origin propagate from the source to a receiving antenna, the amplitude of the interference is reduced with increasing propagation distance. The interference is also delayed in time because of the finite velocity of propagation. Thus, if the interference at its source is denoted as $S_U(t)$, at the receiving antenna it will be of the form $M\ S_U(t - T)$, where M is an amplitude reduction factor and T is a time delay. In general, both M and T depend on the propagation path distance of the interference. If the reference signal for an interference canceller is obtained from a sensor placed close to the receiving antenna, the signals received by the receiving antenna and the reference sensor will differ by amplitude factor K and time delay T', where neither K nor T' is *a priori* known, in most cases. For a narrowband interference and, in particular, for a continuous wave (CW) interference, the time delay will correspond to a phase delay. The signal controller referred to above, for such a case, will have to adjust the amplitude and phase of the reference signal to synthesize the counterinterference required for cancellation. For a wideband interference, however, the signal controller will be required to adjust an amplitude and a time delay. The feasibility of such a signal controller suggests a much broader utilization of the interference cancellation concept.

Various means of controlling the amplitude and time delay in a signal controller have been investigated by Ghose et al. [11]. A mathematical model of one such controller assembly, including a prediction of its performance, was also formulated [13]. This assembly consists of two separate signal controllers operating in parallel, each capable of controlling the amplitude and phase of the signal flowing through it. A fixed time delay is introduced in one of the signal controller paths. With this assembly, it is possible to synthesize an effective variable time delay, in addition to an amplitude control. Other means for controlling an amplitude and time delay of the reference signal to synthesize the required counterinterference and to accommodate a wide dynamic range of interference levels are also feasible.

A radio or radar-receiving antenna may encounter more than one interference simultaneously, some interferences being much stronger than the desired signal. These interferences may arrive at the receiver from arbitrary directions, both characteristics and directions of arrival of the interferences not being *a priori* known. The possibility of a phased array that can place nulls at the radiation pattern of the array and along the directions of arrival of interferences, thereby eliminating such interferences from the receiving antenna line, has been investigated by many researchers [14]–[19]. Weights for the phased array elements, in such cases, are adaptively adjusted to create nulls at the receiving antenna pattern in the directions of unwanted signals or interferences, subject to any one of a number of constraints applied to the array response. The minimum number of array elements needed to eliminate N number of arbitrary interferences arriving at the receiving antenna from different directions is governed by the theory of algebraic equations. This number is usually $N+1$, unless some characteristics of the interferences, other than their directions of arrival, can be exploited to differentiate one interference from another. There are, however, algorithms, commonly used for adaptive adjustments of weights in a phased array to eliminate multiple simultaneous interferences, that utilize fewer array elements than are needed according to the theory of equations.

As will be apparent, the adaptive interference cancellation concept and associated technologies are very powerful tools to not only solve many obvious interference problems, but also to solve problems that do not necessarily relate to interferences. The possibility of devising adaptive filters with specified filter characteristics, adaptive notch filters [20], automatic direction finding [21], same frequency repeater [22], and separations of multiple signals occupying the same frequency bands [23], are only a few examples of the application of adaptive interference cancellation concepts.

1.2 COMMON INTERFERENCE MODES AND SOURCES

Interferences, as addressed in this book, can be classified as collocated or remote, depending on locations of their sources. Sometimes they are also classified as cochannel or adjacent channel interference, based on their frequency spectra. A collocated interference could be a cochannel or an adjacent channel interference. The same is true for a remote interference. These classifications are useful to formulate remedies and approaches needed to rid the interference from the receive line or receiver. By way of definition, such classifications can be further illustrated with examples of common interference modes and their sources.

Collocated Interferences

Interferences from collocated sources are encountered in a variety of scenarios. Two typical scenarios are described here, although one can easily identify many other similar scenarios encountered in real life.

In one scenario, shown in Figure 1-1, there is an unacceptable interference that occurs when communication is attempted from a source close to a broadcast transmitting antenna or a similar radio-frequency-radiating source. If the broadcast antenna is too close, communication can be severely affected. Even when the broadcast signal and communication frequencies are not the same, the broadcast signal can overwhelm and saturate the receiver front end. No conventional frequency-domain filtering is effective under these circumstances.

In another scenario, shown in Figure 1-2, interferences are encountered when more than one simultaneous communication channel is needed, to and from the airplane or helicopter and no adequate frequency separation is possible. Similar problems also arise in vehicular communications where more than one radio with transmitters and receivers operate simultaneously. These interferences could be cochannel or adjacent channel interferences, or both.

Another scenario, in which a collocated interference suppression is necessary involves the same or nearly the same frequency repeater as used in a long-distance

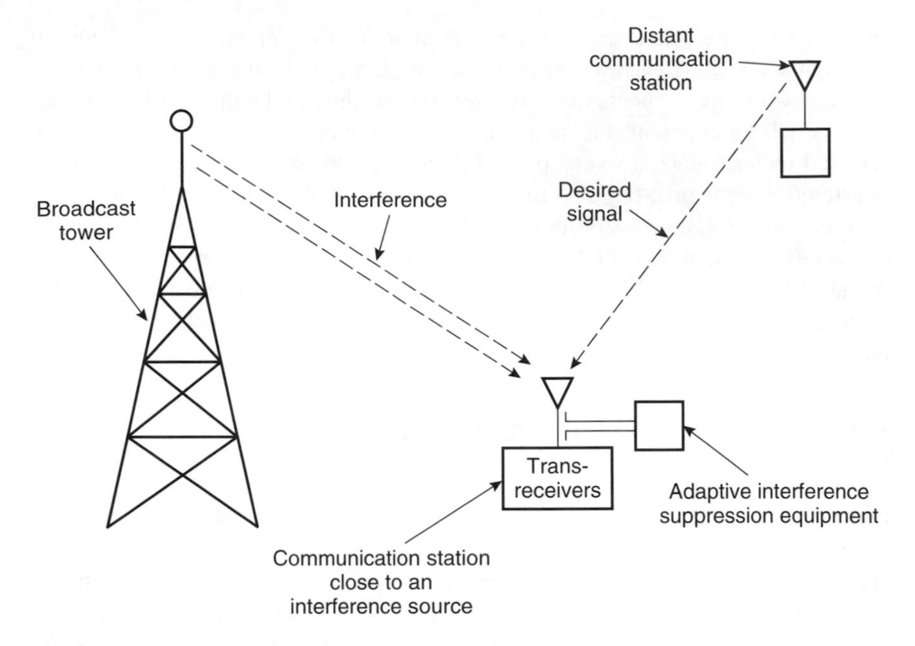

Figure 1-1 A scenario depicting a collocated interference in a communication receiver from an adjacent broadcast tower.

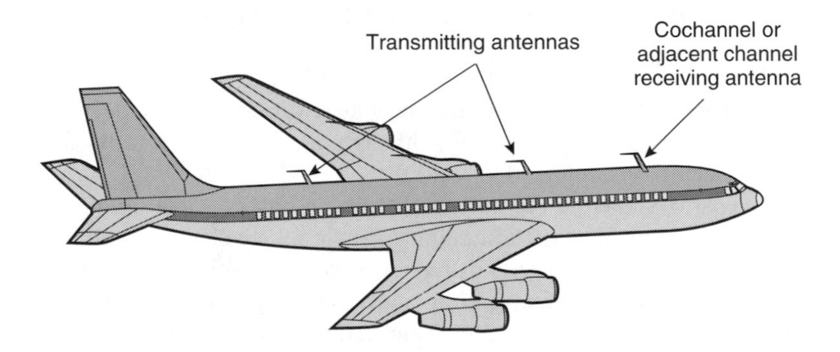

Figure 1-2 Schematic representation of simultaneous transmission and reception of signals from and to an aircraft. Cochannel or adjacent channel interferences, or both, can be suppressed to restore uninterrupted communication.

terrestrial microwave link, shown schematically in Figure 1-3. For such a repeater, a weak signal received at one side is amplified and transmitted by the antenna at the other side. Often a leakage from the transmitted signal couples into the receiving antenna and causes a possible self-oscillation unless the repeater gain is substantially reduced. If, however, the undesired leakage signal can be automatically suppressed at the receiving antenna, one can increase the repeater gain, thus perhaps requiring fewer repeater stations.

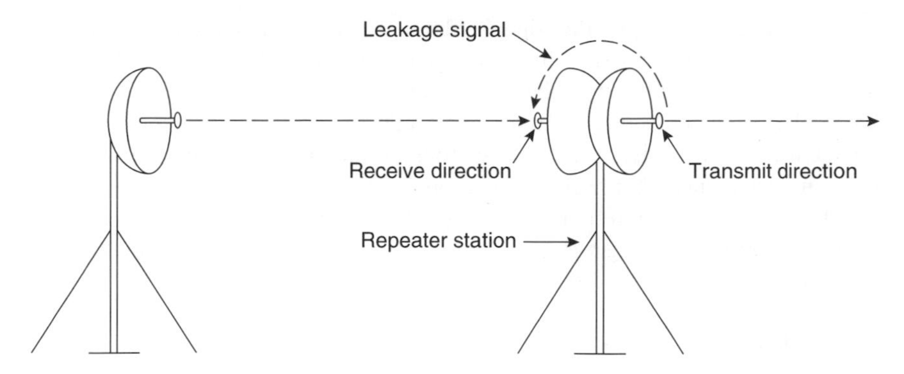

Figure 1-3 Interference in long-distance terrestrial microwave link that leads to a self-oscillation. A suppression of leakage signal restores repeater operation, even with added repeater gain.

Remote Interferences

A remote interference refers to an interference for which the source is not readily accessible to the receiver or receive line, so as to obtain a direct sample of the interference to synthesize a counterinterference. An example of such an interference is shown in Figure 1-4. Here, an intentional or an unintentional interference

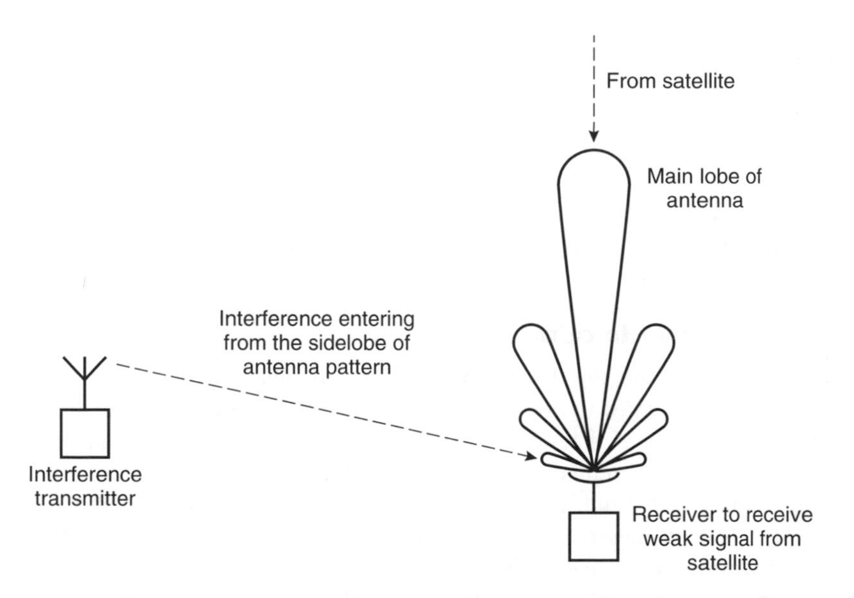

Figure 1-4 Intentional or unintentional interference in a ground station receiver. Here, the interference couples into the receiver through a sidelobe, the gain of which is several decibels less than that of the main lobe. The interference transmitter, however, can be much closer to a ground station, and hence the interference may have a range advantage.

source interferes with a satellite signal. Although such an interference may couple into the ground station receiver line, through the antenna sidelobe, and the antenna directional gain along the sidelobe is considerably lower than that of the main lobe that tracks the desired signal from the satellite, the interference source may have a range advantage when it is nearer to the ground station receiver than is the satellite. Such an interference, then, cannot be ignored.

Another example of remote interference is encountered in communications to and from an airplane flying over a region containing an intentional or unintentional interference source on the ground.

Cochannel Interference

A cochannel interference refers to an interference at the assigned or expected information bandwidth of the desired signal, regardless of the location of the interference source. The same or very nearly the same frequency repeater, shown in Figure 1-3, is an example of a cochannel interference. Another example of the cochannel interference may be found when a television signal for the same channel, that is, the same carrier frequency and allotted bandwidth, originating from an adjoining city, is received with its program being different from that of the desired signal. As noted earlier, a cochannel interference problem cannot be remedied by a conventional frequency-domain filter since the desired signal and the interference occupy the same frequency band.

Adjacent Channel Interference

Usually an adjacent channel interference is a signal in an adjacent channel on either side of the frequency band of the desired signal. Often, an adjacent channel interference occurs because of the poor selectivity of the receiver to receive only the desired signal. However, spurious signals or modulations from the adjacent channel sometimes extend beyond their allotted frequency bands and into the frequency band of the desired signal. Usually, no receiver selectivity can cure such interference problems.

Multipath Interference

A desired signal, traveling through one or more propagation paths other than the direct signal path, may, in effect, create an inadvertent interference leading to a signal fading, as the signals in multipaths tend to cancel or reduce one another because of their different phase relations. Multipath fading is a very common phenomenon at HF (high-frequency) communication, as shown in Figure 1-5, when signals from the ground and sky-wave paths become 180° or nearly 180° out of phase with respect to each other.

Another scenario in which multipath signals can cause mutual interferences and fading is illustrated in Figure 1-6. Here, a reflected signal from a hill in an otherwise line-of-sight communication link causes fading. Other similar scenarios are often encountered in real life.

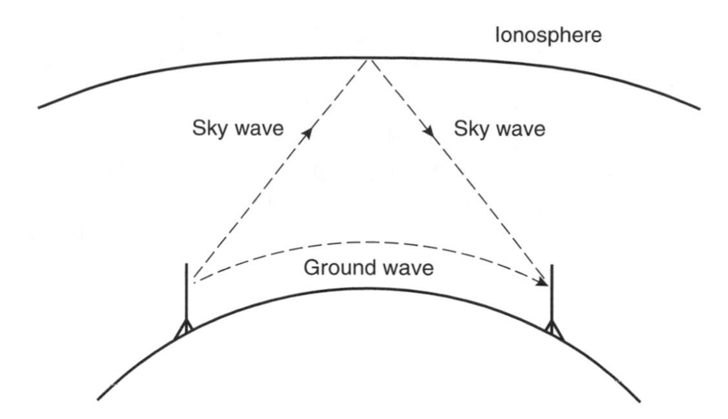

Figure 1-5 In HF communication signals from the ground wave and ionosphere reflected sky wave may interfere with each other, causing fading.

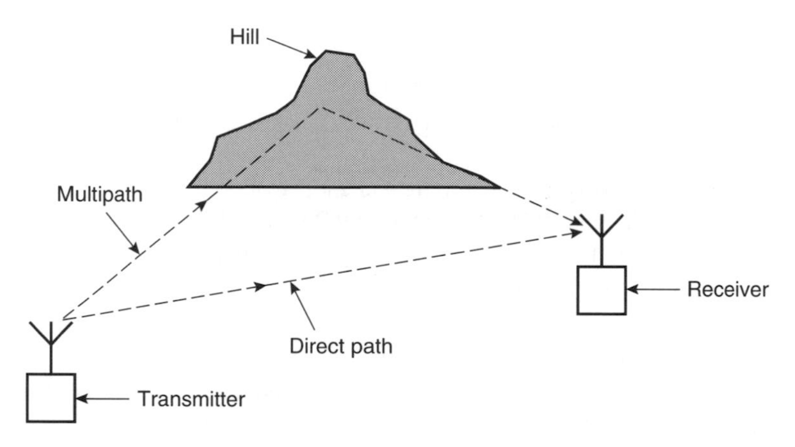

Figure 1-6 Illustration of interference due to a propagation multipath resulting from a reflection from a hill or a similar obstacle.

These scenarios are only a few examples of commonly-occurring interference modes and their sources. Remedies for these and other similar interferences that use adaptive interference cancellers are discussed in the following chapters of the text.

REFERENCES

[1] N. Wiener, *Extrapolation, Intrapolation and Smoothing of Stationary Time Series, with Engineering Applications*. New York: Wiley, 1949.

[2] R. Kalman, "On the general theory of control," in *Proc. 1st IFAC Congr.* London: Butterworth, 1960.

[3] R. Kalman and R. Bucy, "New results in linear filtering and prediction theory," *Trans. ASME, Ser. D, J. Basic Eng.*, vol. 83, pp. 95–107, Dec. 1961.

[4] T. Kailath, "A view of three decades of linear filtering theory," *IEEE Trans. Inform. Theory*, vol. IT-20, pp. 145–181, Mar. 1974.

[5] B. Widrow, J. R. Glover, J. M. McCool, J. Kaunitz, C. S. Williams, R. H. Hearn, J. R. Zeidler, E. Dong, and R. C. Goodlin, "Adaptive noise cancelling: Principles and applications," *Proc. IEEE*, vol. 63, pp. 1692–1716, Dec. 1975.

[6] P. Howells, "Intermediate frequency side-lobe canceller, U.S. Patent 3,202,990, Aug. 24, 1965.

[7] B. Widrow and M. Hoff, Jr., "Adaptive switching circuits," in *IRE WESCON Conv. Rec.*, pt. 4, pp. 96–104, 1960.

[8] F. Rosenblatt, "The Perceptron: A perceiving and recognizing automation," Project PARA, Cornell Aeronaut. Lab., Rep. 85-460.1, Jan. 1957.

[9] D. Gabor, W. P. L. Wilby, and R. Woodcock, "A universal nonlinear filter predictor and simulator which optimizes itself by a learning process," *Proc. Inst. Elec. Eng.*, vol. 108B, July 1960.

[10] R. Lucky, "Automatic equalization for digital communications," *Bell Syst. Tech. J.*, vol. 44, pp. 547–588, Apr. 1965.

[11] R. Ghose and W. Sauter, "Interference cancellation system," U.S. Patent 3,699,444, Oct. 17, 1972.

[12] R. Ghose, "Theory and mathematical model of the adaptive cancellation system," Tech. Note 160004, Technology Research International, June 1986.

[13] R. Ghose, "Analysis of variable time delay system," Tech. Note 150003, Technology Research International, Feb. 1985.

[14] S. Applebaum and D. Chapman, "Adaptive array with mainbeam constraints," *IEEE Trans. Antennas Propagat.*, vol. AP-24, pp. 650–662, Sept. 1976.

[15] B. Widrow, P. Mantly, L. Griffiths, and B. Goode, "Adaptive antenna systems," *Proc. IEEE*, vol. 55, pp. 2143–2159, Dec. 1967.

[16] O. Frost, "An algorithm for linearly constrained adaptive array processing," *Proc. IEEE*, vol. 60, pp. 926–935, Aug. 1972.

[17] S. Haykin, Ed., "Array processing applications to radar," in *Benchmark Papers in Electrical Engineering and Computer Sciences, Vol. 22*. New York: Dowden, Hutchinson and Ross, 1980.

[18] Special Issue on Adaptive Arrays, *IEEE Trans. Antennas Propagat.*, Sept. 1976.

[19] B. Widrow, K. Duvall, R. Gooch, and W. Newman, "Signal cancellation phenomena in adaptive antennas: Causes and cures," *IEEE Trans. Antennas Propagat.*, vol. AP-30, pp. 469–478, May 1982.

[20] W. Sauter and D. Martin, "HF/UHF/VHF adjacent channel measurements," American Nucleonics Corp., Final Report under Contract F30602-69-C-0137, Rome Air Development Center, Griffiss Air Force Base, Rome, NY.

[21] R. Ghose, W. Sauter, and W. Foley, "Automatic direction finder," U.S. Patent 4,486,757, Dec. 4, 1984.

[22] R. Ghose, "Same frequency microwave amplification with adaptive input-output decoupling," *IETE J.*, Oct. 1989.

[23] R. Ghose and W. Sauter, "Automatic separation system," U.S. Patent 4,466,131, Aug. 14, 1984.

2

FILTERS FOR INTERFERENCE CANCELLATION

Notwithstanding various means and technologies advanced to date to rid interferences from electronic communications, radars, and navigation, filters continue to be of great value as interference reduction remedies. Even when conventional filters are not adequate for interference elimination exclusively, they are used along with other means, perhaps because their uses and usefulness are relatively easy to understand, and perhaps because they are often the least expensive remedies. A vast reservoir of information is now available on the theory and implementation of filters and on the application of modern network theory toward the filter design. Of necessity, only a glimpse of the advancement of filter theories and concepts will be feasible in this chapter. More specifically, we will consider only those aspects of filter theories and characteristics as they relate to interference reduction.

2.1 CHARACTERISTICS OF FILTERS, TRANSFER FUNCTIONS, AND OUTPUT SPECTRUM

An electrical filter is a network that has a nonuniform frequency response characteristic in the frequency range of interest. In other words, if a constant voltage e_i is maintained across the input terminal of an electric filter, shown in Figure 2-1(a), and the frequency of the input signal is varied, the response of the network, expressed as a ratio of the output voltage e_o to e_i, will take any one of the forms shown in Figure 2-1(b), (c), (d), or (e). If the response is similar to that shown in Figure 2-1(b), the network is said to be a low-pass filter since low-frequency electrical signals can flow through the network while high-frequency signals are highly attenuated. Similarly, the filters whose frequency response characteristics are shown in Figure 2-1(c), (d), and (e), respectively, are high-pass filters, bandpass filters, and bandstop filters. Filters of all types, such as the low-pass, high-pass, bandpass, etc., are frequently used for interference cancellation, depending on the circumstance. They are used, for example, to prevent unwanted signals to come in or come out of a circuit or system of concern.

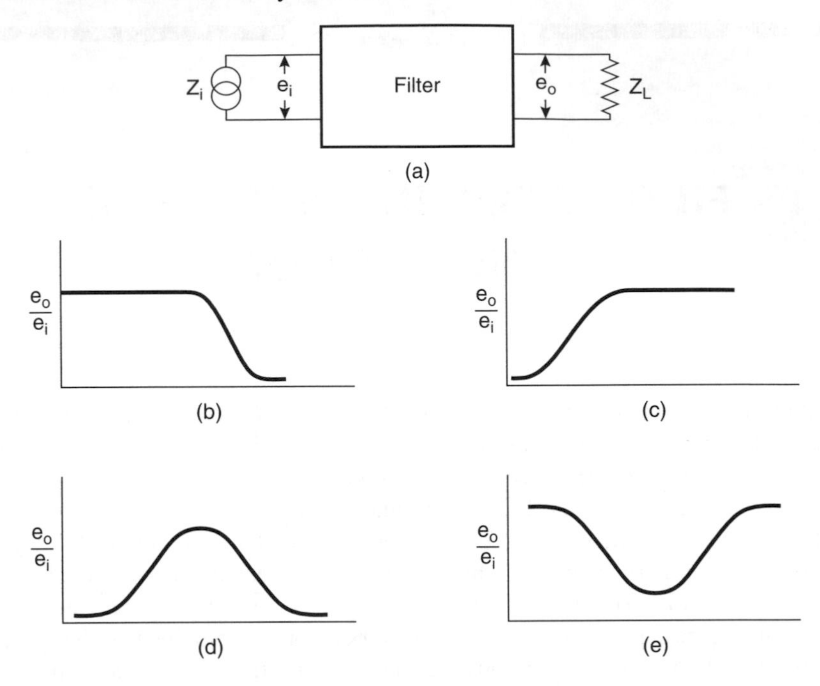

Figure 2-1 (a) A filter network with a nonuniform frequency response characteristic. (b) Characteristic of a low-pass filter. (c) Characteristic of a high-pass filter. (d) Characteristic of a bandpass filter. (e) Characteristic of a bandstop filter.

The design of a filter having specified frequency-response characteristics is a problem of network synthesis. A number of filter synthesis procedures [1]–[6] have been developed over the years, and many such procedures are helpful for designing a filter to reduce or eliminate one or more unwanted interferences from the receive line or receiver. The application of modern network theory to the design of an electric wave filter has introduced new ways to design filter networks based on the actual requirements for signal transmission, in contrast with earlier methodologies to design spectrum shaping networks. While filter design by image-parameter methods of earlier years is relatively simple, only very limited approximations of the specific requirements can be achieved by such methods. Recent approaches using modern network theory are not so straightforward, although more complex signal transmission requirements can be accommodated with such approaches.

A generalized study of filter synthesis is beyond the scope of this section. An approach often followed for the design of a filter is to design a reference low-pass filter, and use known transformation to convert the low-pass filter parameters into a corresponding high-pass, bandpass, or bandstop filter as needed. For the design of a reference low-pass, three types of design parameters are frequently used. They are the Cauer parameters, Chebyshev parameters, and Butterworth parameters. A Cauer parameters low-pass filter is chosen when equal attenuation maximum A_P in the passband and equal attenuation minimum A_R in the stopband region, as shown in Figure 2-2(a), are needed. Similarly, Chebyshev parameters low-pass filters are preferred when a variation of attenuation between zero and maximum is required in the passband and a monotonically increasing attenuation in the stopband, as shown in Figure 2-2(b), are required. On the other hand, a Butterworth parameters low-pass filter is selected when a monotonically increasing attenuation from zero to infinity, as shown in Figure 2-2(c), is needed near the region of passband and stopband interface. Cauer, Chebyshev, and Butterworth parameters for the reference low-pass filters of various degrees or stages are provided in the *Reference Data for Engineers* [6]. Corresponding transformations to convert the low-pass design parameters to high-pass, bandpass, and bandstop filters are also provided in this handbook.

From the viewpoint of interference reduction, the primary concern for designing a filter is how the desired signal will be affected in amplitude and phase due to the filter while it eliminates the interference. Spectral characteristics of the desired signal and the interference and the transfer function of the filter determine the effect of the interference-eliminating filter on the desired signal. By definition, the transfer function of a filter relates the variables at one terminal pair of a two-terminal filter network, such as the filter output, to those at the other terminal pair, such as the input of the filter. The ratio of voltages e_o/e_i, illustrated in Figure 2-1, is one such transfer function. Another transfer function convenient to analyze filter characteristics is the ratio of currents at the two terminal pairs. Thus, one may define

$$\frac{I_1(S)}{I_2(S)} = G(S) \tag{2.1}$$

where

$I_1(S)$ = the Laplace transform of the time-varying input current $I_1(t)$ such that

$$I_1(S) = \int_0^{\infty} I_1(t)e^{-st}\,dt \tag{2.2}$$

and $I_2(S)$ = the Laplace transform of the output or load current.

From Eq. (2.2)

$$I_1(t) = \frac{1}{2\pi j}\int_{c-j\infty}^{c+j\infty} I_1(S)e^{st}\,ds \tag{2.3}$$

where

c = a small real number.

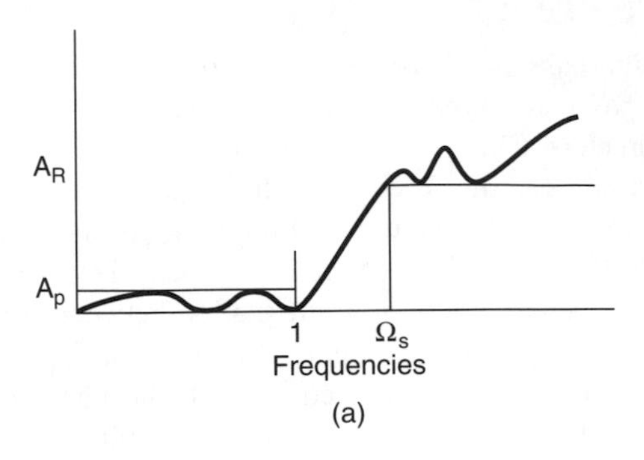

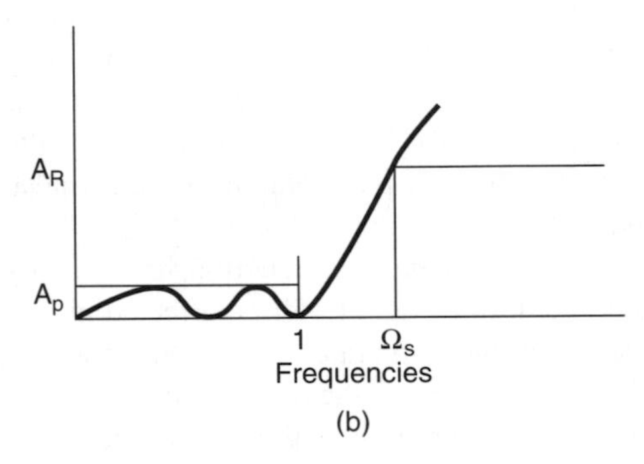

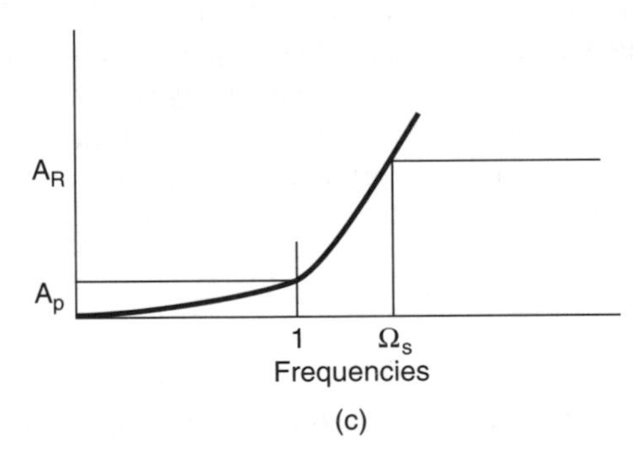

Figure 2-2 Typical characteristics of low-pass filters. (a) Cauer parameters. (b) Chebyshev parameters. (c) Butterworth parameters.

In the sinusoidal steady state, the transfer function becomes a complex number which may be expressed as a magnitude $|G(j\omega)|$ and a phase angle $\arg \cdot G(j\omega)$ such that

$$\begin{aligned} G(j\omega) &= |G(j\omega)|e^{j\arg \cdot G(j\omega)} \\ &= e^{\alpha}e^{j\beta} \\ &= e^{\gamma} \end{aligned} \tag{2.4}$$

where
α = the attenuation constant
β = the phase shift introduced by the filter.

In Eq. (2.4), γ is called the image transfer function.

To illustrate the attenuation and phase shift introduced to any input signal by a filter, one may consider a simple symmetrical T-section filter network as shown in Figure 2-3. Here, the series impedance is Z_1 and the shunt impedance is Z_2. Inductances and capacitances corresponding to Z_1 and Z_2 for the low-pass, high-pass, and bandpass filters are shown in Figure 2-4.

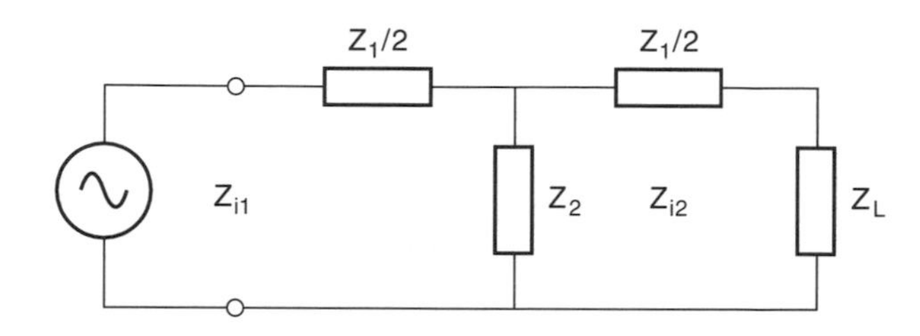

Figure 2-3 A symmetrical T-section filter.

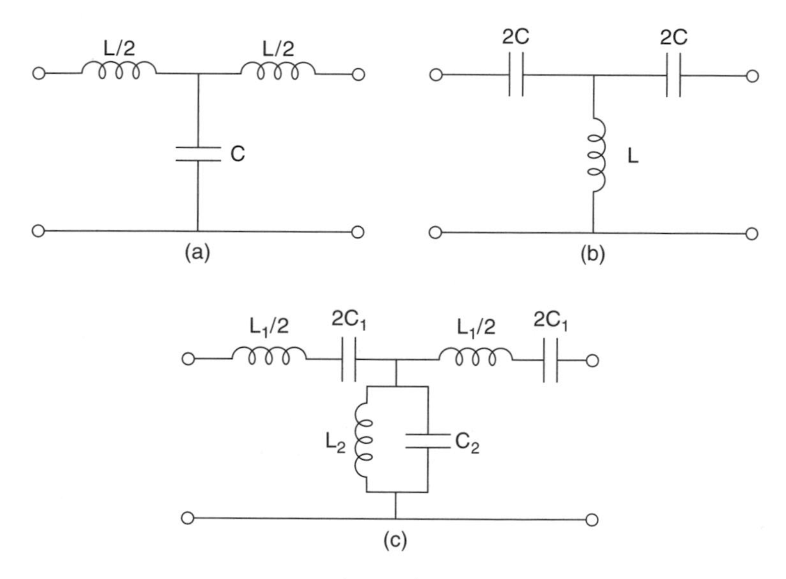

Figure 2-4 Filters using T-sections. (a) Low-pass filter. (b) High-pass filter. (c) Bandpass filter.

The image impedance of this filter network is

$$Z_i = \left[Z_{op} Z_{sc}\right]^{1/2} \tag{2.5}$$

where
Z_{op} = open-circuit impedance $= (Z_{1/2}) + Z_2$
Z_{sc} = short-circuit impedance $= (Z_{1/2}) + Z_1 Z_2/(2Z_2 + Z_1)$.

Thus, the image impedance for the symmetrical T-section filter is

$$Z_i = \left[Z_1^2/4 + Z_1 Z_2\right]^{1/2}. \tag{2.6}$$

The image transfer function of the symmetrical T-section filter, terminated by the image impedance Z_i at both ends, can be expressed as

$$I_1 = e^{\gamma} = 1 + \frac{Z_1}{2Z_2} + \frac{Z_i}{Z_2}. \tag{2.7}$$

For the same symmetrical T-section filter, one may also write [5]

$$\begin{aligned} \cosh\gamma &= 1 + \frac{Z_1}{2Z_2} \\ \sinh\gamma &= Z_i/Z_2 \\ \tanh\gamma &= Z_i/(Z_1/2 + Z_2) \\ &= \sqrt{\frac{Z_{1s}}{Z_{1o}}} = \sqrt{\frac{Z_{2s}}{Z_{2o}}} \end{aligned} \tag{2.8}$$

where
suffixes s and o are short- and open-circuit cases, respectively.

When only reactive elements, such as capacitances and inductances, are used as series and shunt impedances to reduce losses in the passband, one may write from (2.5) and (2.8)

$$\begin{aligned} Z_i &= \sqrt{(\pm jX_{1o})\,(\pm jX_{1s})} \\ \tanh\gamma &= \sqrt{(\pm jX_{1o})\,/\,(\pm jX_{1s})} \end{aligned} \tag{2.9}$$

where
X_{1o} and X_{1s} = the open-circuit and short-circuit impedances, respectively, as measured at terminals 1, 1′.

It is evident from (2.9) that $\tanh\gamma$ can be either real or imaginary for such a case. When $\tanh\gamma$ is real

$$\tanh\gamma = \tanh\alpha,\ \text{and}\ \alpha \neq 0,\ \beta = 0,\ \text{or}\ \pm n\pi,\ n\ \text{being equal to}\ 1, 2 \cdots. \tag{2.10}$$

Similarly, when $\tanh\gamma$ is imaginary

$$\tanh\gamma = j\tan\beta,\ \text{and}\ \alpha = 0, \beta \neq o. \tag{2.11}$$

Thus, if in a filter, X_{1o} and X_{1s} have the same sign at some frequencies, α will be real, the output current I_2 will be attenuated, and these frequencies will be the stop frequencies for the filter. If, on the other hand, X_{1o} and X_{1s} have the opposite sign for some other frequency region, there will be little or no attenuation of the output current I_2. The latter frequencies will be the pass frequencies for the filter. The critical frequency at which transition (real to imaginary values of γ or vice versa) takes place is called the cutoff frequency. The cutoff frequencies are, in general, the poles and zeros of the transfer function, and the distribution of these cutoff frequencies in the entire frequency range determines the frequency response characteristic of a filter.

If the desired signal and interference occupy two distinctly separate frequency bands, the design of a filter becomes straightforward since X_{1o} and X_{1s}, as referred to above, for example, can be designed for such a case to have the same sign for interference frequencies and the opposite sign for the desired signal frequencies. In most cases of practical interest, such an ideal situation seldom occurs. More specifically, the transfer function $G(j\omega)$ for a practical filter seldom has a unity gain uniformly at all passband frequencies, as may be seen in Figure 2-2. The phase of $G(j\omega)$ is also not zero uniformly in the passband. Consequently, the desired signal $S(t)$ at the output of the filter becomes

$$\bar{S}(t) = \frac{1}{2\pi j} \int_{c-j\infty}^{c+j\infty} S(j\omega)G(j\omega)e^{j\omega t} d(j\omega) \tag{2.12}$$

where

$S(j\omega)$ = transform of the input desired signal as given by

$$S(j\omega) = \int_{0}^{\infty} S(t)e^{-j\omega t}\, dt. \tag{2.13}$$

Thus, if $G(j\omega)$ is the transfer function of the filter designed to eliminate the interference, it will necessarily affect the desired signal from $S(t)$ to $\bar{S}(t)$. It should be noted that the contribution toward $\bar{S}(t)$ arises from the entire frequency band for which either $S(j\omega)$ or $G(j\omega)$ is not zero.

When the desired signal and the interference occupy the same frequency band, clearly no conventional filters as discussed above will be adequate to reduce interference. In fact, more difficult problems of interference reduction in a receiver or receive line usually involve interferences which not only occupy the same frequency band, but also are comparable in amplitude with respect to the desired signal, the only distinction between them being that they are not correlated with each other. In some cases, it is still possible to substantially reduce the interference even when it is in the same frequency band as the desired signal. For example, if the desired signal is an audio signal and the interference is in the form of a 60 Hz hum, one may seek a filter with an infinite transfer impedance to 60 Hz and zero transfer impedance to all other audio frequencies of interest. Such a filter is unrealizable, but a filter with an impedance characteristic sharply peaked at 60 Hz would be satisfactory. If, however, the interference is spread over the entire frequency band of the desired signal and beyond, an interference cancellation from the receive line, as will be discussed later in this text, may be the only practical remedy.

Before we examine the interference cancellation approach in detail, however, a brief discussion of a few other means that are useful for reducing the interference or reducing the effect of interference may be in order. Important among such means are the matched filter and the adaptive filter. The concept of "matching" in a matched filter denotes a form of optimization which enhances in some appropriate sense the reception of the desired signal in the undesired noise background, the noise being the interference for such a case. An adaptive filter, on the other hand, adaptively shapes the frequency response of a network that favors the desired signal or reduces or nulls some specific undesired signals. Further discussions on these means for combating interferences are provided in the following subsections.

2.2 MATCHED FILTER—CONCEPT AND APPLICATION

As noted earlier, the concept of matching in a matched filter denotes a process of optimization of the desired signal. Perhaps this characteristic of desired signal optimization can be better explained by an example. Suppose that we have received a signal of waveform $x(t)$ which consists solely of either a white noise $n(t)$ of power density $N_0/2$ W/Hz or of $n(t)$ plus the desired signal $S(t)$ of known form. We wish to determine which of these contingencies is true by operating $x(t)$ with a linear filter in such a way that if $S(t)$ is present, the filter output at some time, $t = \Delta$, will be considerably greater than if $S(t)$ is not present. Since the filter is assumed to be linear, its output $y(t)$, as shown schematically in Figure 2-5, will be composed of a noise component $y_n(t)$ due to $n(t)$ only. If $S(t)$ is present, there will be a signal component $y_s(t)$ due to $S(t)$ only. One way to quantify the filter requirement that $y(\Delta)$ be considerably greater when $S(t)$ is present than when $S(t)$ is absent as stated above is to design the filter such that the instantaneous power in $y_s(\Delta)$ is as large as possible compared to the average power in $n(t)$ at time Δ.

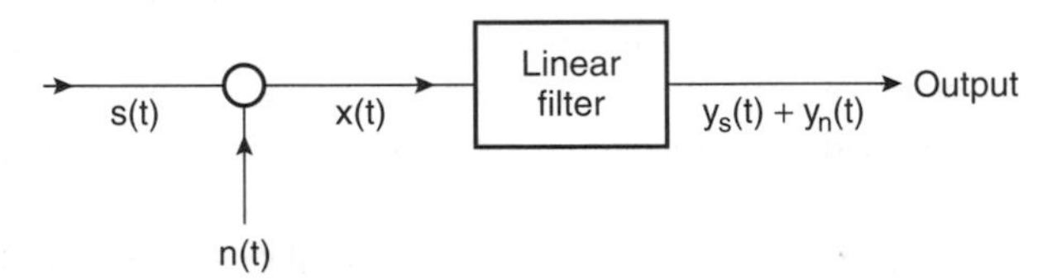

Figure 2-5 Illustration of the requirement of a matched filter.

Let us assume that the noise $n(t)$ of concern is stationary in a statistical sense. The average power of $n(t)$ at any instant is the integrated power under the noise power density spectrum at the filter output. Let $G(j\omega)$ be the transfer function of the filter. The output noise power density then is $(N_0/2)|G(j\omega)|^2$. The output noise power therefore is

$$P_N = \frac{N_0}{2}\int_{-\infty}^{\infty} |G(j\omega)|^2 d\omega. \tag{2.14}$$

Now, if $S(j\omega)$ is the input desired signal spectrum, the output signal spectrum of the filter will be $S(j\omega)G(j\omega)$. Thus, $y_s(\Delta)$ will be the inverse Fourier transform of $S(j\omega)G(j\omega)$ evaluated at $t = \Delta$. We may then write

$$y_s(\Delta) = \int_{-\infty}^{\infty} S(j\omega)G(j\omega)e^{j\omega\Delta} d\omega. \tag{2.15}$$

The ratio R of the square of $y_s(\Delta)$ to P_N is the power we wish to maximize. Here

$$R = \frac{2\left[\int_{-\infty}^{\infty} S(j\omega)G(j\omega)e^{j\omega\Delta} d\omega\right]^2}{N_0\int_{-\infty}^{\infty} |G(j\omega)|^2 d\omega}. \tag{2.16}$$

Now, from the Schwarz inequality, it is known that

$$\left|\int f(x)g(x)dx\right|^2 \le \int |f(x)|^2 dx \int |g(x)|^2 dx. \tag{2.17}$$

If we identify $G(j\omega)$ with $f(x)$ and $S(j\omega)e^{j\omega\Delta}$ with $g(x)$ in Eq. (2.17), we obtain

$$R \le \frac{2}{N_0}\int_{-\infty}^{\infty} |S(j\omega)|^2 d\omega. \tag{2.18}$$

Since $|S(j\omega)|^2$ is the energy density spectrum of $S(t)$, the integral in Eq. (2.18) is the total energy E in $S(t)$. Thus

$$R \le \frac{2E}{N_0}. \tag{2.19}$$

It is clear on inspection that the equality in Eq. (2.17) holds when $f(x) = Kg^*(x)$, K being a constant and $g^*(x)$ being the conjugate of $g(x)$. Similarly, the equality in Eqs. (2.18) and (2.19) holds when

$$G(j\omega) = KS^*(j\omega)e^{-j\omega\Delta}. \tag{2.20}$$

Thus, when the filter is matched to $S(t)$, a maximum of R is obtained. Further, it turns out that the equality in Eq. (2.17) holds only when $f(x) = g^*(x)$. Thus, the matched filter in Eq. (2.20) represents the only type of linear filter that maximizes R.

Thus far, we have assumed nothing about the statistics of $n(t)$ except that it is stationary and white with a power density $N_0/2$. If it is not white but has some arbitrary power density spectrum $|N(j\omega)|^2$, the derivation similar to that given in Eq. (2.18) leads to the solution

$$G(j\omega) = \frac{KS^*(j\omega)e^{-j\omega\Delta}}{|N(j\omega)|^2}. \tag{2.21}$$

Equation (2.20) results directly from the definition of the matched filter. More specifically, if $S(t)$ is any physical waveform, then a filter which is matched to $S(t)$ is, by definition, one with an impulse response $h(\tau)$ such that

$$h(\tau) = KS(\Delta - \tau) \tag{2.22}$$

where
K and Δ are arbitrary constants.

Since the transfer function $G(j\omega)$ is the Fourier transform of the impulse response, one may write

$$\begin{aligned} G(j\omega) &= \int_{-\infty}^{\infty} h(\tau)e^{-j\omega\tau}\,d\tau \\ &= K\int_{-\infty}^{\infty} S(\Delta-\tau)e^{-j\omega\tau}\,d\tau \\ &= Ke^{-j\omega\Delta}\int_{-\infty}^{\infty} S(\tau')e^{j\omega\tau'}\,d\tau' \\ &= Ke^{-j\omega\Delta}S^*(j\omega), \end{aligned} \tag{2.23}$$

the last result in Eq. (2.23) being the same as in Eq. (2.20).

The effect of a matched filter on the desired signal $S(t)$ and the noise $n(t)$ is shown in Figure 2-6. Here, at $t = \Delta$, the filter output $y_n(t)$ due to the noise is not substantially different from that at any other time. But $y_s(\Delta)$ due to $S(t)$ is maximum at $t = \Delta$, and it is, in fact, considerably greater at $t = \Delta$ than if $S(t)$ is not present. Thus, in this case, a matched filter does not reduce or eliminate the interference or noise, but reduces the adverse effect of noise while deciding the presence or absence of $S(t)$, which might not have been possible without the matched filter.

Although the possibility of using a matched filter is tempting based on Figure 2-6, there are two major problems which need to be overcome. First, the spectrum of the desired signal must be known, as the necessary transfer function of the matched filter is directly proportional to the complex conjugate of the spectrum of the signal being addressed by the matched filter. Second, the synthesis of a physical filter for the signal to be addressed by the matched filter may not be easy. For a

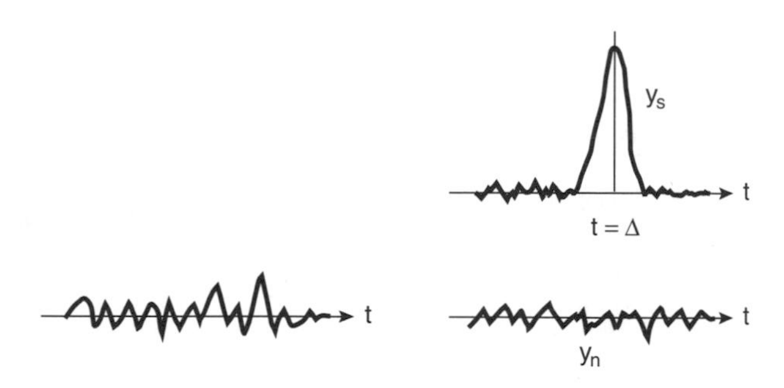

Figure 2-6 Characteristics of a matched filter.

radar where the desired signal is the radar return, the spectrum of the desired signal is usually known. Thus, only the synthesis of a physically realizable filter based on the radar signal spectrum will remain to take advantage of the signal processing similar to that shown in Figure 2-6. For most communication systems, however, the desired signal spectrum at a given time will not be known. Hence, the design or use of a matched filter is not feasible. When there is flexibility of designing the desired signal, one may design such a signal based on a physically realizable matched filter, and then use the combination of desired signal and matched filter to realize the advantage shown in Figure 2-6.

Sometimes, for communication systems, the interference and its spectrum may be known, although the desired signal is not known. For such cases, the matched filter may be built for the interference spectrum to cause the interference energy to peak at a given time, say $t = \Delta$, and then a limiting device may be used to "clip off" the peak to reduce the interference energy in the receiving system. Finally, an inverse filter can be used to restore the desired signal. This process is shown schematically in Figure 2-7. Again, the success of such an approach for interference reduction will depend on how well the interference spectrum is known, and how well the matched and inverse matched filter can be built within the constraints of physical realizability. In any event, one has to conclude that when both the desired signal and interference are not known from the viewpoint of their spectra, or their spectra are complex, the "state of the art" on matched filter for interference cancellation or reduction is, perhaps, still not very satisfactory.

Before concluding the discussion on matched filters, we may note that sometimes a useful application of the matched filter is found in the case of a multipath propagation environment. When, for example, the same signal arrives at the receiving antenna through more than one propagation path, a matched filter may be designed for one of the paths such that the contributions of signals traveling through other paths may be discarded. A common example of multipath propagation is experienced in long-range, high-frequency communications or broadcasts when the signal from its transmitter travels through ground waves and through ionosphere-reflected "sky" waves. These two signal components often cause a

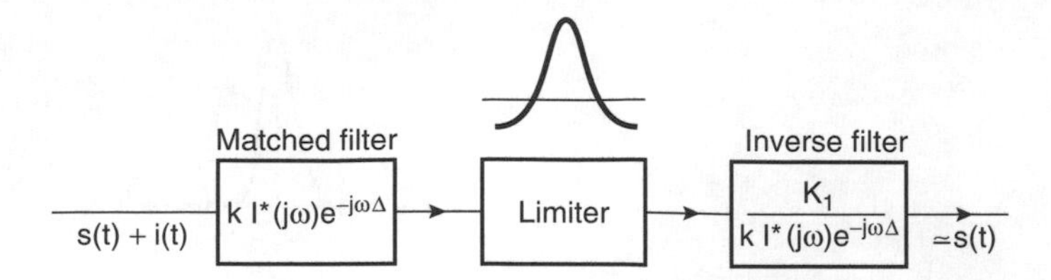

Figure 2-7 Interference reduction approach by employing a filter matched to interference and an inverse filter. Here, k and K_1 are constants.

fading which could be avoided by a matched filter. More about the propagation multipaths and associated interference mitigations is discussed in Chapter 7.

2.3 ADAPTIVE FILTERS AND INTERFERENCE REDUCTION

Usually, an adaptive filter refers to that whose frequency response is shaped by a closed-loop control. The schematic concept of an adaptive filter is shown in Figure 2-8. Here, an input signal $x(t)$ flows through the adaptive filter, and the filter output $y(t)$ is compared with the desired response $d(t)$ of the filter. If the response of the filter is not the same as the desired response $d(t)$, an error signal is generated. This error signal $\epsilon(t)$ is used to change the filter response until such time when the error signal approaches zero. For such a condition, the response of the filter becomes the desired response.

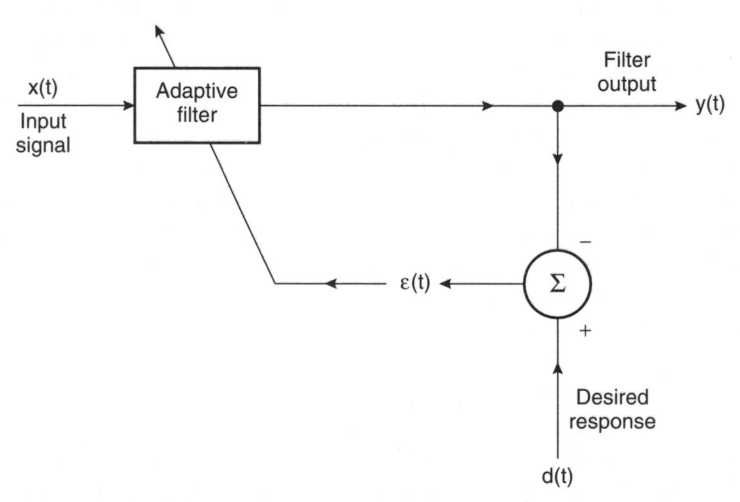

Figure 2-8 Schematic concept of an adaptive filter.

In most cases, the input $x(t)$ contains the sum of the desired signal $S(t)$ and noise $n(t)$. The adaptive filter then shapes its own frequency response in such a way that the filter output approaches the desired signal, for example, as closely as possible or acceptable. The filter is called adaptive because the exact knowledge of $x(t)$ is not necessary in advance, and the filter response adapts to $x(t)$ with the goal of making the filter output $y(t)$ as close as the desired response $d(t)$.

A simplified symbolic representation of an adaptive filter is shown in Figure 2-9. Here, the input signal $x(t)$ flows through a series of delay lines or taps of a delay line, each unit delay being τ. The output of each delay line is weighted or adjusted in amplitude by a factor W_1 or W_2 or W_3, etc. The weighted outputs are then summed to constitute the filter output $y(t)$. As in the case shown in Figure 2-8, the filter response $y(t)$ is compared with the desired response $d(t)$. If the filter response is not the same as the desired response, an error signal is used to control the adjustments of W_1, W_2, W_3, etc., until $y(t)$ approaches $d(t)$.

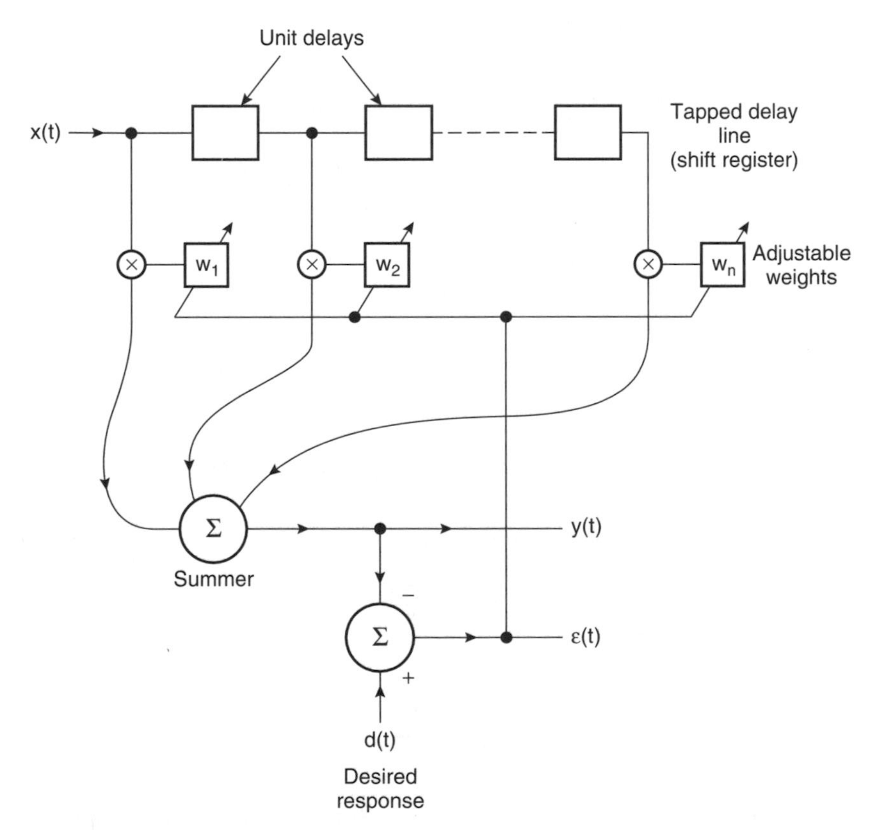

Figure 2-9 A simplified symbolic representation of an adaptive filter.

To illustrate the characteristic of the adaptive filter shown in Figure 2-9, let $x(t)$ contain the desired signal $S(t)$ and noise $n(t)$. The output $y(t)$, then, can be written as

$$y(t) = \sum_{m=0}^{N-1} W_m S(t - m\tau) + \sum_{m=0}^{N-1} W_m n(t - m\tau) \tag{2.24}$$

where
N = the number of adjustable weights, W_0, W_1, W_2, etc.

Let the transform of $S(t)$ and $n(t)$ be defined such that

$$S(j\omega) = \int_{-\infty}^{\infty} S(t)e^{-j\omega t}dt$$
$$N(j\omega) = \int_{-\infty}^{\infty} n(t)e^{-j\omega t}dt. \tag{2.25}$$

The transform of $y(t)$, then, is

$$\begin{aligned} Y_m(j\omega) &= \sum_{m=0}^{N-1} W_m \int_{-\infty}^{\infty} S(t - m\tau)e^{-j\omega t}dt \\ &\quad + \sum_{m=0}^{N-1} W_m \int_{-\infty}^{\infty} n(t - m\tau)e^{-j\omega t}dt \\ &= \sum_{m=0}^{N-1} W_m e^{-jm\tau} \int_{-\infty}^{\infty} S(t - m\tau)e^{-j\omega(t-m\tau)}d(t - m\tau) \\ &\quad + \sum_{m=0}^{N-1} W_m e^{-jm\tau} \int_{-\infty}^{\infty} n(t - m\tau)e^{-j\omega(t-m\tau)}d(t - m\tau) \\ &= S(j\omega)\sum_{m=0}^{N} W_m e^{-jm\tau} + N(j\omega)\sum_{m=0}^{N} W_m e^{-jm\tau} \\ &= S(j\omega)G(j\omega) + N(j\omega)G(j\omega) \end{aligned} \tag{2.26}$$

where
the filter transfer function $G(j\omega)$ is given by

$$G(j\omega) = \sum_{m=0}^{N-1} W_m e^{-jm\tau}. \tag{2.27}$$

Thus, if one knows the filter transfer function $G(j\omega)$ that will yield an optimum response for the desired signal or minimum response for the noise or both, one can synthesize that $G(j\omega)$ as closely as is physically realizable or practicable from Eq. (2.27) by choosing W_m and the number of delay line taps N. Adaptive filters can be built also to null sinusoidal interferences.

Filters other than a tapped delay line can also be made to work as an adaptive filter. For example, adaptive systems[1] have been made that combine the output of analog *RC* filters of different time constants, driven by a common signal source.

From the viewpoint of using an adaptive filter for interference reduction, one has to note that:

(a) the desired response of the signal being addressed by the adaptive filter must be known, and

(b) the filter must be physically realizable and can be implemented in real time.

In many cases, the desired response is not known since neither the desired signal nor the interference is known sufficiently to rid the interference without affecting the desired signal. Moreover, if the actual desired response $d(t)$ were known, one might not need the adaptive filter in most cases.

The more complex problem for the adaptive filter is its physical realizability. It is shown in Figure 2-9 that the error signal $\epsilon(t)$ is used to simultaneously adjust various weights W_1, W_2, W_3, etc. In many cases when the weights have different values and their values with respect to one another are not related in any specific way, it may not be easy to converge to their required values in a reasonable time, particularly when the number of filter taps is large and there is only one error signal.

Adaptive algorithms [14] have been developed to derive individual weights W_1, W_2, W_3, etc., based on making the average value of the square of the error signal $\epsilon(t)$ a minimum. A theory [11] of adaptive filtering involving signal extraction for time-discrete data with limited assumptions on the distribution of signal and noise was advanced by Davisson. Notwithstanding the establishment of a sound theoretical basis for adaptive filters, the problem of practical implementation of an adaptive filter based on the concept shown in Figure 2-9 has not been lessened, particularly for arbitrary desired signals and interferences.

An interference cancellation system, to be discussed in the next section and in other chapters of this text, can be used as an adaptive filter where the desired signal is a broadband signal and the interference is periodic and no external reference of the interference is available. Examples of such problems include the playback of speech or music from a tape or turntable record in the presence of tape hum or turntable rumble. Alternatively, an adaptive filter using an interference cancellation system can be used to extract a periodic desired signal in the presence of a broadband noise. Such applications, including implementation of physical filters to achieve the desired objectives, are discussed later in this text.

An important difference between the filter, regardless of whether its response is fixed or shaped by adaptive means, and an interference cancellation system is

[1] J. Kaunitz, "General purpose hybrid adaptive signal processor," SEL-71-023 (TR No. 6793-2), Stanford Electronics Laboratories, Stanford, CA, Apr. 1971.

noteworthy here. Let the Laplace transform of the desired signal $S(t)$ be $S(j\omega)$ and that of the interference $i(t)$ be $I(j\omega)$. If the filter transfer function is $G(j\omega)$, then the output of the filter $E_0(j\omega)$ is always given by

$$E_0(j\omega) = G(j\omega)[S(j\omega) + I(j\omega)]. \tag{2.28}$$

The corresponding filter output in the time domain is

$$\begin{aligned} e_0(t) = & \frac{1}{2\pi} \int_{c-j\infty}^{c+j\infty} G(j\omega) S(j\omega) e^{+j\omega t} d\omega \\ & + \frac{1}{2\pi} \int_{c-j\infty}^{c+j\infty} G(j\omega) I(j\omega) e^{+j\omega t} d\omega. \end{aligned} \tag{2.29}$$

When the filter is designed to reject the interference, the second integral in Eq. (2.29) is made zero or minimum. However, the transfer function $G(j\omega)$ that makes the interference contribution at the filter output zero or minimum necessarily also affects the desired signal, as shown by the first integral in Eq. (2.29), unless there are no overlapping frequencies for the desired signal and the interference. If there are no such overlapping frequencies, $G(j\omega)$ can be designed such that $G(j\omega)$ approaches unity or a constant for the entire spectrum of the desired signal. Similarly, when the filter is designed to maximize the desired signal output, as is common for a matched filter discussed in Section 2.2, the contribution of the second integral is not necessarily zero or minimum unless the spectral characteristics of the desired signal and interference are significantly different. Thus, when the desired signal and the interference have overlapping frequencies or their spectral characteristics and magnitudes are comparable, no conventional filters discussed so far in this chapter will be helpful for avoiding the harmful or intolerable interference.

In the interference cancellation system, on the other hand, one subtracts from the sum of the desired signal and the interference as it appears at the input of a system or a receive line a counterinterference which is made to approach the interference in an adaptive manner. Thus, there is no multiplication product similar to that shown in Eq. (2.28) involved. The process, shown schematically in Figure 2-10, is more akin to an arrangement where the output is given by

$$E_0(j\omega) = S(j\omega) + I(j\omega) - I_c(j\omega) \tag{2.30}$$

where
$I_c(j\omega)$ = a countersignal which is made to approach $I(j\omega)$.

Thus, ideally, whatever is done to eliminate or cancel the interference need not necessarily affect the desired signal, except for a nominal insertion loss. The basic undesired signal cancellation concept and the necessary and sufficient criteria to implement such a concept are discussed in the following section.

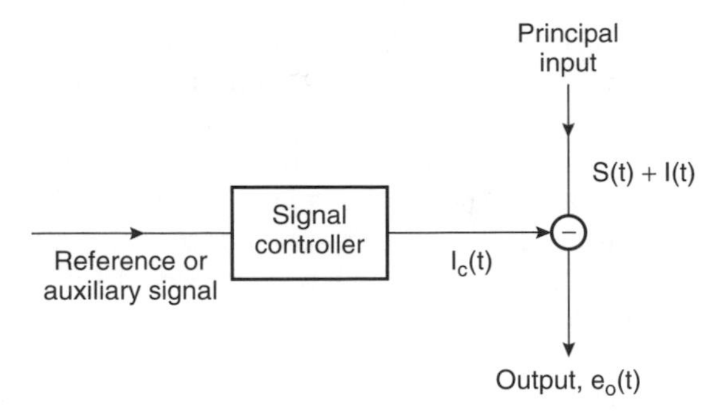

Figure 2-10 Concept of interference or undesired signal cancellation.

2.4 UNDESIRED SIGNAL CANCELLATION CONCEPT

The basic concept of undesired signal cancellation by adaptive means may be illustrated by Figure 2-11. Here, the principal input or receive line consists of a desired signal $S(t)$ and an undesired signal $U(t)$. There is an auxiliary or reference input consisting of the undesired signal $U_1(t)$ only or with a weak or undetectable desired signal. The undesired signals at the principal input and the reference input are not the same, although they are correlated. By definition, the desired signal is not correlated with the undesired signal at either input.

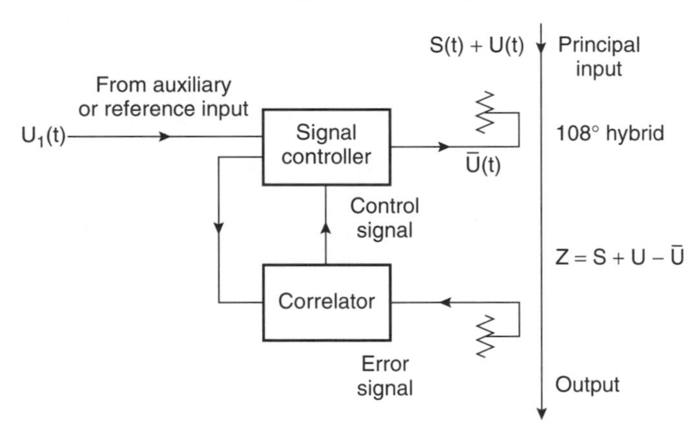

Figure 2-11 Illustration of undesired signal cancellation concept.

The object of the undesired signal cancellation scheme is to synthesize a cancelling undesired signal $\bar{U}(t)$ from $U_1(t)$ and subtract $\bar{U}(t)$ from the principal input. If the synthesis is appropriate, $\bar{U}(t)$ becomes equal to $U(t)$, and nothing further remains to obtain $S(t)$ almost exclusively following the subtraction. If, however, the synthesis is not perfect, the difference $U(t) - \bar{U}(t)$ is used as an error signal to control the synthesis until it becomes perfect. Thus, the adaptive control process drives the error signal to zero or a minimum. We shall discuss the operation of the adaptive control process in further detail in the following chapters.

At first glance, subtracting an undesired signal, which could be like noise from another noise-like signal, seems a risky operation since, if it is not done properly, the undesired signal may even increase following the subtraction. This will be the case, for example, when one subtracts one random noise from another random noise. We will see, however, that if the adaptive control process is appropriate, there is little or no risk of enhancing the level of the undesired signal.

If one knows the exact relationship between $U(t)$ and $U_1(t)$, all that will be needed to synthesize $\bar{U}(t)$ from $U_1(t)$ will be a fixed network that simulates the known relation. Such relationships are seldom known exactly, and more often, the relationship is not fixed or a constant function of time. Moreover, even when the relationship is known, the adjustment during the synthesis has to be very precise since the slightest error could prevent the undesired signal cancellation.

Given that $U_1(t)$ and hence $\bar{U}(t)$ are correlated with $U(t)$, and $S(t)$ is not correlated with either $U(t)$ or $\bar{U}(t)$, one can examine the criterion for cancelling the undesired signal. Thus, for example, the resultant signal Z, following the subtraction, may be written as

$$Z = S + U - \bar{U}. \tag{2.31}$$

Squaring both sides yields

$$Z^2 = S^2 + (U - \bar{U})^2 + 2S(U - \bar{U}). \tag{2.32}$$

Taking the expectation of both sides of Eq. (2.32) and recalling that S is uncorrelated with U and $\bar{U}$, so that the expectations of SU and $S\bar{U}$ are zero, one obtains

$$E[Z^2] = E[S^2] + E[U - \bar{U})^2]. \tag{2.33}$$

The signal power $E[S^2]$ will be unaffected as the adaptive process minimizes $E[Z^2]$. Accordingly, the minimum output power following the subtraction is

$$\min E[Z^2] = E[S^2] + \min E[(U - \bar{U})^2]. \tag{2.34}$$

It is evident from Eq. (2.34) that when $E[Z^2]$ is minimized, $E[(U - \bar{U})^2]$ is also minimized. The synthesized cancelling signal $\bar{U}$, then, is the best least-square estimate of the undesired signal $U(t)$. Moreover, when $E[(U - \bar{U})^2]$ is minimized, $E[(Z - S)^2]$ is also minimized, since from Eq. (2.31)

$$Z - S = (U - \bar{U}). \tag{2.35}$$

The adaptive control process to minimize the total output power following the subtraction is thus equivalent to causing the output Z to be the best least-square estimate of the desired signal S for the given cancellation system and the reference input $U_1(t)$.

In general, the output Z will contain the desired signal and some undesired signal, although substantially reduced in magnitude. Since minimizing $E[Z^2]$ minimizes $E[(U - \bar{U})^2]$, minimizing the total output power minimizes the power of the residual undesired signal. Also, since the signal at the output following the subtraction or cancellation remains constant, minimizing the total output power maximizes the output signal-to-undesired-noise ratio.

From Eq. (2.33), it is seen that the smallest possible output power is $E[Z^2] = E[S^2]$. When this is achievable, $E[(U - \bar{U})^2]$ becomes zero, $\bar{U}$ becomes equal to U, and Z becomes equal to S. For such a case, minimizing the output power causes the output signal, following the subtraction or cancellation, to be free of undesired signal.

The conclusions from the above discussion can be extended to the case where the principal input or the receive-line input and the reference input contain additive noise in addition to U and U_1. They can also be extended to the case where U and U_1 are deterministic rather than stochastic.

The above discussion establishes the sufficiency condition for the cancellation of an undesired signal. The implementation of an adaptive signal cancellation or the mechanization of the adaptive control that minimizes the output power will be considered in further detail in the following chapters.

2.5 DIFFERENT ROLES OF FILTERS AND INTERFERENCE CANCELLERS

In a crowded signal and interference or noise environment, the primary objective to maintain an electromagnetic compatibility is to reduce interference or noise or to enhance "desired signal-to-noise" ratio by whatever practical and economical means available or feasible. Clearly, when the desired signal and interference spectra are distinctly different, conventional frequency-domain filters with fixed or adaptive filter elements deserve first consideration. Thus, if the interference frequencies are always higher or lower than those of the desired signal, a suitable low-pass or high-pass or bandpass filter can be designed to rid the interferences, depending on the characteristics of the desired signal and interference spectra. Even when there is a common frequency band for the desired signal and interferences, it is possible sometimes to design and implement a conventional filter to enhance the "desired signal-to-interference" ratio without seriously affecting or compromising the quality of desired signal.

In general, when the desired signal and interference occupy the same frequency band or the same principal frequency band, the use of a conventional filter to maintain an electromagnetic compatibility becomes discouraging. As we have seen earlier, in Eq. (2.12) for example, the primary reason for the difficulty is that the desired signal is necessarily modified at the filter output, even when the filter is designed to rid the interference exclusively. Thus, the filter output of the desired signal is the inverse transform of the product of $S(j\omega)$ and the filter transfer function $G(j\omega)$, and as long as this product is nonzero, the filter will affect the desired signal, although it is supposed to address the interference only.

The matched filter, on the other hand, is capable of, at least conceptually, separating the desired signal and the interference even when the desired signal and interference occupy the same frequency band. From the viewpoint of filtering, the means of discrimination for such a case is the correlation of either the desired signal or the interference to enhance the effective signal-to-noise ratio. Unfortunately, the required filter transfer function to achieve this objective assumes knowledge of the transform of the desired signal or the interference, and such knowledge is seldom available in most interference environments of concern. Also, as in the case of a frequency-domain filter, the output of this matched filter is the inverse transform of the product of the desired signal transform and the filter transfer function or the product of the interference transform and filter transfer function, and as long as this product is not zero, the complete elimination of one or the other, not intended to be addressed by the matched filter, is not possible.

In contrast, there is no multiplication product of the interference transform, and the filter transfer function is involved for the interference cancellation system, although, like the matched filter, the fact that the desired signal and interference are not correlated is utilized for discrimination of the desired signal and the interference. Instead, for the interference cancellation concept, the synthesized interference is subtracted from the receive-line interference until the net interference at the receive line approaches zero without affecting the desired signal. This subtraction process, in contrast with the multiplication process as discussed earlier, permits the feasibility of, at least conceptually, eliminating the interference completely when the interference and the desired signal occupy the same or the same principal frequency band.

Figures 2-12–2-14 show, for example, the ability of an adaptive canceller at VHF and UHF bands to unmask a desired signal overwhelmed in each case by an interference. As shown in the figures, the desired signal and interference share a common frequency band. Figures 2-12(a), 2-13(a), and 2-14(a) show the receive-line signal and interference from the receiving antenna before interference cancellation. Figures 2-12(b), 2-13(b), and 2-14(b) show the corresponding residual and desired signals when the interferences are subtracted from the receive-line signals. Clearly, such results are not obtainable through the use of frequency-domain or matched filters.

Sometimes, in real life, one encounters an interference problem where the interferences are at fixed or predictable frequencies, but the desired signal frequency changes at a fast rate as in a frequency-hopping radio. During this frequency change, the desired signal spectrum moves through the spectra of interferences, and the signal reception is adversely affected as a consequence, even when the interferences are intermittent in character. This is a common occurrence, for

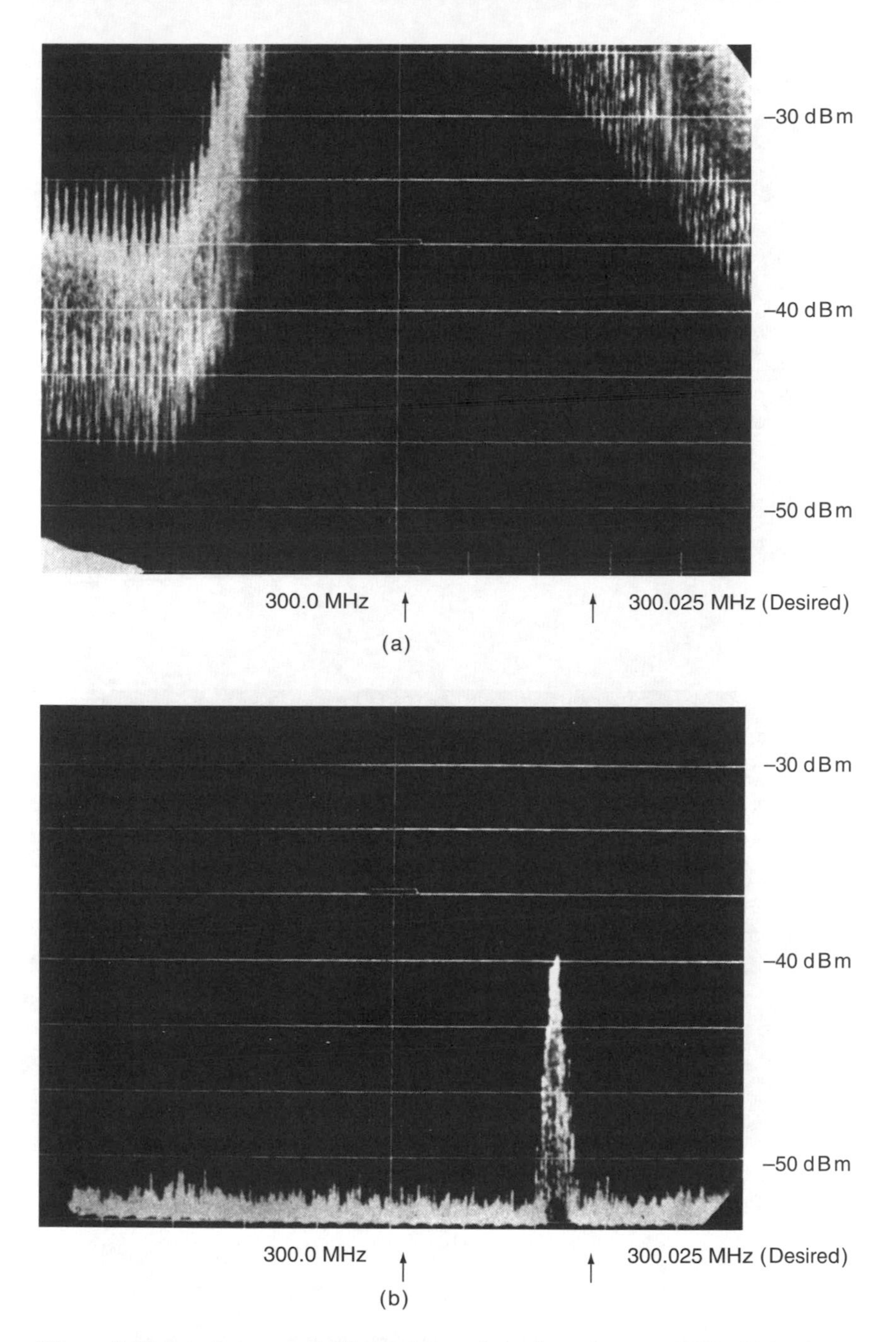

Figure 2-12 Interference cancellation by a subtraction process at UHF. (a) Desired signal masked by an overwhelming interference at a receiving antenna line before cancellation. (b) Desired signal after interference cancellation.

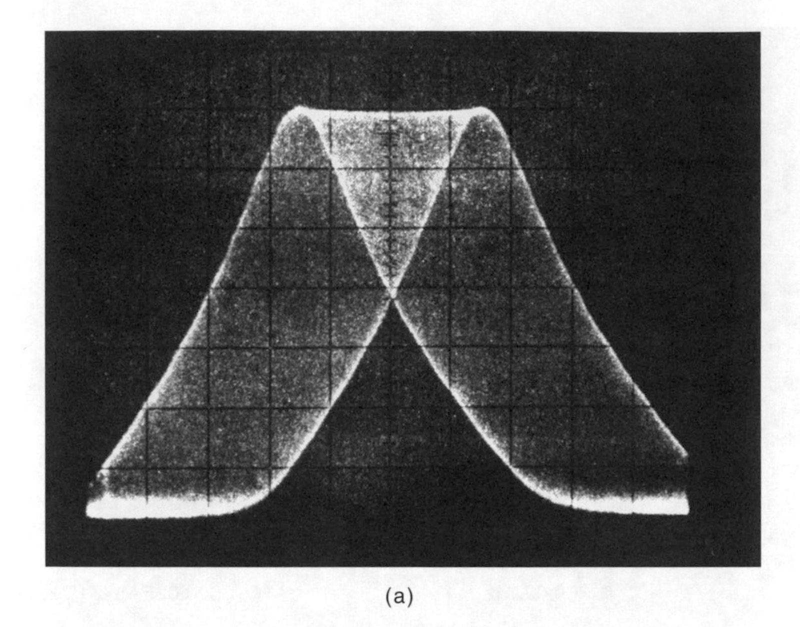

(a)

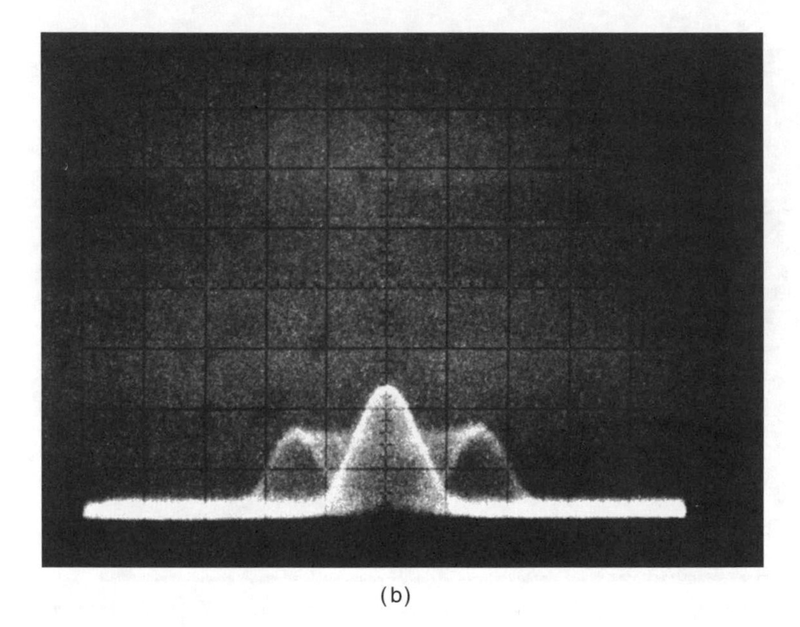

(b)

Figure 2-13 Frequency-modulated interference cancellation by a subtraction process at VHF. Center frequency: 105 MHz. Desired signal: amplitude modulated at −66 dBm. Undesired signal: −10 dBm. (a) Before cancellation. (b) After cancellation.

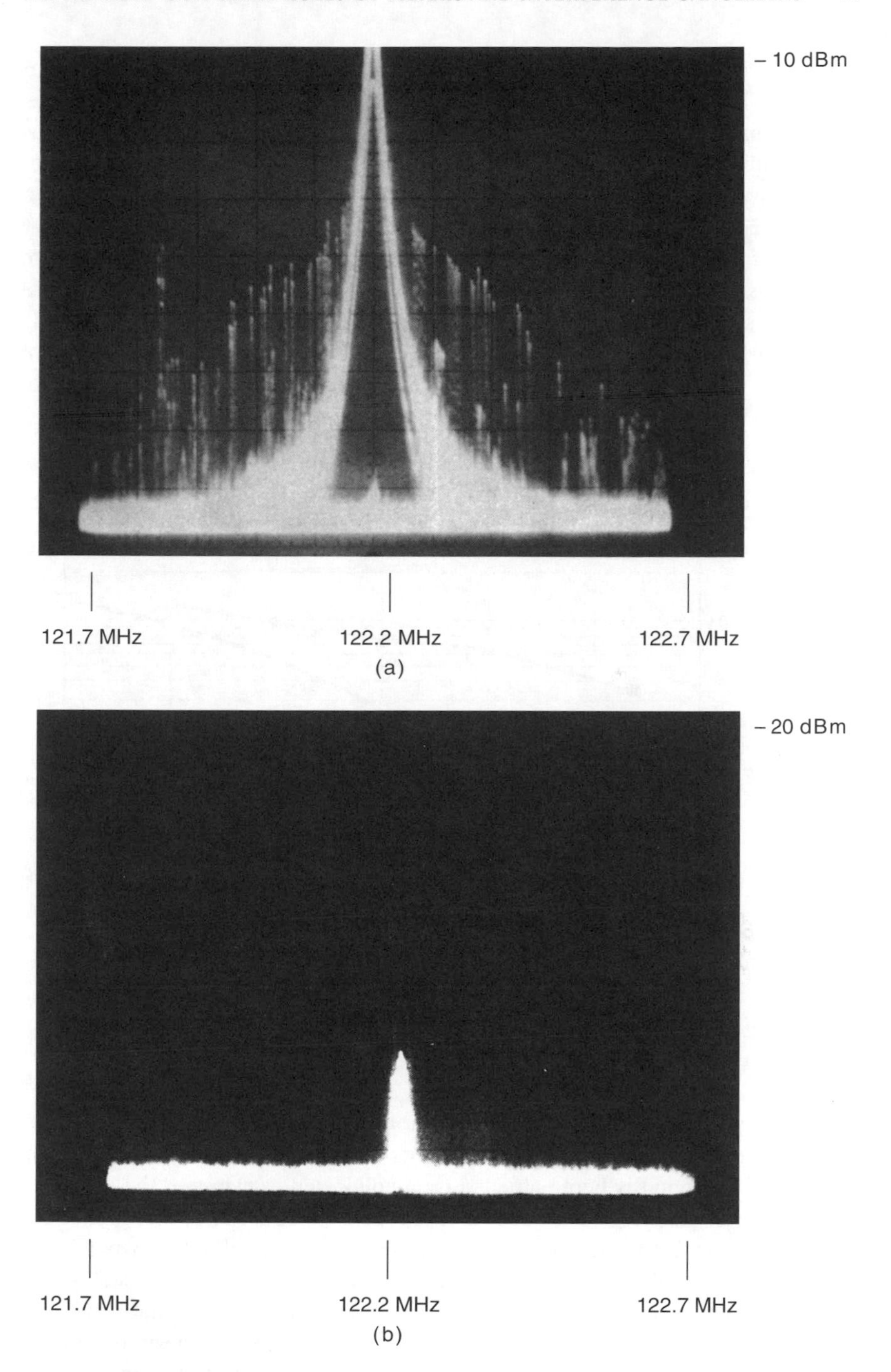

Figure 2-14 VHF interference cancellation by a subtraction process. (a) Before cancellation. (b) After cancellation, showing desired signal.

example, when a frequency-hopping radio and a single-channel radio are collocated on a ship. The bit error rate, which is an important parameter that characterizes the communication effectiveness and which is averaged over a period, deteriorates significantly, even for the intermittent interferences. An adaptive canceller for each receiver line significantly improves the bit-error-rate performance, as shown in Figure 2-15, for example. Here, the measured bit error rate for various relative transmitter powers at HF is shown with and without an adaptive canceller. Figure 2-15 shows that one needs a significantly increased transmitter power to maintain a specified or permissible signal-to-noise ratio or a bit error rate in the presence of the interference, while such an increase will not be necessary if an appropriate adaptive canceller is used at the receive line. Again, such a performance improvement cannot be expected with a frequency-domain or matched filter.

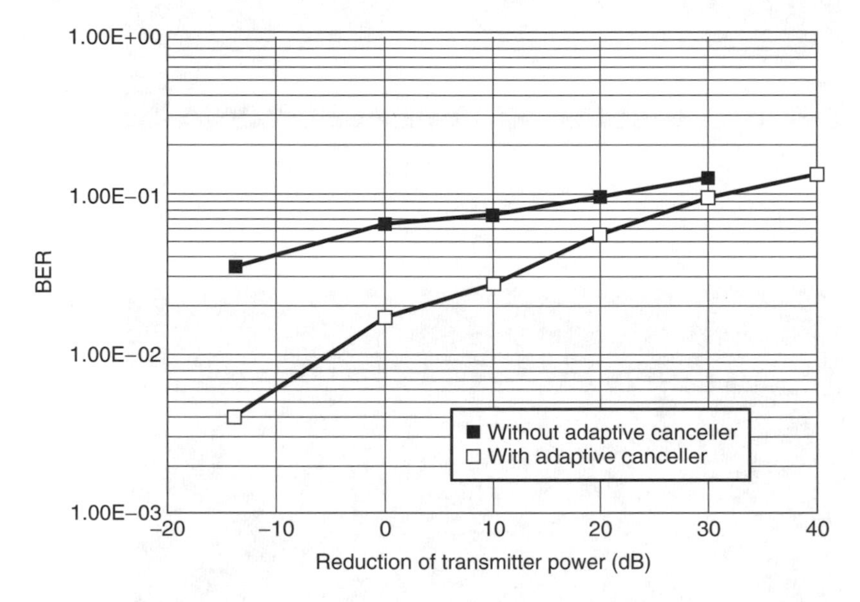

Figure 2-15 Improvement of bit error rate in a communication receiver with an adaptive canceller.

Perhaps the differences among the various remedies for combatting interferences can be illustrated by an example. Let us assume that the reception of a weak distant desired signal is being affected by a collocated transmitter, the source of interference in this case. The spectra of the desired signal and interference are shown in Figure 2-16. The interference in this case is one of a narrowband, and hence it is conceivable to design a frequency-domain notch filter to reject the interference. However, the characteristic of the notch filter is such that it will not only attenuate the undesired signal or interference, but also the desired signal at the notchband frequencies and their vicinity because the multiplication product of the notch filter transfer function and the transform of the desired signal is not zero for such frequencies. If an interference cancellation system is used instead, one can synthesize a cancelling signal for the interference from a sample from the interference source and subtract this synthesized signal from the receive line. This process is adaptive in the sense that no advance knowledge of the interference frequency, its spectrum, or its amplitude is necessary for the operation of the system. Also, because this subtraction process does not necessarily affect the desired signal, it is conceptually possible to eliminate the interference completely. Thus, unlike the frequency-domain filter, the interference cancellation system can eliminate the interference completely when the desired signal and interference occupy the same or common frequency band, and unlike the matched filter, the interference cancellation system does not require the knowledge of the spectral characteristics of the desired signal or interference. More about the characteristics and practical implementation of the interference cancellation concept is discussed in the following chapters.

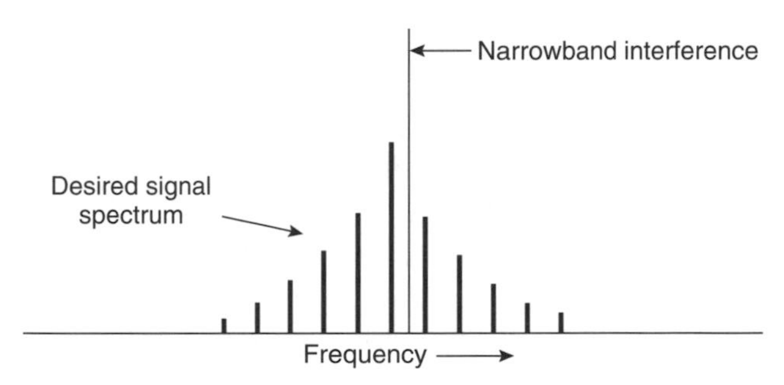

Figure 2-16 Desired signal and a narrowband interference occupying a common frequency band.

A discussion on the reduction of an arbitrary interference of concern at the receive line would not be complete without consideration of the role of an adaptive array for such a purpose. When an array comprises a number of antenna elements, each connected to a weighting network that can introduce an amplitude and phase change in signals flowing through the network, and the outputs of the weighting networks are summed to constitute the array output, one may change the radiation or receiving array pattern by changing the weights. For a receiving antenna array, if the weights are changed so that a null occurs along the direction of an interference source, one may eliminate or substantially reduce the interference at the array output, and hence at the receiver. Since in a multielement array more than one null can be created in desired directions, such an array may accommodate multiple interferences simultaneously. Additionally, when the directions of arrival of the interferences are not *a priori* known, the weights can be changed in an adaptive manner until appropriate nulls are formed at the array pattern. Such an array, then, adaptive array. Although an adaptive array may not always be practical :ific interference of concern, it is a very useful tool. More about the nplementation of an adaptive array is considered in Chapters 5 and 6.

REFERENCES

[1] W. Cauer, *Synthesis of Linear Communication Networks.* New York: McGraw-Hill, 1958.

[2] S. Darlington, "Synthesis of reactance 4-poles which produce prescribed insertion loss characteristics," *J. Math. Phys.*, pp. 257–353, 1939.

[3] R. Saal and E. Ulbrich, "On the design of filters by synthesis," *IRE Trans. Circuit Theory*, pp. 284–327, Dec. 1958.

[4] E. Wetherhold, "Additional modern filters and selected filter bibliography," *1984 Radio Amateur's Handbook*, 61st ed. Newington, CT: American Radio Relay League, pp. 2-40–2-46, 1984.

[5] R. Ghose, *Microwave Circuit Theory and Analysis.* New York: McGraw-Hill, pp. 237–287, 1964.

[6] E. Jordan, Ed.-in-Chief, *Reference Data for Engineers: Radio, Electronics, Computer, and Communications*, 7th ed. Indianapolis, IN: Howard W. Sams, 1985.

[7] G. Turin, "An introduction to matched filter," *IRE Trans. Inform. Theory*, pp. 311–329, June 1960.

[8] D. Brennan, "On the maximum signal-to-noise ratio realizable from several noisy signals," *Proc. IRE*, vol. 43, p. 1530, Oct. 1955.

[9] R. Fano, "On matched-filter detection in the presence of multipath propagation," unpublished paper, M.I.T., Cambridge, MA, 1956.

[10] S. Sussman, "A matched-filter communication system for multipath channels," *IRE Trans. Inform. Theory*, p. 367, June 1960.

[11] L. Davisson, "A theory of adaptive filtering," *IEEE Trans. Inform. Theory*, pp. 97–102, Apr. 1966.

[12] J. McCool, "The basic principles of adaptive systems with various applications," Naval Undersea Center, San Diego, CA, Sept. 1972.

[13] B. Widrow, "Adaptive filters," in *Aspects of Network and System Theory*, R. Kalman and N. Declaris, Eds. New York: Holt Reinhart and Winston, 1971.

[14] B. Widrow and S. Stearns, *Adaptive Signal Processing*. Englewood Cliffs, NJ: Prentice-Hall, NJ 1985.

[15] R. Ghose and W. Sauter, "Interference cancellation system," U.S. Patent 3,699,444, Oct. 17, 1972.

3

CHARACTERIZATION OF ADAPTIVE CANCELLERS

Adaptive interference cancellation usually refers to an automatic cancellation of interference at a receive line when no advance knowledge of the interference, such as its amplitude and waveform, is available, leaving the desired signal almost unaffected at the receive line. Thus, if the receive line contains a sum of the desired signal $S(t)$ and an undesired signal $U(t)$, an adaptive cancellation involves the subtraction of a cancelling or counterundesired signal $U_1(t)$ from the sum $S + U$ and making the error signal $(U - U_1)$ approach zero by a closed-loop control. As it is possible in some cases to make the counterundesired signal U_1 very nearly the same as U, the enhancement of the S/U ratio following the cancellation can approach a very large value as U approaches zero.

3.1 BASIC CONCEPT OF ADAPTIVE CANCELLING

The mechanism of adaptive undesired signal cancellation may be further explained by considering a practical example illustrated in Figure 3-1. Here, the undesired signal $U(t)$ at the receive antenna line is due to a collocated transmitter operating at the frequency band of the desired signal for the receiver, and arrives at the receiver antenna through a leakage path. Such a situation is very common when more than one radio is used on a ship or in an aircraft, as shown in Figure 1-2. A similar problem also arises when one attempts to collocate a radio receiver at the transmitter station of a broadcasting or radio-relay station, as in a Voice of America system, even when the transmitter frequency is slightly different from the receiver frequency. Let the undesired signal transmitted by the collocated transmitter antenna, as shown in the figure, be $U_1(t)$. The leakage path introduces a time delay and a reduction of amplitude, both depending primarily on the leakage path distance. Thus, if there is only a single leakage path, the signal arriving at the receiving antenna may be expressed as $KU_1(t - T)$, where K denotes the reduction of amplitude and T denotes the time delay introduced by the propagation path. If a second path, identical to the leakage path, can be created through a signal controller as shown in Figure 3-1, and the signal controller path introduces an attenuation equal to K and a time delay T, the same as the leakage path, then the undesired signal arriving at the 180° hybrid through the leakage and signal controller paths will be the same, making $KU_1(t - T) = U(t)$. The output of the 180° hybrid that subtracts the counterundesired signal or the cancelling signal created at the output of the signal controller from the sum of the desired and undesired signals $S + U$ present at the receiving antenna line will be free of the undesired signal $U(t)$. The desired signal $S(t)$ at the receiving antenna line will not be materially affected by the subtraction process, except for some through-line insertion losses.

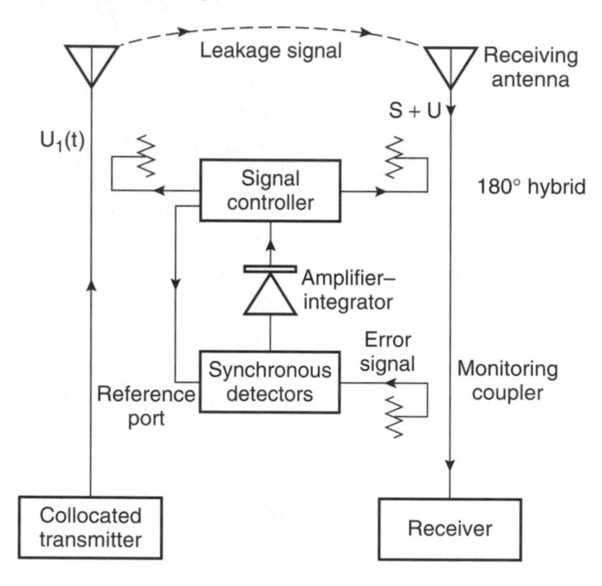

Figure 3-1 Cancellation of an undesired signal due to a collocated transmitter from the receive line.

The exact values of K and T for the leakage path will usually not be known, and they may change even when they are known at one time. Thus, one cannot set fixed values of K and T at the signal controller. Now, if the attenuation and time delay introduced by the signal controller are not the same as those due to the leakage path, the undesired signal $U_1(t)$ flowing through the two paths will not be the same. Consequently, there will be a residual undesired signal at the output of the 180° hybrid. Previously, we have termed this residual signal as the error signal. As shown in Figure 3-1, the error signal becomes the input signal of the synchronous detector which is referenced to the same undesired signal $U_1(t)$ as at the input of the signal controller. The principal characteristic of the synchronous detector is that if there is any signal at its input which has the same waveform as that of its reference signal, there will be a dc signal at the output of the synchronous detector. We shall discuss other characteristics of the synchronous detector when used for the adaptive interference cancellation system later in the text. Since the reference signal in this case is $U_1(t)$, there will be a dc signal as long as there is any nonzero error signal present at the synchronous detector input port. The dc signals from the output of the synchronous detector control and change the values of K and T of the signal controller until such values are adjusted so that the undesired signals arriving at the 180° hybrid through the leakage and the signal controller paths are identical. For this equilibrium condition of the control loop, comprised of the signal controller, 180° hybrid, the synchronous detector, and the amplifier–integrator assembly as shown in the figure, there will be no error signal, and hence no $U(t)$ at the receive antenna line following the 180° hybrid. The implementation of the adaptive interference cancellation shown in Figure 3-1, then, becomes similar to that shown in Figure 2-11, except for the synchronous detector, which plays the role of a correlator. Quantitative aspects of this correlation by the synchronous detector will be further discussed later in this chapter.

The effectiveness[1] of the adaptive canceller for the Voice of America problem referred to earlier is illustrated in Figure 3-2(a) and (b). Here, the inner and outer traces of the spectrum analyzer records show the suppression of interferences due to collocated transmitters at the receive line at 9.7 MHz in Figure 3-2(a) and at 11.8 MHz in Figure 3-2(b), respectively, when the receiver is collocated with the transmitters. As expected, the overwhelming transmitter signals from the adjacent transmitters are substantially reduced to enable the receiver to be operative again.

An examination of a few important characteristics of adaptive cancelling as illustrated by Figure 3-1 may now be in order. First, the cancelling or counterundesired signal is synthesized by the signal controller which is characterized by the parameters K and T, the same as those of the leakage path. Once the values of K and T are correctly synthesized by the signal controller, the undesired signal is cancelled at the output of the 180° hybrid, regardless of the amplitude, frequency, or waveform of the undesired signal since it is equally affected by the two identical paths. Thus, the operation of adaptive cancelling is independent of the characteristics of the undesired signal.

It is also evident from Figure 3-1 that the adaptive cancellation process involves more of the synthesis of an equivalent propagation path, identical in characteristics to the leakage path, than the synthesis of a counterundesired signal. Since the characteristic parameters K and T of the leakage path are seldom known, it is very likely that at the beginning, when the cancellation system is just turned on, the K-T parameters for the leakage and signal controller paths will not be the same. Consequently, the undesired signals arriving at the 180° hybrid through the

[1]R. N. Ghose, "Collocation of receivers and high power broadcast transmitters," *IEEE Trans. Broadcasting*, vol. 34, no. 2, June 1988.

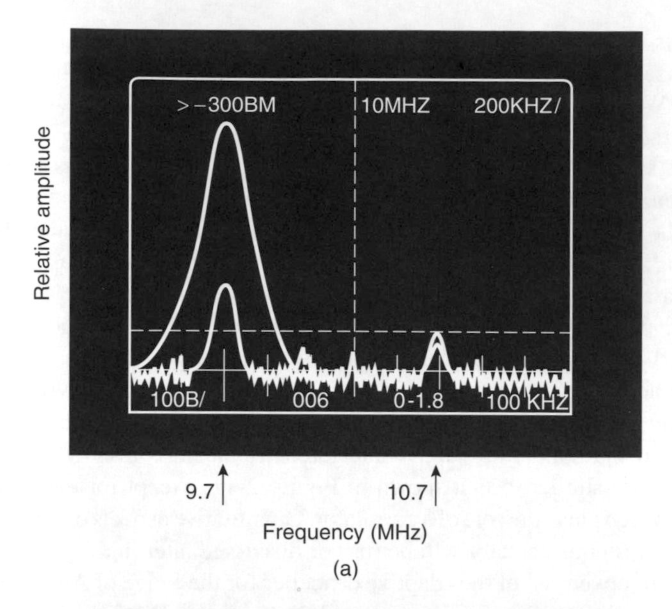

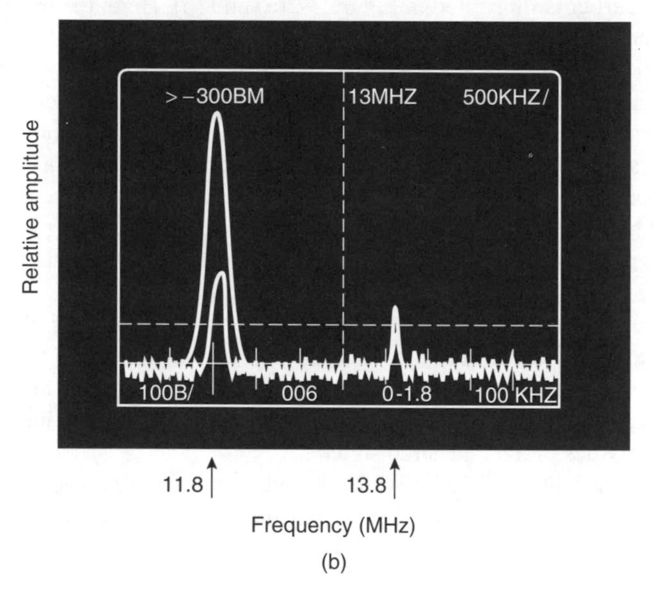

Figure 3-2 (a) Adaptive interference suppression at 9.7 MHz. Outer and inner traces show interference spectra without and with adaptive canceller, respectively. (b) Adaptive interference suppression at 11.8 MHz. Outer and inner traces show interference spectra without and with adaptive canceller, respectively.

two paths will not be the same, and there will be a residual undesired signal or error signal present at the receive line following the signal subtraction at the 180° hybrid. This error signal will, in turn, generate the control signals for the signal controller at the output of the synchronous detectors. And the control signals, in turn, will change the values of K and T. This process will continue until there is no error signal, and the closed-loop control arrives at an equilibrium condition. For this condition, the K-T parameters of the leakage and signal controller paths will be the same, and the undesired signal will be eliminated from the receive line following the 180° hybrid. If for any reason the characteristics of the leakage path, rather than the characteristics of the undesired signal, change, an error signal will be generated again, and the closed-loop control will drive the error signal to zero, causing a new equilibrium. Thus, the operation of the adaptive cancelling process is adaptive with a continuous tracking provision.

The third important characteristic of the adaptive cancellation process is that the degree of undesired signal cancellation is governed more by the closed-loop control characteristics than by the deviation of the values of K and T of the signal controller path from the corresponding values of the leakage path. This is because any deviation will cause an error signal which will be driven to zero automatically by the closed-loop control, and the absence of the error signal will imply the elimination of the undesired signal from the receive line, the only objective of the adaptive cancellation process. Thus, the permissible gain of the closed-loop control becomes the dominant factor that determines the degree of cancellation of the undesired signal.

In addition to the degree of cancellation of the adaptive cancelling process, one is often concerned with the response time for the cancellation. Again, from Figure 3-1 and the operation of the closed-loop control described above, it is seen that this response time is a function of the bandwidth of the closed-loop control, and the response time has nothing to do with the bandwidth of the undesired signal. In general, the bandwidth of the closed-loop control is set by design to accommodate the rapidity with which the equivalent K and T parameters of the leakage path is likely to change with time. Once this bandwidth of the closed-loop control is fixed, the choice of the permissible gain of the closed-loop becomes limited, as the upper bound of the product of loop–gain bandwidth is usually a severe constraint in the loop design.

It should be noted here that, unlike many closed-loop operations, the response time for adaptive cancellation will depend not only on the loop bandwidth, but also on the initial condition of the signal controller path. Thus, for example, when an undesired signal is cancelled once by the adaptive cancellation process, but disappears and reappears with some periodicity such as a pulsed signal, or even irregularly with time, the response time can be zero since the correct values of the K-T parameters for the signal controller are held fixed by the amplifier–integrator assembly, at least for a short duration. A nonzero response time will be involved, if at all, only when the K-T parameters change during the interval of disappearance and reappearance of the undesired signal.

From Figure 3-1, it is also seen that the subtraction of the counterundesired signal at the through receive line is a linear process which usually generates no harmonics or intermodulation products of the desired and undesired signals. This is important since any nonlinearity introduced at the through receive line will create one or more separate undesired signals that do not exist in the absence of the cancelling system and cannot be removed from the receive line, particularly when such signals fall within the bandwidth of the desired signal.

Perhaps the characterization of the adaptive cancellation process will not be complete without the consideration of its power-handling capability for the undesired signal. More specifically, it is seen from Figure 3-1 that at the 180° hybrid, the undesired signal $U(t)$ as received through the leakage path is cancelled by the counterundesired signal generated at the signal controller path. Such a cancellation implies that the two undesired signals must have the same amplitudes and phases at any instant of time. This also means that the power level of the undesired signal, as generated at the output of the signal controller, must be the same as that of the undesired signal $U(t)$ received through the leakage path. Now, the signal controller, like most subsystems or components, introduces a loss for its through-line signal. Thus, the input to the signal controller must be at a power level higher than that of $U(t)$. We will see in the following section on signal controllers that the loss in the signal controller path can be on the order of 7–10 dB or more in some cases. Thus, there exists a power level of $U(t)$ for which a given signal controller will not be able to operate and generate the counterundesired signal as needed, the components of the signal controller usually being solid-state devices. In many practical situations, the power level of $U(t)$ from a collocated transmitter could be overwhelmingly high. For such cases, the constraints on the power-handling capability of the signal controller often results in an insurmountable problem.

To summarize, the adaptive cancellation process to suppress or cancel an undesired signal may be characterized as follows:

(a) Its operation is independent of the characteristics of the undesired signal.

(b) Its operation is adaptive in the sense that no *a priori* knowledge of the undesired signal characteristics is necessary.

(c) The degree of cancellation of the undesired signal is governed by the characteristics of the closed-loop control, such as the loop gain, instead of the K-T parameters of the signal controller path.

(d) The response time for cancellation depends not only on the loop bandwidth of the closed-loop control, but also on the initial condition of the signal controller path.

(e) The cancellation operation is a linear process not likely to generate harmonics and intermodulation products of the desired and undesired signals.

(f) The maximum power level of the undesired signal at the receive line becomes, perhaps, the most limiting constraint in the design of an adaptive cancelling system, particularly when solid-state devices are used at the signal controller.

To understand the implications of the above-mentioned characteristics of the adaptive cancellation further, one needs to study the essential system elements that effect such a cancellation such as the signal controller, synchronous detector, amplifier–integrator assembly, etc. We begin with the characteristics and implementation of the signal controller, the most important of such system elements, in the following section.

3.2 SIGNAL CONTROLLERS

For adaptive cancelling, a signal controller refers to a device that can effect changes in the amplitude and time delay of a signal at the output of the signal controller as compared with the same signal at the input. The changes in amplitude and time delay must be controllable independently, and the control for such changes are usually effected by direct-current control signals. The conceptual diagram of a signal controller is shown in Figure 3-3. Here, the input signal is assumed to be of the form

$$S_i = A(t)\sin[\omega t + \phi(t)] \tag{3.1}$$

where

A = a time-varying amplitude, similar to that of an amplitude-modulated signal

ω = the angular carrier frequency

ϕ = is a time-varying phase.

Figure 3-3 Conceptual diagram of a signal controller that introduces an amplitude gain factor K and a time delay T.

The output of the signal controller, then, assumes the form

$$S_0 = KA(t-T)\sin[\omega(t-T) + \phi(t-T)] \tag{3.2}$$

denoting that the signal controller has introduced an amplitude gain factor K and a time delay T.

If the time delay is so small for the given rate of variations of A and ϕ such that

$$A(t-T) \simeq A(T) \quad \text{and} \quad \phi(t-T) \simeq \phi(t) \tag{3.3}$$

then

$$\begin{aligned} S_0 &\simeq KA(t)\sin[\omega(t-T)+\phi(t)] \\ &= KA(t)\sin[\omega t+\phi(t)+\theta] \end{aligned} \tag{3.4}$$

where
$\theta = -\omega T$.

The required function of the signal controller for such a case is to introduce an amplitude gain factor K and a phase delay θ. The physical implementation of such a signal controller is very straightforward, as is shown schematically in Figure 3-4. Here, the input signal is branched into two paths. At the first path, there is an attenuator with a gain (loss) factor K_1. At the second path, there is also an attenuator with a gain (loss) factor K_2, but there is a phase delay of 90° with respect to the first branch in addition. The outputs of the attenuators at the two branches are summed, and the summed signal becomes the output of the signal controller.

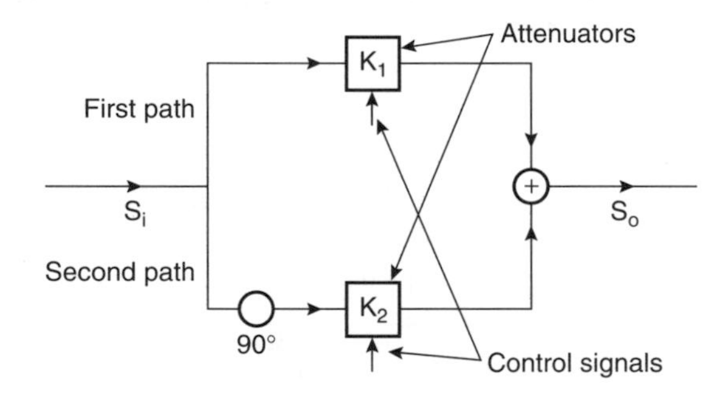

Figure 3-4 Physical implementation of a signal controller introducing an amplitude factor K and a phase delay $\theta = \text{arc tan}\ (k_2/k_1)$.

From the expression of the signal input as given in Eq. (3.1), the output of the signal controller shown in Figure 3-4 can be written as

$$\begin{aligned} S_0 &= K_1A(t)\sin[\omega t+\phi(t)] + K_2A(t)\sin[\omega t+\pi/2+\phi(t)] \\ &= K_1A(t)\sin[\omega t+\phi(t)] + K_2A(t)\cos[\omega t+\phi(t)] \\ &= KA(t)\sin[\omega t+\phi(t)+\theta] \end{aligned} \tag{3.5}$$

where

$$K = \sqrt{K_1^2 + K_2^2}$$
$$\theta = \text{arc tan}\,(K_2/K_1).$$

Thus, a signal controller shown in Figure 3-4 can introduce any arbitrary amplitude factor K and any arbitrary phase delay θ as long as one has independent control to obtain the necessary values of the gain (loss) factors K_1 and K_2. Although K_1 and K_2 denote only amplitude gain (loss) factors, any arbitrary amplitude gain factor K as well as any arbitrary phase delay θ can be synthesized by the arrangement shown in Figure 3-4. It should be noted, however, that in Eq. (3.5), K_1 and K_2 are assumed to be real; that is, the attenuators in Figure 3-3 were assumed to have introduced no phase shifts. If K_1 is associated with a phase shift $(-\Delta\phi_1)$ and K_2 is associated with a different phase shift $(-\Delta\phi_2)$, then Eq. (3.5) becomes

$$S_0 = K' A(t) \sin[\omega t + \phi(t) + \theta'] \tag{3.6}$$

where

$$K' = [K_1^2 + K_2^2 + 2K_1K_2 \sin(\Delta\phi_2 - \Delta\phi_1)]^{1/2}$$
$$\theta' = \text{arc tan}\left[\frac{K_2 \cos\Delta\phi_2 - K_1 \sin\Delta\phi_1}{K_2 \sin\Delta\phi_2 + K_1 \cos\Delta\phi_1}\right].$$

Thus, at least one way to implement a signal controller as needed for adaptive cancelling is to arrange two attenuators and a 90° phase shifter as shown. Each attenuator, however, must have means to control its gain (loss) to provide an independent control. Also, if K_1 and K_2 assume only positive values, the value of the phase delay θ, obtainable from the signal controller, will be confined to $0 \leq \theta \leq 90°$ since $\tan\theta = (K_2/K_1)$. Such a restriction is not permissible in a signal controller which must accommodate a value of θ ranging from 0° to 360°. Therefore, the attenuators in a signal controller, arranged in the form shown in Figure 3-4, must be capable of providing both negative and positive values of K_1 and K_2.

Let us consider a simple potentiometer-like arrangement as shown in Figure 3-5. Here, the RF input signal is connected to the "primary" of a transformer, its "secondary" being connected to a resistive load with a slide. The ratio of output-to-input signal, which constitutes the value of K_1 or K_2, changes as the slide moves up and down. Also, as the transformer "secondary" is connected to the ground at the midpoint, the value of K_1 or K_2 assumes a positive value with respect to the ground point as the slide moves upward from the midpoint of the resistive load. Similarly, the value of K_1 or K_2 becomes negative with respect to the ground as the slide moves downward from the midpoint. The slide, in turn, can be moved up or down by an electric motor driven by a control signal. A pair of attenuators, as shown in Figure 3-5, and a 90° phase shifter, then, will constitute a complete signal controller as needed for adaptive cancellation. A similar arrangement for the output–input ratio control using two variable capacitances is shown in Figure 3-6.

When the resistive load in Figure 3-5 is a pure resistance, or when the capacitive load in Figure 3-5 is a pure capacitive reactance, there will be no phase shift between the output and input of the attenuator.

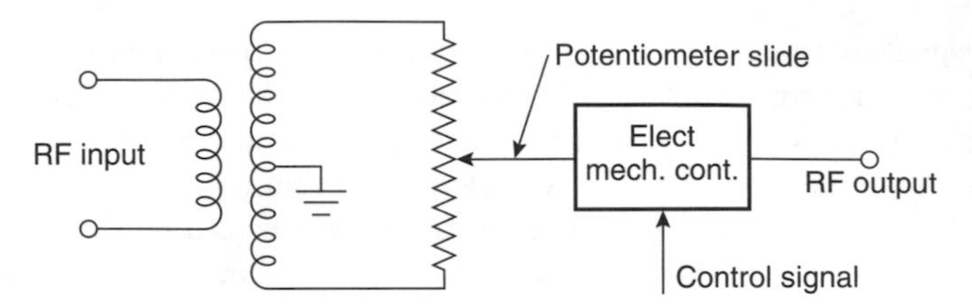

Figure 3-5 Conceptual arrangement for a signal controller employing a variable resistance for signal amplitude control.

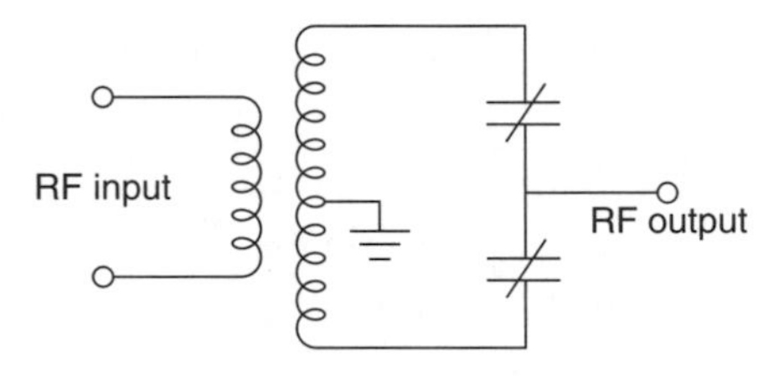

Figure 3-6 Another conceptual arrangement for a signal controller employing two variable capacitors.

It is not necessary to use an electromechanical device such as shown in Figure 3-5 to obtain a signal control function. Solid-state devices, such as p-i-n diodes, with appropriate biases, are often used to construct an attenuator and thus a signal controller. A typical characteristic of a p-i-n diode attenuation with dc control voltage is shown in Figure 3-7(a). The attenuation increases monotonically in this case with increased control voltage. For the attenuators useful for a signal controller, however, it is often desirable to have a linear relationship of the attenuation and control voltage. Such a relationship is easily obtained by biasing the p-i-n diodes appropriately to linearize the attenuation characteristics as shown in Figure 3-7(b). In addition to a linearized characteristic, an attenuator for a signal controller must be capable of assuming both positive and negative values, as otherwise the controllable phase, θ, will be restricted to $0 \leq \theta \leq 90°$, as discussed above. The p-i-n diodes can be connected in a complex network to provide a signal controller function. Figure 3-8 shows an example of such a p-i-n diode network. Here, the radio frequency input signal is fed to a 90° hybrid, A, with a matched termination at one of its ports. Each of the remaining two ports of this 90° hybrid, corresponding to 0° or 90°, is coupled to another 90° hybrid, B or C. The hybrids B and C have a common lead, *CL*, that provides the signal controller output. The remaining two ports of each of the hybrids B or C, corresponding to 0° and 90°, are connected to p-i-n diodes D, E, F, and G, as shown in Figure 3-8. Each of the p-i-n diodes

D and E has an independent dc bias B_1 or B_2 and a variable control voltage C_1 or C_2, constituting the diode power supply. The bias and control voltages for the p-i-n diode D are connected to its input through dropping resistors R_1 and R_2. Similarly, the bias and control voltages for the p-i-n diode E are connected to its input through resistors R_3 and R_4. The p-i-n diodes F and G are similarly connected with their respective power supplies of bias and control voltages.

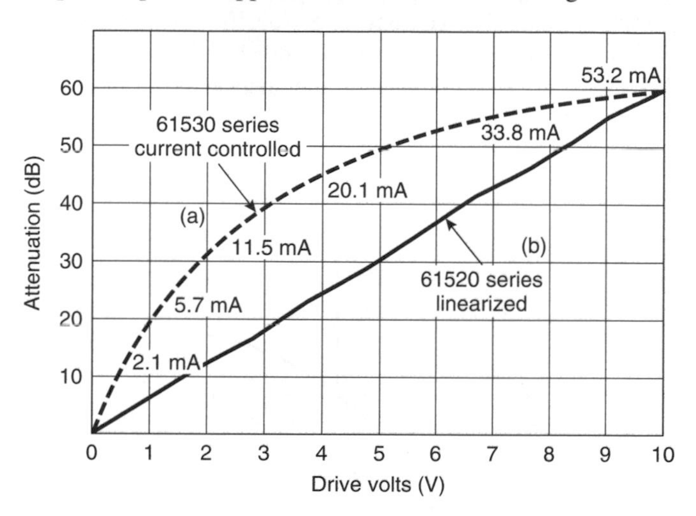

Figure 3-7 Typical attenuation characteristic of a p-i-n diode (a) without linearization bias, (b) with linearization bias.

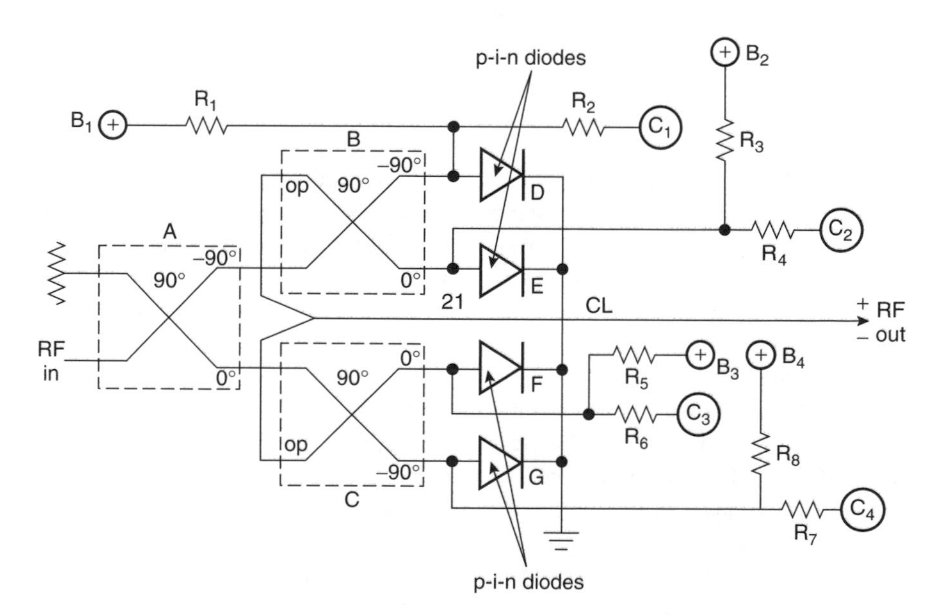

Figure 3-8 A signal controller employing p-i-n diodes and a complex biasing network.

Operating characteristics of each attenuator in Figure 3-8 are shown in Figure 3-9. The bias supply may be varied continuously from a positive to a negative voltage. The effect of the control voltage C_1 or C_2 is shown in Figure 3-9, with the RF output voltage in millivolts as ordinate and the dc control voltage as abscissa. It should be noted that the RF output versus control voltage graph is asymmetrical, although the RF output passes through the origin without any discontinuity, thus providing a smooth control from the negative to positive values of RF output signal. As noted earlier, linearizing circuits may be added to the control circuitry shown in Figure 3-8 to obtain a virtually linear graph, instead of the asymmetrical one as shown in Figure 3-9.

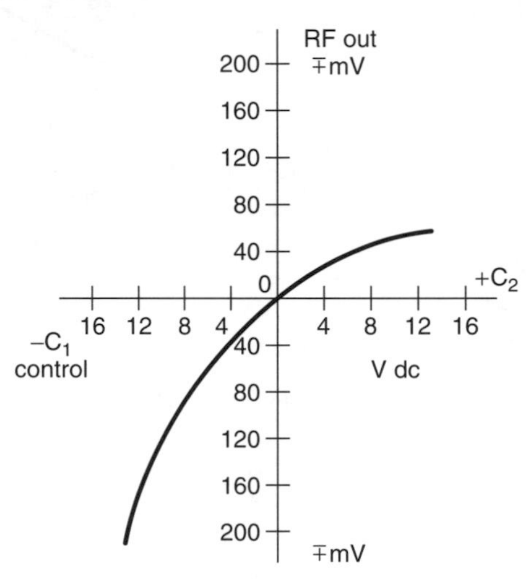

Figure 3-9 Operating characteristic of a p-i-n diode signal controller.

The effect of the control voltage C_1 or C_2 on the amplitude of RF output of the attenuator without regard to its polarity is shown in Figure 3-10. As may be seen from this figure, the RF output varies from almost 0 to 0.5 mW, almost linearly on either side of the zero control voltage, as the control voltage varies from 12 V dc to −12 V dc, although at a different rate. The linear range from 0 to 0.5 mW is a typical operating range of the attenuators used for typical signal controllers for adaptive cancelling.

Another form of a signal controller [4] with some variation from what is shown in Figure 3-8 is illustrated in Figure 3-11. Here, also, the RF input signal is fed to a 90° hybrid *A* with a matched termination at one of the ports. Each of the remaining two ports of the hybrid *A*, corresponding to 0° and 90°, is coupled to

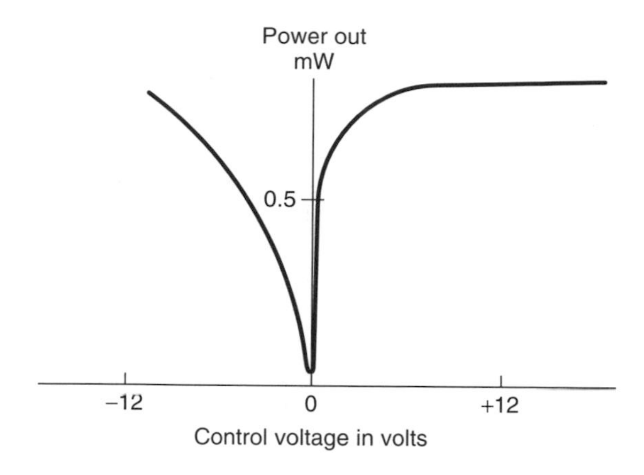

Figure 3-10 The effect of control voltage on the output power of a p-i-n diode signal controller.

another hybrid B or C. The adjacent ports of the hybrids B and C are connected to a 0° summer which sums the signals from hybrids B and C without introducing a phase shift. The output of the summer becomes the output of the signal controller. The remaining two ports of each of the hybrids B and C are connected to p-i-n diodes P_1, P_2, P_3, and P_4. Each p-i-n diode output is connected to the ground, and each diode is assumed to have an internal capacitance and inductance, as shown in Figure 3-11.

The inputs of the p-i-n diodes P_1 and P_2 are controlled in a manner similar to the comparable inputs of the p-i-n diodes in Figure 3-8 by a dc supply network S_1, including the fixed dc supply S_{1F} connected by dropping resistors R_1 and R_2 to the inputs of the p-i-n diodes P_1 and P_2. Active control voltages are supplied from the source C_1 via transistor networks T_1 and T_2 as well as respective resistors R_3 and R_4. Therefore, similar to the signal controller arrangement shown in Figure 3-8, a fixed bias as well as a variable bias are supplied to the input of each of the two control p-i-n diodes P_1 and P_2. A similar bias control network S_2 is provided for the p-i-n diodes P_3 and P_4. As noted earlier, the RF output of the signal controller is obtained from the output of the 0° summer.

The resistance values of R_1 and R_2 are chosen so that when the control voltage C_1 is 0, the current flowing through each of the diodes P_1 and P_2 causes the diode to have a resistance equal to the characteristic impedance of the hybrid B such that, when $C_1 = 0$, the RF output is zero. For this condition, the entire signal applied to p-i-n diodes P_1 and P_2 is dissipated. If the resistances of the specific p-i-n diodes P_1 and P_2 are adjusted so that they are less than the characteristic impedance of the hybrid B, some of the RF input signal would be reflected to the output port

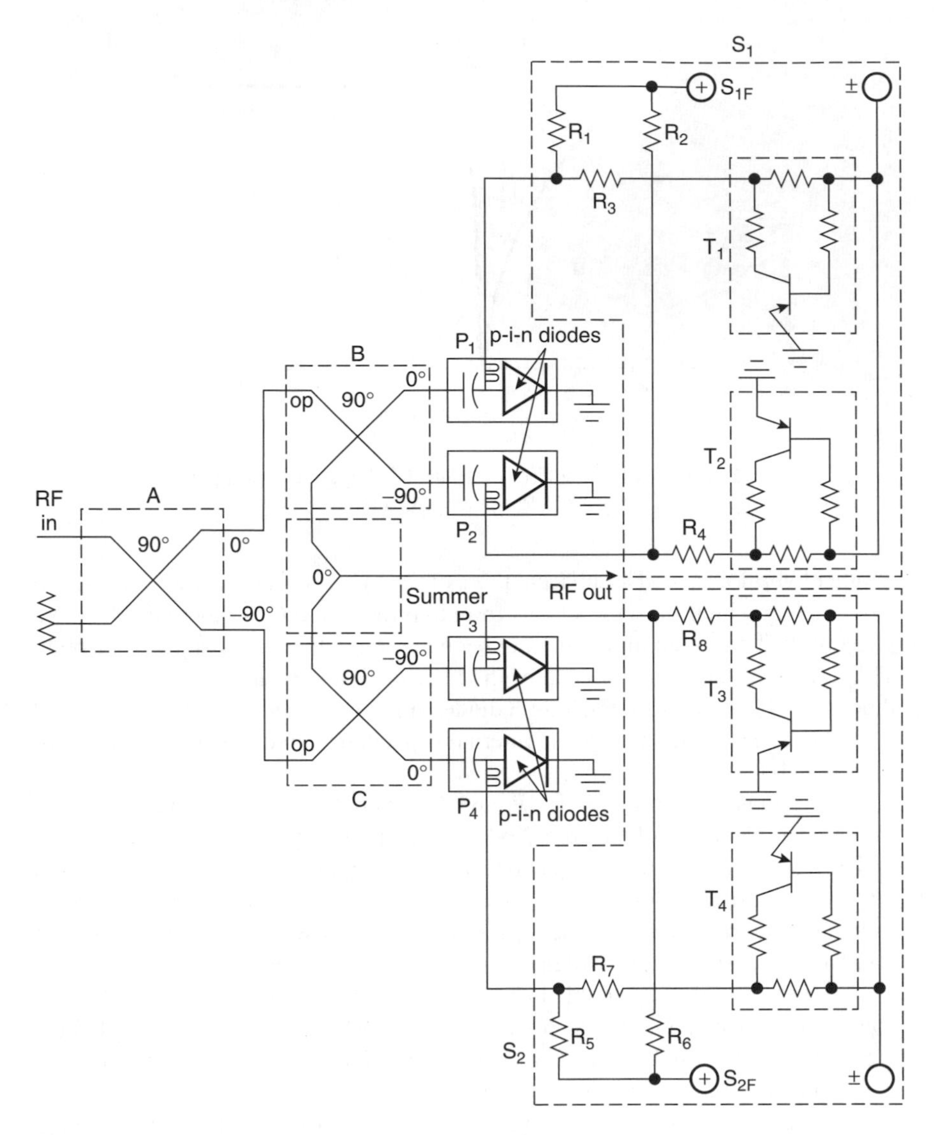

Figure 3-11 Another form of a signal controller with some variation from that shown in Figure 3-8.

of the hybrid B with the same relative polarity as the RF input signal. Similarly, if the resistances of the p-i-n diodes P_1 and P_2 are adjusted so as to have values greater than the characteristic impedance of the hybrid B, some of the RF input signal would be reflected to the output port of the hybrid B with a relative phase which is 180° out of phase with respect to the RF input signal. Thus, the purpose of the fixed source S_{1F} is to supply bias currents to each of the diodes P_1 and P_2 via the resistors R_1 and R_2 at a level that causes the reflected RF output signal to be zero when $C_1 = 0$. The same function is performed by the fixed source S_{2F} in the dc supply network S_2.

As is evident from the above discussion, the signal controllers shown in Figures 3-8 and 3-11 are very similar as they both operate by applying a controllable variable impedance mismatch to the second and third hybrids B and C, depending on the level of the control voltage C_1 and C_2. When the impedance of the p-i-n diodes matches that of the line and its associated components, no energy is reflected to its associated hybrid, and, in turn, no energy is coupled to the output line of the signal controller. When the energy is reflected to the second and third hybrids B and C, the reflected energy is directed toward the RF output line. Thus, signals in signal controllers shown in Figures 3-8 and 3-11 are attenuated to yield variable values of K_1 and K_2 by providing two matching terminations rather than by a series attenuation. Such signal controllers, then, may be called reflective mode signal controllers.

3.3 VARIABLE TIME-DELAY CONTROL

We have noted earlier that for an ideal adaptive cancelling, the signal controller must effect changes in amplitude gain (or loss) and time delay of a signal between its output and input ports. Means for controlling the amplitude and electrical phase as discussed in the previous section may not be adequate at times, since for an undesired signal having a nonzero bandwidth, electrical phases at different frequencies within the band will be different, and a single set of values of K_1 and K_2, as shown in Figure 3-3, cannot provide different phase controls for different frequencies. A variable gain control and a variable time delay in tandem, as shown schematically in Figure 3-12, for example, could, on the other hand, be adequate for an ideal signal controller. The variable gain control for such an arrangement could be obtained by an attenuator that can change the ratio of its output-to-input signals as a function of a control signal. The variable time delay could be achieved by switching in and out different sections of delay lines. The advantages of such an arrangement are that it not only provides the necessary signal controller function for a broadband undesired signal or interference, but also it can accommodate a higher power level of the undesired signal than what is permissible in a p-i-n diode controller. A high-power potentiometer can be used, for example, instead of a p-i-n diode controller with a restricted power-handling capability. Also, no polarity reversal of the signal within the signal controller, as is necessary for an electric phase control over a 360° range, will be needed for the arrangement shown in Figure 3-12. Difficulties, however, are experienced to derive appropriate control signals to change the gain and time-delay controls, particularly for adaptive cancelling where a rapid response time for cancellation is necessary. A frequency-independent time-delay control [5] can be obtained by a pair of gain and phase controlling signal controllers, as discussed in the previous section, along with a delay line.

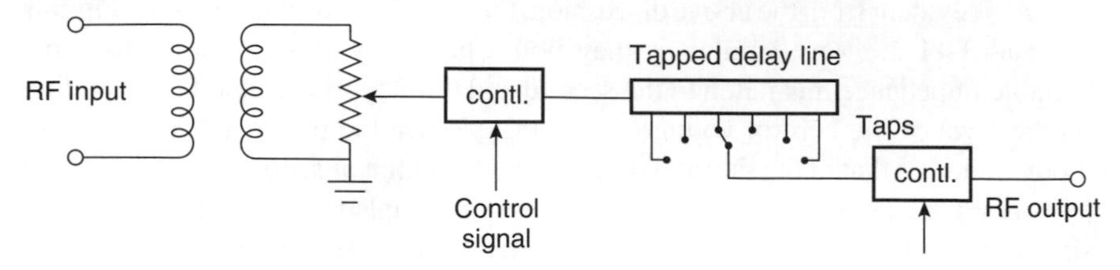

Figure 3-12 A variable gain and a variable time delay control in tandem.

To illustrate the operation of such a signal controller, let $S(t)$ be an arbitrary undesired signal at the input, and let the desired output of the controller be $KS(t - \delta t)$ where the amplitude factor K and the delay δt are independent of frequency. The signal controller then will be a network characterized by K and δt. Let it be assumed that such a network can be designed with a pair of signal controllers, each similar to that shown in Figure 3-4, separated by a fixed delay line providing a time delay T, as shown in Figure 3-13.

For most undesired signals, which may even be desired signals for others, one may write

$$S(t) = A(t)\sin[\omega t + \phi(t)] \tag{3.7}$$

where
$A(t)$ = some arbitrary time-varying amplitude
$\phi(t)$ = some arbitrary time-varying phase.

The signal e_1, shown in Figure 3-13, resulting from the first signal controller, can be written as

$$\begin{aligned} e_1 &= A(t)K_1\sin[\omega t + \phi(t)] \\ &\quad + A(t)K_2\sin[\omega t + \phi(t) + \pi/2] \\ &= \alpha_1 A(t)\sin[\omega t + \phi(t) + \beta_1] \end{aligned} \tag{3.8}$$

where
$\alpha_1 = [K_1^2 + K_2^2]^{1/2}$
$\beta_1 = \tan^{-1}(K_2/K_1)$.

Similarly, the signal e_2, also shown in Figure 3-13, at the output of the second signal controller can be expressed as

$$e_2 = A(t-T)\alpha_2\sin[\omega t - \omega T + \phi(t-T) + \beta_2] \tag{3.9}$$

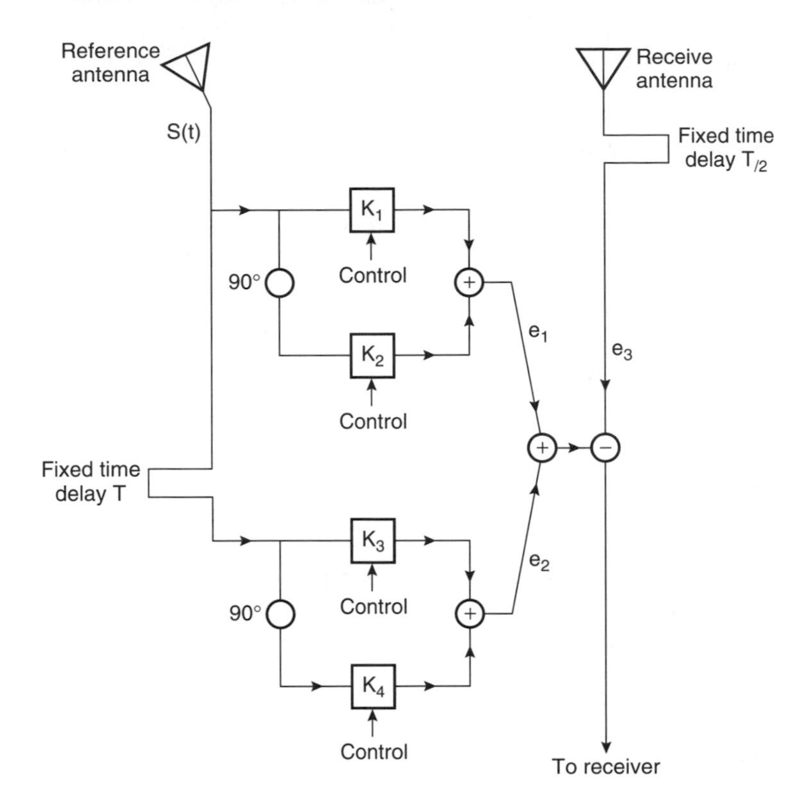

Figure 3-13 Schematic arrangement for a variable time delay signal controller.

where

$\alpha_2 = [K_3^2 + K_4^2]^{1/2}$

$\beta_2 = \tan^{-1}(K_4/K_3).$

Expressing

$$\begin{aligned} A(t) &= A(t - T/2 + T/2) \\ \phi(t) &= \phi(t - T/2 + T/2) \\ A(t - T) &= A(t - T/2 - T/2) \\ \phi(t - T) &= \phi(t - T/2 - T/2) \end{aligned} \tag{3.10}$$

one may write from Eqs. (3.8) and (3.10)

$$e_1 = A(t' + T/2)\alpha_1 \left\{ \sin[\omega t' + \phi(t' + T/2)] \cos\left(\frac{\omega T}{2} + \beta_1\right) + \cos[\omega t' + \phi(t' + T/2)] \sin\left(\frac{\omega T}{2} + \beta_1\right) \right\} \tag{3.11}$$

where
$t' = t - T/2$.

Similarly, from Eqs. (3.9) and (3.10)

$$e_2 = A(t' - T/2)\alpha_2 \left\{ \sin[\omega t' + \phi(t' - T/2)] \cos\left(-\frac{\omega T}{2} + \beta_2\right) + \cos[\omega t' + \phi(t' - T/2)] \sin\left(-\frac{\omega T}{2} + \beta_2\right)\right\}. \quad (3.12)$$

If $T/2$ is assumed so small that

$$A(t' + T/2) \simeq A(t') \simeq A(t' - T/2)$$

and

$$\phi(t' + T/2) \simeq \phi(t') \simeq \phi(t' - T/2) \quad (3.13)$$

then, from Eqs. (3.11)–(3.13), one obtains

$$e_1 + e_2 = A(t')\bar{K} \sin[\omega t' + \phi(t') + \gamma] \quad (3.14)$$

where

$$\bar{K}^2 = \alpha_1^2 + \alpha_2^2 + 2\alpha_1\alpha_2 \cos(\bar{\beta}_1 - \bar{\beta}_2)$$

$$\tan\gamma = \frac{\alpha_1 \sin\bar{\beta}_1 + \alpha_2 \sin\bar{\beta}_2}{\alpha_1 \cos\bar{\beta}_1 + \alpha_2 \cos\bar{\beta}_2}$$

$$\bar{\beta}_1 = \frac{\omega T}{2} + \beta_1$$

$$\bar{\beta}_2 = \frac{-\omega T}{2} + \beta_2.$$

Since $(e_1 + e_2)$ is the output of the network shown in Figure 3-12, one obtains

$$\bar{K} = K$$
$$\gamma = -\omega\delta t \quad (3.15)$$

to achieve the desired K, δt characteristic for the signal controller.

So far, we have kept the signal-controller parameters K_1, K_2, K_3, and K_4 completely arbitrary. Now, let us suppose that each signal controller pair (K_1, K_2) and (K_3, K_4) can be adjusted so that Eq. (3.15) is satisfied at the angular frequencies ω_1 and ω_2 which are at the two extreme ends of the bandwidth of the undesired signal which needs to be cancelled by adaptive cancelling.

From Eqs. (3.14) and (3.15), then, we may write

$$A(t - T/2)\bar{K}' \sin\left[\omega_1 t - \frac{\omega_1 T}{2} + \phi(t - T/2) + \gamma_1\right] = A(t - T/2)K \sin\left[\omega_1 t - \frac{\omega_1 T}{2} + \phi(t - T/2) - \omega_1 \delta t\right]$$

and

$$A(t - T/2)\bar{K}'' \sin\left[\omega_2 t - \frac{\omega_2 T}{2} + \phi(t - T/2) + \gamma_2\right] = A(t - T/2)K \sin\left[\omega_2 t - \frac{\omega_2 T}{2} + \phi(t - T/2) - \omega_2 \delta t\right] \quad (3.16)$$

since $t' = t - T/2$, and since the relations of $A(t')$ and $\phi(t')$ as given in Eq. (3.13) are assumed.

In Eq. (3.16), $\bar{K}'$ and γ_1 are assumed to correspond to $\bar{K}$ and γ in Eq. (3.14) for the angular frequency ω_1. Similarly, $\bar{K}''$ and γ_2 are assumed to correspond to $\bar{K}$ and γ for the angular frequency ω_2. Recalling that

$$\begin{aligned}
\sin\left[\tan^{-1}\frac{K_2}{K_1}\right] &= K_2/\alpha_1 \\
\cos\left[\tan^{-1}\frac{K_2}{K_1}\right] &= K_1/\alpha_1 \\
\sin\left[\tan^{-1}\frac{K_4}{K_3}\right] &= K_4/\alpha_2 \\
\cos\left[\tan^{-1}\frac{K_4}{K_3}\right] &= K_3/\alpha_2
\end{aligned} \quad (3.17)$$

and equating the amplitude and phase on both sides of Eq. (3.16), one obtains

$$\bar{K}' = \bar{K}'' = K \quad (3.18)$$

and

$$\gamma_1 = \tan^{-1}\left\{\frac{\cos\frac{\omega_1 T}{2}[K_2 + K_4] + \sin\frac{\omega_1 T}{2}[K_1 - K_3]}{\cos\frac{\omega_1 T}{2}[K_1 + K_3] + \sin\frac{\omega_1 T}{2}[K_4 - K_2]}\right\} = -\omega_1 \delta t \quad (3.19)$$

and

$$\gamma_2 = \tan^{-1}\left\{\frac{\cos\frac{\omega_2 T}{2}[K_2 + K_4] + \sin\frac{\omega_2 T}{2}[K_1 - K_3]}{\cos\frac{\omega_2 T}{2}[K_1 + K_3] + \sin\frac{\omega_2 T}{2}[K_4 - K_2]}\right\}$$
$$= -\omega_2 \delta t. \tag{3.20}$$

Further simplification of the expressions for γ_1 and γ_2 as given in Eqs. (3.19) and (3.20) yields

$$\gamma_1 = \tan^{-1}\left\{B\frac{\sin\left(\frac{\omega_1 T}{2} + \gamma_1'\right)}{\cos\left(\frac{\omega_1 T}{2} + \gamma_2'\right)}\right\} = -\omega_1 \delta t \tag{3.21}$$

$$\gamma_2 = \tan^{-1}\left\{B\frac{\sin\left(\frac{\omega_2 T}{2} + \gamma_1'\right)}{\cos\left(\frac{\omega_2 T}{2} + \gamma_2'\right)}\right\} = -\omega_2 \delta t \tag{3.22}$$

where

$$B^2 = \frac{(K_1 - K_3)^2 + (K_2 + K_4)^2}{(K_1 + K_3)^2 + (K_2 - K_4)^2}$$

$$\gamma_1' = \tan^{-1}\frac{K_2 + K_4}{K_1 - K_3}$$

$$\gamma_2' = \tan^{-1}\frac{K_2 - K_4}{K_1 + K_3}.$$

From Eqs. (3.21) and (3.22)

$$-\tan(\omega_2 - \omega_1)\delta t = \frac{B\left\{\frac{\sin S_3}{\cos S_4} - \frac{\sin S_1}{\cos S_2}\right\}}{1 + B^2\frac{\sin S_1 \sin S_3}{\cos S_2 \cos S_4}} \tag{3.23}$$

where

$$S_1 = \frac{\omega_1 T}{2} + \gamma_1'$$

$$S_2 = \frac{\omega_1 T}{2} + \gamma_2'$$

$$S_3 = \frac{\omega_2 T}{2} + \gamma_1'$$

$$S_4 = \frac{\omega_2 T}{2} + \gamma_2'.$$

Further simplification of the right-hand side of Eq. (3.23) yields

$$\frac{2B\cos(\gamma_1' - \gamma_2')\sin q}{\cos(p + 2\gamma_2') - B^2\cos(p + 2\gamma_1') + (1 + B^2)\cos q} = -\tan(\omega_2 - \omega_1)\delta t \tag{3.24}$$

where

$$p = \frac{\omega_2 + \omega_1}{2}T$$

$$q = \frac{\omega_2 - \omega_1}{2}T.$$

Further, from Eq. (3.22)

$$\tan\gamma_1' = \frac{K_2 + K_4}{K_1 - K_3} \qquad \tan\gamma_2' = \frac{K_2 - K_4}{K_1 + K_3}$$

$$\sin\gamma_1' = \frac{K_2 + K_4}{\sqrt{(K_1 - K_3)^2 + (K_2 + K_4)^2}}$$

$$\cos\gamma_1' = \frac{K_1 - K_3}{\sqrt{(K_1 - K_3)^2 + (K_2 + K_4)^2}}$$

$$\sin\gamma_2' = \frac{K_2 - K_4}{\sqrt{(K_2 - K_4)^2 + (K_1 + K_3)^2}}$$

$$\cos\gamma_2' = \frac{K_1 + K_3}{\sqrt{(K_2 - K_4)^2 + (K_1 + K_3)^2}}. \tag{3.25}$$

Substituting the results of Eq. (3.25) into Eq. (3.24), one obtains

$$-\tan(\omega_2 - \omega_1)\delta t = \frac{(K_1^2 + K_2^2 - K_3^2 - K_4^2)\sin q}{2\cos p(K_1K_3 + K_2K_4) - 2\sin p(K_2K_3 - K_1K_4) + (K_1^2 + K_2^2 + K_3^2 + K_4^2)\cos q}. \tag{3.26}$$

Thus far, the two pairs of signal controllers are assumed to be independent. That is, the values of the pair K_1 and K_2 are assumed to be unrelated to the values of the pair K_3 and K_4. However, if the operation of each pair of signal controllers, which is capable of adjusting the amplitude and phase of the signal $S(t)$ as given in Eq. (3.7), is considered independently, it is seen that at the midfrequency, $(\omega_2 + \omega_1)/2$, of the frequency band under consideration, both pairs of signal controllers must synthesize the same signal phase at their outputs. Thus, at the midband frequency $(\omega_2 + \omega_1)/2$, $\bar{\beta}_1$ and $\bar{\beta}_2$, as given in Eq. (3.14), will be equal. This requirement relates the values of the pairs (K_1, K_2) and (K_3, K_4).

One may thus write

$$\tan^{-1}\frac{K_2}{K_1} = -\frac{\omega_2+\omega_1}{2}T + \tan^{-1}\frac{K_4}{K_3}. \tag{3.27}$$

But, from Eq. (3.24)

$$\frac{\omega_2+\omega_1}{2}T = p. \tag{3.28}$$

Thus, from Eqs. (3.27) and (3.28)

$$\tan p = -\frac{K_2K_3 - K_1K_4}{K_1K_3 + K_2K_4}. \tag{3.29}$$

Substitution of Eq. (3.29) into (3.26) yields

$$-\tan(\omega_2-\omega_1)\delta t = \frac{(\alpha_1^2-\alpha_2^2)\sin q}{(\alpha_1^2+\alpha_2^2)\cos q + 2\alpha_1\alpha_2}. \tag{3.30}$$

Finally, when $q = (\omega_2 - \omega_1/2)T \ll 1$ and is on the order of $\pi/8$ or less

$$\cos q \simeq 1, \qquad \sin q \simeq q. \tag{3.31}$$

For such cases

$$\begin{aligned}-\tan(\omega_2-\omega_1)\delta t &\simeq \frac{\omega_2-\omega_1}{2}T\cdot\frac{\alpha_1^2-\alpha_2^2}{(\alpha_1+\alpha_2)^2}\\ &\simeq -(\omega_2-\omega_1)\delta t\end{aligned} \tag{3.32}$$

and

$$-\delta t = \frac{T}{2}\left[\frac{\alpha_1-\alpha_2}{\alpha_1+\alpha_2}\right]. \tag{3.33}$$

Thus, when $q \ll 1$ and Eq. (3.31) is satisfied for the requirement, the variable time delay δt synthesized by the arrangement shown in Figure 3-12 becomes independent of the frequency and the values of ω_1 and ω_2. This is precisely what is needed for the signal controller that controls the amplitude and time delay instead of a phase delay. As noted earlier, this characteristic is important for an adaptive cancellation system as the propagation path for the undesired signal or the signal to be cancelled introduces a time delay independent of frequency.

It should be noted from Eq. (3.33) that δt, realizable for the arrangement shown in Figure 3-13, has a range of $\pm T/2$, and that δt can assume all values within this range. This realizability, however, does not mean that an arbitrarily large value of δt could be obtained by merely increasing the fixed time delay T. More specifically, the validity of Eq. (3.33) depends on the validity of Eq. (3.31), that is,

$$\sin q \simeq q = \frac{\omega_2 - \omega_1}{2} T.$$

Since the value of $\sin q$ is approximately the same as that of q when q is on the order of $\pi/8$ radian or less, the maximum value of realizable δt is

$$\delta t_{\max} = \frac{T}{2} \simeq \frac{\pi}{8} \frac{1}{(\omega_2 - \omega_1)}. \tag{3.34}$$

Thus, when the instantaneous bandwidth of the undesired signal or $(\omega_2 - \omega_1)$ is large, the value of $T/2$, and hence the maximum value of variable time control, become limited in accordance with Eq. (3.34). The effect of instantaneous bandwidth $(\omega_2 - \omega_1)$ on the variable time control δt is further illustrated in Figure 3-14. Here, the normalized δt with respect to the fixed delay $T/2$ is plotted against

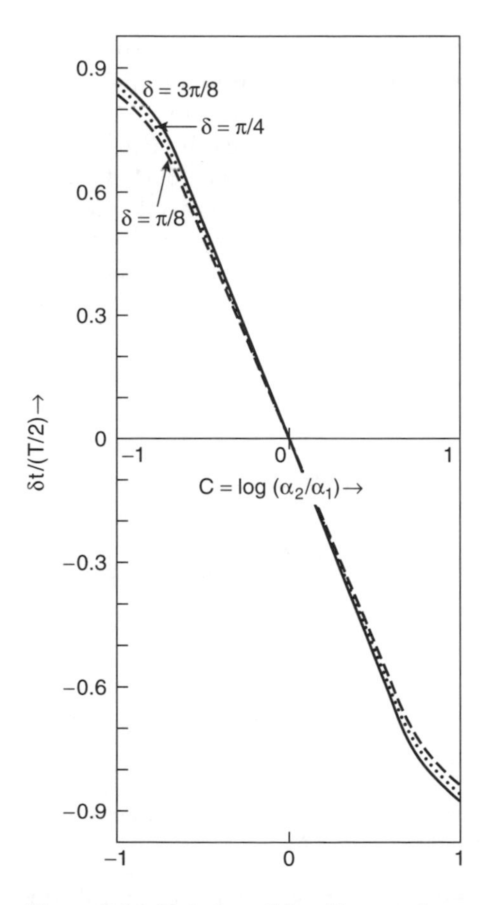

Figure 3-14 Variation of δt with α_1 and α_2.

log C where C is α_2/α_1. For a given value of $T/2$, q varies with the interference bandwidth $(\omega_2 - \omega_1)$. As shown in the figure, δt varies linearly with log C, except when α_1 and α_2 assume a value approaching zero. For such cases, the variation of δt is no longer linear, and the nature of variation depends on the value of q.

When the assumption in Eq. (3.33) holds, that is, when the product of the interference bandwidth $(\omega_2 - \omega_1)$ and $T/2$ is less than or equal to $\pi/8$, the variable time that can be synthesized at the signal controller is independent of interference frequency. In contrast, for a signal controller which is capable of adjusting the phase and of providing one phase delay only, the equivalent time delay for this fixed phase ϕ changes as ϕ/ω, which is different for different frequencies within the interference frequency band.

3.4 SYNCHRONOUS DETECTORS

An important element of the closed-loop control associated with adaptive interference cancellation is the synchronous detector that provides the necessary control signals for the signal controller. Such control signals are created by correlating the error signal with the interference to be cancelled. A schematic representation of a synchronous detector is shown in Figure 3-15. Here, $f(t)$ and $g(t)$ denote, respectively, the reference port and error signals. In one type of synchronous detector, the error signal is switched by the zero crossing of the reference port signal, which is the interference to be cancelled. For such a case, a simple mathematical model, that relates the synchronous detector output $e(t)$ with $f(t)$ and $g(t)$, may be written as

$$\begin{aligned} e(t) &= g(t) \qquad \text{when } f(t) \geq 0 \\ &= -g(t) \qquad \text{when } f(t) < 0. \end{aligned} \tag{3.35}$$

This relation may also be expressed as

$$e(t) = g(t)\frac{f(t)}{|f(t)|}. \tag{3.36}$$

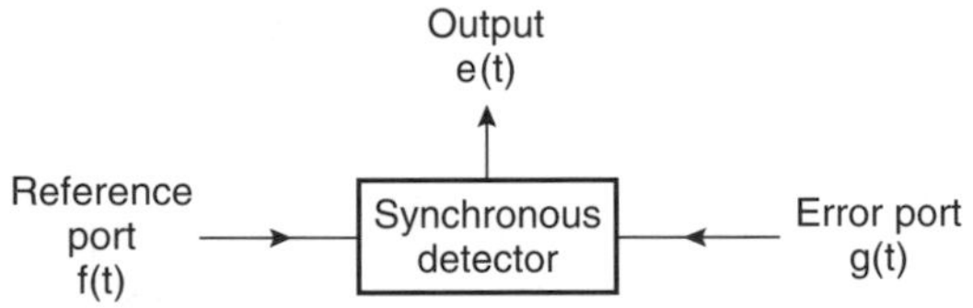

Output
e(t)
Reference
port
f(t)
Synchronous
detector
Error port
g(t)

Figure 3-15 Schematic representation of a synchronous detector.

In general, the reference port signal $f(t)$, which is also the interference sought to be cancelled, is kept at a much higher level than the maximum expected magnitude of the error signal, that is, $g(t)$.

To further examine the operation of a synchronous detector, let us assume that the reference port signal is an amplitude-modulated signal of the form

$$f(t) = A(t)\sin\omega t \tag{3.37}$$

where
A = the time-varying amplitude
ω = is the angular frequency.

Let the error port signal $g(t)$ denote the difference of the two interferences of the same origin as $f(t)$ and a desired signal such that

$$\begin{aligned} g(t) &= KA(t)\sin(\omega t - \phi) - K_1 A(t)\sin(\omega t - \theta) \\ &\quad + K_D B(t)\sin(\omega_D t - \phi_D) \end{aligned} \tag{3.38}$$

where the amplitude factor K and phase delay ϕ characterize one interference, while the amplitude factor K_1 and phase delay θ characterize the second interference, the angular frequency ω for both interferences being the same as that of $f(t)$. Also, the desired signal, which is not to be cancelled, is assumed to be uncorrelated with either interference, having a different time-varying amplitude $B(t)$ and a different angular frequency ω_D.

Let us consider the RF cycle of $f(t)$, as given in Eq. (3.37), beginning with $t = 0$. During the period when ωt changes from 0 to π, $f(t) \geq 0$, and hence from Eq. (3.35), the output signal of the synchronous detector $e(t)$ is the same as $g(t)$. Similarly, during the period when ωt changes from π to 2π, $f(t) < 0$, and hence $e(t) = -g(t)$. For the period covering the first RF cycle, the output of the synchronous detector which represents the integral of $e(t)$ over this period may be written as

$$\begin{aligned} \bar{e}_1(t) &= \int g(t)\frac{f(t)}{|f(t)|}\,dt \\ &= \omega K \int_0^{\pi/\omega} A(t)\sin(\omega t - \phi)\,dt - \omega K_1 \int_0^{\pi/\omega} A(t)\sin(\omega t - \theta)\,dt \\ &\quad + K_D\omega \int_0^{\pi/\omega} B(t)\sin(\omega_D t - \phi_D)\,dt \\ &\quad - \omega K \int_{\pi/\omega}^{2\pi/\omega} A(t)\sin(\omega t - \phi)\,dt + \omega K_1 \int_{\pi/\omega}^{2\pi/\omega} A(t)\sin(\omega t - \theta)\,dt \\ &\quad - K_D \int_{\pi/\omega}^{2\pi/\omega} B(t)\sin(\omega_D t - \phi_D)\,dt. \end{aligned} \tag{3.39}$$

Assuming that the time-varying amplitudes of the interference $A(t)$ and that of the desired signal $B(t)$ do not change within one RF cycle, one may write from Eq. (3.39)

$$\bar{e}_1(t) = 4A(t)[K\cos\phi - K_1\cos\theta] + \frac{\omega}{\omega_D}K_D B(t)\left[\cos\phi_D - 2\cos\left(\frac{\omega_D}{\omega}\pi - \phi_D\right) + \cos\left(\frac{2\omega_D}{\omega}\pi - \phi_D\right)\right] \quad (3.40)$$

Similarly, for the next RF cycle when ωt varies from 2π to 4π, the integrated output of the synchronous detector can be written as

$$\bar{e}_2(t) \simeq 4A(t)[K\cos\phi - K_1\cos\theta] + \frac{\omega}{\omega_D}K_D B(t)\left[\cos\left(2\frac{\omega_D}{\omega}\pi - \phi_D\right) - 2\cos\left(\frac{3\pi\omega_D}{\omega} - \phi_D\right) + \cos\left(\frac{4\pi\omega_D}{\omega} - \phi_D\right)\right]. \quad (3.41)$$

For the first M number of RF cycles where $A(t)$ and $B(t)$ do not change appreciably so as to permit the same approximation used in Eqs. (3.40) and (3.41), one may write

$$\begin{aligned}\bar{e}(t) &= \sum_M \bar{e}_1(t) \\ &= 4MA(t)[K\cos\phi - K_1\cos\theta] \\ &\quad + \frac{\omega}{\omega_D}K_D B(t)\sum_{n=0}^{2M}\epsilon_n\cos\left(n\pi\left[1+\frac{\omega_D}{\omega}\right] - \phi_D\right) \end{aligned} \quad (3.42)$$

where

$$\begin{aligned}\epsilon_n &= 1 \quad \text{for } n = 0 \\ &= 2 \quad \text{for } n \neq 0.\end{aligned}$$

Thus, it is seen from Eq. (3.42) that if the output of the synchronous detector is integrated over a number of RF cycles of the interference, the first part of Eq. (3.42) involving the factor $(K\cos\phi - K_1\cos\theta)$ increases linearly with M, but the second part averages toward zero. For the closed-loop control, the first part of the integrated synchronous detector output is used to generate the control signals for the signal controllers.

3.5 CLOSED-LOOP CONTROL

An essential aspect of adaptive interference cancelling is the closed-loop control, the signal controller and the synchronous detector, as discussed in the preceding sections, being the essential elements of this closed loop. To further illustrate the operation of the closed-loop control, we may refer to Figure 3-16, where e_1 denotes a sample of an interference. To simplify the illustration of the closed-loop control, we assume that this interference is amplitude modulated and is of the form

$$e_1 = A(t) \sin \omega_1 t. \tag{3.43}$$

A being a time-varying amplitude. We further assume that the input signal e_i of the receiving antenna contains the interference and a desired signal such that

$$e_i = K_i A(t) \sin(\omega_1 t - \phi_{i1}) + R_i B(t) \sin(\omega_2 t - \phi_{i2}) \tag{3.44}$$

where

K_i = an arbitrary amplitude
ϕ_{i1} = an arbitrary phase factor of the interference
R_i and ϕ_{i2} = an abritrary amplitude and an arbitrary phase factor of the desired signal, respectively, the desired signal being of the form $e_D = B(t) \sin \omega_2 t$
B = a time-varying amplitude.

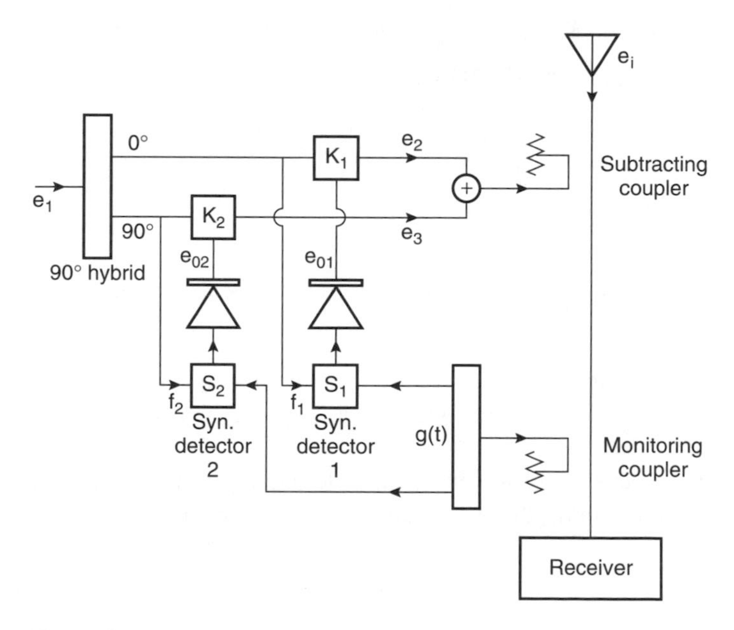

Figure 3-16 Illustration of closed-loop control to cancel the interference e_i automatically even when the characteristics of e_1 and e_i are not *a prior* known.

As shown in Figure 3-16, the sampled interference e_i is led to a 90° hybrid to obtain its two quadratured components at the 0° and 90° output ports of the hybrid. Each of the hybrid outputs is led to a signal controller, capable of altering the amplitude of the input signal at its output, as discussed in Section 3.2. The outputs of the two signal controllers may then be written as

$$\begin{aligned} e_2 &= K_1 A(t)\sin(\omega_1 t) \\ e_3 &= K_2 A(t)\sin\left(\omega_1 t + \frac{\pi}{2}\right) \\ &= K_2 A(t)\cos(\omega_1 t) \end{aligned} \tag{3.45}$$

where
K_1 and K_2 = the two amplitude factors introduced by the two signal controllers.

The synchronous detector S_1, similar to the one discussed in Section 3.3, is referenced to the 0° component, while the second synchronous detector S_2 is referenced to the 90° component of e_1, as shown in the figure.

The outputs of the signal controllers are subtracted from the received signal e_i such that the residual signal becomes

$$\begin{aligned} e_e = {} & K_i A(t)\sin(\omega_1 t - \phi_{i1}) - K A(t)\sin(\omega_1 t - \phi) \\ & + R_i B(t)\sin(\omega_2 t - \phi_{i2}) + n(t) \end{aligned} \tag{3.46}$$

where

$K^2 = K_1^2 + K_2^2$

$\tan\phi = \dfrac{K_2}{K_1}$

$n(t)$ = some additive noise uncorrelated with both the desired signal and the interference.

The residual signal is often referred to as an error signal, and a sample of the error signal becomes the input signal of each synchronous detector S_1 or S_2, as shown in the figure.

If we use the suffix 1 to denote this relevant signal input and output of the synchronous detector S_1, then based on the notation used in Eqs. (3.37) and (3.38)

$$\begin{aligned} f_1 &= A(t)\sin\omega_1 t \\ g_1 &= K_c e_e \end{aligned} \tag{3.47}$$

where

K_c = a coupling factor introduced while sampling the error signal.

From the characteristic of the synchronous detector as given in Eq. (3.36), the integral of output e_{01} of the synchronous detector S_1 when $\omega_1 t$ varies from 0 to 2π

becomes

$$
\begin{aligned}
e_{01} &= \omega_1 \int_0^{\pi/\omega_1} g_1\, dt - \omega_1 \int_{\pi/\omega_1}^{2\pi/\omega_1} g_1\, dt \\
&= \omega_1 K_c \left\{ A(t) \int_0^{\pi/\omega_1} [K_i \sin(\omega_1 t - \phi_{i1}) - K \sin(\omega_1 t - \phi)]\, dt \right. \\
&\quad + B(t) \int_0^{\pi/\omega_1} R_i \sin(\omega_2 t - \phi_{i2})\, dt + \int_0^{\frac{\pi}{\omega_1}} n(t)\, dt \\
&\quad - A(t) \int_{\frac{\pi}{\omega_1}}^{2\frac{\pi}{\omega_1}} [K_i \sin(\omega_1 t - \phi_{i1}) - K \sin(\omega_1 t - \phi)]\, dt \\
&\quad \left. - B(t) \int_{\frac{\pi}{\omega_1}}^{2\frac{\pi}{\omega_1}} R_i \sin(\omega_2 t - \phi_{i2})\, dt - \int_{\frac{\pi}{\omega_1}}^{2\frac{\pi}{\omega_1}} n(t)\, dt \right\}.
\end{aligned}
\tag{3.48}
$$

Similarly, for the synchronous detector S_2, the reference interference is f_2, the error signal is g_2, and the integral of the output e_{02}, when $\omega_1 t$ varies from 0 to 2π, can be written as

$$
\begin{aligned}
f_2 &= A(t) \cos \omega_1 t \\
g_2 &= g_1
\end{aligned}
\tag{3.49}
$$

and

$$
e_{02} = \omega_1 \int_0^{\frac{\pi}{2\omega_1}} g_2\, dt - \omega_1 \int_{\frac{\pi}{2\omega_1}}^{\frac{\pi}{\omega_1}} g_2\, dt - \omega_1 \int_{\frac{\pi}{\omega_1}}^{\frac{3\pi}{2\omega_1}} g_2\, dt + \omega_1 \int_{\frac{3\pi}{2\omega_1}}^{\frac{2\pi}{\omega_1}} g_2\, dt \tag{3.50}
$$

based on the characteristic of the synchronous detector as given in Eqs. (3.35) and (3.36).

When the integrals of the output of the synchronous detectors S_1 and S_2 are carried out over several RF cycles of $\omega_1 t$, the integrals involving $R_i B(t)$ $\sin(\omega_2 t - \phi_{i2})$ vanish because of the randomness of the values of such integrals over several cycles, as noted in Section 3.3.

The integrals involving $n(t)$ also vanish because of the same reason. Thus, when the integrals in Eq. (3.48) are evaluated over M number of RF cycles of $\omega_1 t$, one obtains for $M \gg 1$

$$
e_{01} = 4MK_c(K_i \cos \phi_{i1} - K \cos \phi). \tag{3.51}
$$

Similarly, when the integrals in Eq. (3.50) are evaluated over M number of RF cycles of $\omega_1 t$, one obtains

$$e_{02} = 4MK_c(K_i \sin\phi_{i1} - K \sin\phi). \tag{3.52}$$

The signal e_{01} becomes the control signal for the signal controller 1 that introduces a gain (loss) factor K_1. Similarly, the signal e_{02} becomes the control signal for the signal controller 2 that introduces the gain (loss) factor K_2. Thus, as long as e_{01} is not zero, the gain (loss) factor K_1 is continuously changed, and as long as e_{02} is not zero, the gain (loss) factor K_2 is continuously changed until the values of K_1 and K_2 are such that e_{01} and e_{02} become zero simultaneously. For such conditions, the operation of the closed loop, comprising the signal controllers 1 and 2, the coupler that subtracts the signal controller outputs from the received signal e_i, the monitoring coupler, and the two synchronous detectors, comes to an equilibrium requiring no further variation of the gain (loss) factors K_1 and K_2. For the same conditions, the interference is eliminated from the receive line. The operation of the closed loop therefore automatically cancels the interference at the receive line, even when the interference parameters K_i and ϕ_{i1} are not *a priori* known, and even when their values change with time. The dynamics of the closed-loop operation and the convergence of solution for cancellation need to be discussed further. Before that, however, some considerations on the physical realizability of the essential components that make up the loop will be desirable.

The physical realizability of a signal controller employing p-i-n diodes and associated bias circuits and a combination of signal controllers and a fixed delay line to provide a variable time delay have been considered in two previous sections. The practical implementation of a synchronous detector, as indicated by the mathematical model of one type of detector, outlined earlier in this section, turns out to be relatively simple. More specifically, a balanced modulator commonly used for amplitude modulation may serve the purpose of a synchronous detector.

A balanced modulator is a three-port nonlinear device where the ports are usually referred to as the local oscillator (LO) port, the radio frequency (RF) port, and the intermediate frequency (IF) port. When the signal voltage at the LO port is fixed, the IF port signal level at the balanced modulator usually increases in magnitude with increasing RF port signal level. Also, the frequency of the IF port signal commonly used is the difference of the RF and LO port signal frequencies. A nonlinearity in the form of multiplication of signals causes this difference-frequency signal to appear at the IF port. Now, if the RF port signal frequency is changed gradually until it becomes the same as the LO port signal frequency, the IF port signal frequency will approach zero, corresponding to dc signals. The magnitude of the dc signals at the IF port also will increase with the increasing RF port signal level. This is exactly the situation that occurs at a synchronous detector where the LO and RF ports of the balanced modulator correspond to the reference and error ports of the synchronous detector, respectively. When the reference port or the LO port signal is kept at a high level, the signal level at the output port of the synchronous detector, corresponding to the IF port of the balanced modulator, will increase with the increase of error or RF port signal level.

It is interesting to note that, physically, a balanced modulator may also serve the purpose of a signal controller where one needs to control the level of the output signal by varying the dc control signal. Again, if one assumes that the frequency of the LO port signal of the balanced modulator is varied until it approaches zero, corresponding to a dc signal, the IF port signal of the balanced modulator must have a frequency which is the difference between the frequencies of the signals at the LO and RF ports. In this case, as the LO port signal frequency approaches zero, the IF port signal frequency will be the same as that of the RF port signal. Also, from the inherent signal multiplication characteristic of the balanced modulator, the IF port signal level will increase as the LO port signal level increases, at least for the linear operating part of the modulator. This is precisely the characteristic needed in a signal controller where the output should be proportional to the dc control voltage, at least for the linear operating part of the controller.

There are indeed other ways to physically construct a synchronous detector or a signal controller that follow the same mathematical models discussed earlier.

Now, returning to the consideration of the control loop operation, it is seen in Figure 3-16 that the control signal e_{01} is used to change the gain (loss) K_1 of the signal controller 1, and the control e_{02} is used to change the gain (loss) K_2 of the signal controller 2. If the closed-loop control corresponds to a velocity-type servo system, one may write the characteristic equations governing the dynamics of K_1 and K_2 as

$$\frac{dK_1}{dt} = M_1 e_{01} \quad \text{and} \quad \frac{dK_2}{dt} = M_2 e_{02} \tag{3.53}$$

where
M_1 = a constant corresponding to the 0° loop gain
M_2 = a constant corresponding to the 90° loop gain.

Thus, from Eqs. (3.51) and (3.53)

$$\begin{aligned} \frac{dK_1}{dt} &= M_1 M'(K_i \cos\phi_{i1} - K\cos\phi) \\ M' &= 4MK_c. \end{aligned} \tag{3.54}$$

Similarly, from Eqs. (3.52) and (3.53)

$$\frac{dK_2}{dt} = M_2 M'(K_i \sin\phi_{i1} - K\sin\phi). \tag{3.55}$$

Now, since

$$K^2 = K_1^2 + K_2^2 \quad \text{and} \quad \tan\phi = \frac{K_2}{K_1}$$
$$K_1 = K\cos\phi \quad \text{and} \quad K_2 = K\sin\phi. \tag{3.56}$$

From Eqs. (3.54) and (3.56)

$$\frac{dK_1}{K_1 - K_i\cos\phi_{i1}} = -M_1 M'\,dt \tag{3.57}$$

and from Eqs. (3.55) and (3.56)

$$\frac{dK_2}{K_2 - K_1\sin\phi_{i1}} = -M_2 M'\,dt. \tag{3.58}$$

It should now be recalled that $K_i\cos\phi_{i1}$ and $K_i\sin\phi_{i1}$ are related to the propagation path geometry of the interference to be cancelled. If they are constants or if they vary slowly with respect to time, as is most often the case, one obtains the solutions for K_1 and K_2 by integrating both sides of Eqs.(3.57) and (3.58) respectively. Thus

$$K_1 = K_i\cos\phi_{i1} + A_1 e^{-M_1 M' t} \tag{3.59}$$

and

$$K_2 = K_i\sin\phi_{i1} + A_2 e^{-M_2 M' t} \tag{3.60}$$

where
A_1 and A_2 = constants depending on the boundary condition.

It is evident from Eqs. (3.59) and (3.60), however, that the steady-state solutions for K_1 and K_2, as obtained when $t \to \infty$, are

$$K_1 = K_i\cos\phi_{i1}$$
$$K_2 = K_i\sin\phi_{i1}. \tag{3.61}$$

Also, for this model of the closed-loop control, the convergence toward the steady-state solutions is assured, particularly when the effective loop gains $M_1 M'$ and $M_2 M'$ are very large.

It may be remarked that for the transient variations of K_1 and K_2, one has to take into account the exact expressions for e_{01} and e_{02} that result from the outputs of the filters that follow the synchronous detectors as shown in Figure 3-16, thus taking into account the characteristics of the filters. Also, to evaluate the performance of the interference cancellation system due to the closed-loop control, the presence of the desired signal, which should remain uneffected, and the noise have to be taken into account. Analyses involving most of the relevant factors for the specific cosite and remote interferences are considered later in this text.

3.6 ADAPTIVE CANCELLATION OF CERTAIN STATISTICAL NOISE

So far, we have considered the problems of adaptive cancellation of one or more discrete interferences. Widrow *et al.* [1] have suggested the cancellation of certain statistical noise from the optimal unconstrained Wiener solutions employing adaptive transversal filters. The idealized solutions do not take into account the issues of finite filter length or causality, which are important in practical applications. However, means approximating optimal unconstrained Wiener performance are available using physically realizable adaptive filters.

As in the case of discrete interference cancellation, fixed filters are also, for the most part, not applicable for the cancellation of statistical noise because the correlation and cross-correlation functions of input and reference signals (or noise) are generally unknown, and they are often variable with time. Adaptive filters can "learn" the statistics initially and track if they vary slowly.

For stationary stochastic inputs, however, the steady-state performance of adaptive filters closely approximates that of fixed Wiener filters. Thus, the Wiener filter theory provides a convenient method of mathematically analyzing certain statistical noise cancellation problems.

Figure 3-17 shows an example of a single-input, single-output Wiener filter where the input and output signals are digital signals denoted by x_j and y_j, respectively, the desired response being d_j. They represent the jth elements in the time series represented by x, y, and d, respectively. Usually, the time series may be assumed to be obtained by sampling continuous signals. Thus, $x_j = x(jT)$, and so on, where T is the time step or interval between samples. The error signal in this case is $\epsilon_j = d_j - y_j$. The filter is linear, discrete, and designed to be optimal in the minimum mean-square sense.

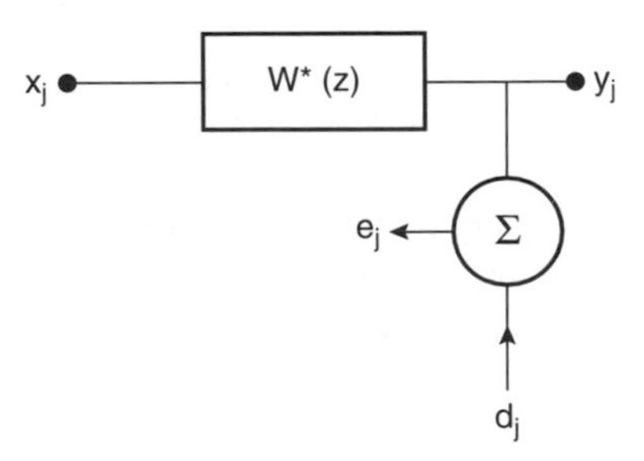

Figure 3-17 Schematic representation of a single-channel Wiener filter.

Let the discrete autocorrelation function of the input signal be defined as

$$\phi_{xx} = E[x_j \, x_{j+k}] \tag{3.62}$$

where E denotes the expected value.

Similarly, let the cross-correlation function between x_j and the desired response be defined as

$$\phi_{xd} = E[x_j \, d_{j+k}]. \tag{3.63}$$

The optimal impulse response $w^*(k)$ of the Wiener filter is obtained from the discrete Wiener–Hopf equation [1].

$$\sum_{\ell=-\infty}^{\infty} w^*(\ell)\phi_{xx}(k-\ell) = \phi_{xd}(k). \tag{3.64}$$

The convolution of Eq. (3.64) can be written as

$$w^*(k) \cdot \phi_{xx}(k) = \phi_{xd}(k). \tag{3.65}$$

The transfer function of this Wiener filter may now be obtained from the ratio of the transform of the output to the transform of the input. With continuous systems, one usually uses the Laplace transform to obtain the transfer function. With digital systems, as is being considered now, one uses the Z-transform. The power-density spectrum of the input signal is the Z-transform of ϕ_{xx}. Thus

$$S_{xx}(z) = \sum_{k=-\infty}^{\infty} \phi_{xx}(k)z^{-k} \tag{3.66}$$

and the transfer function of the Wiener filter is

$$\bar{w}(z) = \sum w^*(k)z^{-k}. \tag{3.67}$$

Transforming Eq. (3.64) yields the optimum unconstrained Wiener transfer function

$$\bar{w}(z) = \frac{S_{xd}(z)}{S_{xx}(z)}. \tag{3.68}$$

Based on the Wiener filter theory outlined above, one may now consider the adaptive noise cancellation problem. Figure 3-18 shows a single-channel noise canceller where the primary input is a desired signal S_j and two noises N_{0j} and $I_j \cdot h(j)$ where $h(j)$ is the impulse response of the channel whose transfer function is $H(z)$. The noises I_j and $I_j \cdot h(j)$ have a common origin. They are correlated with each other, but are uncorrelated with the desired signal S_j. Additionally, they are assumed to have a finite power spectrum at all frequencies. The noises N_{0j} and N_{1j} are uncorrelated with each other and with the desired signal S_j. They are also uncorrelated with the noises I_j and $I_j \cdot h(j)$. For the purpose of analysis, all noise propagation paths are assumed to be equivalent to linear time-variant filters.

Since the input of the filter x_j consists of the two noises, N_{1j} and $I_j \cdot h(j)$, the spectrum of the filter's input can be expressed as the sum of the spectra of the

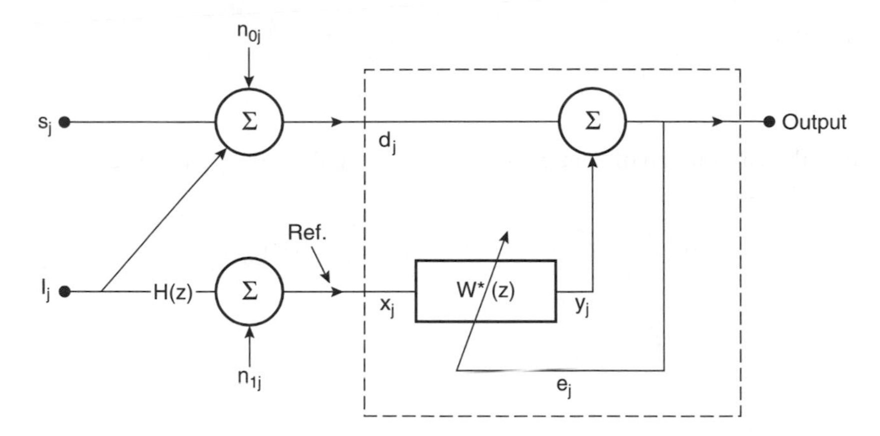

Figure 3-18 Illustration of the concept of adaptive noise canceller.

two mutually uncorrelated components. Thus, the filter's input spectrum S_{xx} is

$$S_{xx} = S_{n_1 n_1}(z) + S_{II}(z)|H(z)|^2. \tag{3.69}$$

The cross-power spectrum between the filter's input and the desired response depends only on the mutually correlated primary and reference components, and is given by

$$S_{xd} = S_{II}(z)H(z^{-1}). \tag{3.70}$$

From Eqs. (3.68)–(3.70), the Wiener transfer function is

$$W^*(z) = \frac{S_{II}(z)H(z^{-1})}{S_{n_1 n_1}(z) + S_{II}(z)|H)z)|^2}. \tag{3.71}$$

It may be noted that this Wiener transfer function is independent of the primary signal spectrum $S_{ss}(z)$ and of the primary suncorrelated noise spectrum $S_{n_o n_o}(z)$. Furthermore, in the absence of the additive noise N_{1j}, the transfer function is

$$W^*(z) = \frac{1}{H(z)}. \tag{3.72}$$

For such a condition, the adaptive filter completely cancels the noise I_j.

One may compare the result in Eq. (3.72), as obtained from the Wiener transfer function, with the transfer function of a signal controller, discussed earlier, for a continuous signal or interference. Figure 3-17 shows such a canceller schematically.

Consistent with the notation used for a signal controller, let $I(t)$ and $KI(t-T)$ denote, respectively, a continuous interference at the reference and a primary or receive line, as shown in Figure 3-17. The autocorrelation function of $I(t)$ can be written as

$$\phi_a = \text{Lim}_{\tau=0} \int_{-\infty}^{\infty} I(t)I(t+\tau)\,dt \tag{3.73}$$

Now, the convolution of the two functions $S(w)$ and $G(\omega)$ as a function of ω can be written as

$$(1/2\pi) \int_{-\infty}^{\infty} S(\omega)G(\omega)e^{-j\omega t}\,d\omega = \int_{-\infty}^{\infty} s(t_1)g(t-t_1)\,dt_1 \tag{3.74}$$

where

$$S(\omega) = \int_{-\infty}^{\infty} s(t)e^{j\omega t}\,dt$$

$$G(\omega) = \int_{-\infty}^{\infty} g(t)e^{j\omega t}\,dt$$

Substituting $(t_1 - \tau)$ for t in Eq. (3.74), one may write

$$G(\omega) = g(t_1 - \tau)e^{j\omega(t_1-\tau)}\,dt_1 \tag{3.75}$$

And, when $g(t)$ is real, the complex conjugate of $G(\omega)$ can be written as

$$G^*(\omega) = \int_{-\infty}^{\infty} g(t_1 - \tau)e^{-j\omega(t_1-\tau)}\,dt_1 \tag{3.76}$$

Again, substituting $-t_1$ for t_1 in Eq. (3.76), one obtains

$$G^*(\omega)e^{-j\omega\tau} = -\int_{-\infty}^{\infty} g(-t_1 - \tau)e^{j\omega t_1}\,dt_1 \tag{3.77}$$

Now, from Eq. (3.74) and (3.77), one may write

$$(1/2\pi) \int_{-\infty}^{\infty} S(\omega)G^*(\omega)e^{-j\omega\tau}e^{-j\omega t}\,d\omega = \int_{-\infty}^{\infty} s(t_1)g(t+t_1+\tau)\,dt_1 \tag{3.78}$$

and, by setting $t = 0$, results

$$(1/2\pi) \int_{-\infty}^{\infty} S(\omega)G^*(\omega)e^{-j\omega\tau}\,d\omega = \int_{-\infty}^{\infty} s(t_1)g(t_1+\tau)\,dt_1 \tag{3.79}$$

Substituting $s(t) = g(t) = I(t)$, one obtains from Eqs. (3.73) and (3.79)

$$\begin{aligned}\phi_a &= \text{Lim}_{\tau=0} \int_{-\infty}^{\infty} I(t)I(t+\tau)\,dt \\ &= (1/2\pi) \int_{-\infty}^{\infty} |I(\omega)|^2\,d\omega,\end{aligned} \tag{3.80}$$

the transform of ϕ_a in the ω-plane being

$$\phi_a(\omega) = |I(\omega)|^2 \tag{3.81}$$

Similarly, the cross-correlation function of the primary and reference line interferences is

$$\begin{aligned}\phi_c &= \mathrm{Lim}_{\tau\to 0} K \int_{-\infty}^{\infty} I(t-T)I(t+\tau)\,dt \\ &= \mathrm{Lim}_{\tau\to 0}(K/2\pi)\int_{-\infty}^{\infty} I(\omega)I^*(\omega)e^{-j\omega(t+T)}\,d\omega \end{aligned} \quad (3.82)$$

the transform of ϕ_c in the ω-plane being

$$\phi_c = K|I(\omega)|^2 e^{-j\omega T} \quad (3.83)$$

The Wiener transfer function for the optimal filter, then, is the ratio $\phi_c(\omega)/\phi_a(\omega)$. Thus,

$$\begin{aligned} W(\omega) &= K[|I(\omega)|^2/|I(\omega)|^2]e^{-j\omega T} \\ &= Ke^{-j\omega T} \end{aligned} \quad (3.84)$$

From Figure 3-17 and Eq. (3.84), or Eq. (3.72), it is seen that the transfer function for the optimal filter is the ratio of the transfer function for the interference-propagation path, from its source to the primary input line, to the transfer function for the propagation path for the same interference from its source to the reference line.

It should be noted that Eq. (3.84) results directly from the optimization of the weighting function $W(\omega)$, based on the least mean square (LMS) of the error signal, which, as defined earlier, is the difference between the interference at the receive line and the synthesized counterinterference obtained from the reference interference flowing through the weighting function network. More specifically, if $I(\omega)$ and $I'(\omega)$ denote, respectively, the transforms of a continuous interference at the reference and receive lines, then, by definition, the error signal ϵ is

$$\epsilon = I'(\omega) - I(\omega)W(\omega) \quad (3.85)$$

Hence

$$\epsilon^2 = I'^2(\omega) + I^2(\omega)W^2(\omega) - 2I(\omega)I'(\omega)W(\omega) \quad (3.86)$$

The optimum value of W, that which results in the least mean value of ϵ^2, is obtained by differentiating ϵ^2 with respect to W and setting the result to zero. Thus,

$$\begin{aligned} (d\epsilon^2/dW) &= 2I^2(\omega)W(\omega) - 2I(\omega)I'(\omega) \\ &= 0 \end{aligned} \quad (3.87)$$

or,

$$W(\omega) = [I(\omega)I'(\omega)]/I^2(\omega) \quad (3.88)$$

Obviously, when $I'(\omega) = KI(\omega)e^{-j\omega T}$, corresponding to the time-domain interference, $KI(t - T)$, as assumed in Figure 3-17, one obtains from Eq. (3.88)

$$W^* = W(\omega) = Ke^{-j\omega T} \tag{3.89}$$

REFERENCES

[1] B. Widrow, J. Glover, Jr., J. McCool, J. Kaunitz, C. Williams, R. Hearn, J. Zeidler, E. Dong, Jr., and R. Goodlin, "Adaptive noise cancelling: Principles and applications," *Proc. IEEE*, vol. 63, pp. 1692–1716, Dec. 1975.

[2] R. Ghose and W. Sauter, "Interference cancellation system," U.S. Patent 3,699,444, Oct. 1972.

[3] J. McCool, "The basic principles of adaptive systems with various applications," Naval Undersea Center Rep., Sept. 1972.

[4] W. Sauter and D. Martin, "Reflective mode signal controller," U.S. Patent 4,016,516, Apr. 1977.

[5] R. Ghose, "Analysis of variable time delay system," Technology Research International Note WP150003, Feb. 1985.

[6] R. Ghose, "Degree of cancellation potential with ICS," Technology Research International Note WP150002, Feb. 1985.

[7] F. Kretschmer, Jr. and B. Lewis, "A digital open-loop adaptive processor," *IEEE Trans. Aerospace Electron. Syst.*, vol. AES-14, Jan. 1978.

CANCELLATION OF INTERFERENCES FROM COLLOCATED SOURCES

The approaches suitable or usable for adaptive interference cancellation differ significantly, depending on whether the source of the interference is collocated with the victim receiver or is one of remote origin. In the case of a collocated interference, a "good" or "pure" reference is readily accessible, and the signal- controller input could be made free of the desired signal, thus avoiding the possibility of degrading the desired signal as received by the receiving antenna due to a contribution from the interference cancellation process. Characteristics of an adaptive canceller for a collocated interference source and the expected degree of cancellation as a function of various relevant canceller parameters are discussed in this chapter.

4.1 BASIC CONCEPT AND CANCELLER CHARACTERIZATION

When an interference cancellation system is used to suppress an interference at the receive line from a collocated source, usually a sample of the interference from its source transmitter is used as an input to the signal controller, the output of which yields the synthesized cancelling signal. We have already considered some examples of collocated interference sources in Chapter 1. In this chapter, we will first examine the factors that influence the degree of cancellation by an adaptive canceller in general, and then in particular consider how the performance may differ for collocated interference sources.

From the principle of adaptive interference cancellation as discussed in the previous chapters, one may expect that the more "pure" the sample of interference at the reference line is, the higher will be the degree of cancellation. The other signals, desired or undesired, or noise present at the reference line will not only be affected by the correlation process at the synchronous detectors, but will cause unwanted signals and noise to be injected at the receive line through the signal controller and the 180° hybrid, as shown in Figure 2-11. When a sample of the interference which needs to be cancelled at the receive line can be obtained by coupling the reference line directly to the interference transmitter line, one obtains a reasonably "pure" interference as a reference as needed for a high degree of cancellation since other signals or noise will not be present at the interference transmitter line. This is possible only for a collocated interference source because of the accessibility of the interference source. Just exactly how much of a gain in the degree of cancellation is realizable for a direct-coupled reference line can be determined by examining the general case first, where the presence of other signals or noise is assumed at the reference line along with the interference of concern.

It is readily recognized from the physical cancellation principle that if the cancellation of interference effected by an adaptive canceller has to be selective without degrading the desired signal, there must be means to discriminate the desired and undesired signals in the control process. Since a reference antenna used for obtaining a sample of the interference can receive both desired and undesired signals, it is to be expected that the performance of the canceller would, in some form, depend on the relative levels of the desired signal and the interference at the reference antenna. To obtain a quantitative measure of the steady-state cancellation for a canceller when the reference line contains both the desired and undesired signals in any arbitrary proportion, one can formulate a model which essentially describes the operation of the canceller under such circumstances for further study.

4.2 OPERATION OF AN ADAPTIVE CANCELLER

Figure 4-1 shows the schematic arrangement of an adaptive canceller, where the sample of interference is obtained by an auxiliary antenna so that it may contain both the desired signal and the interference. Such a situation occurs frequently in real life where a receiver is collocated with an interfering transmitter operating at the same frequency band, but it is not convenient or desirable to obtain a direct sample from the interference transmitter as considered in Figure 3-1. For the purpose of simplification of the analysis, it is assumed that only one desired signal and one interference are present, and the impact of both external and internal noise is ignored. Let S_1 and S_2 denote, respectively, the undesired and desired signals at the reference antenna. With no loss of generality, the corresponding signals at the receiving antenna may be expressed as

$$S_R = L_1 S_1 + L_2 \bar{S}_1 + L_3 S_2 + L_4 \bar{S}_2 \tag{4.1}$$

where

$L_1, L_2, L_3, L_4,$ = arbitrary constants

$\bar{S}_1, \bar{S}_2$ = the quadrature components of S_1 and S_2, respectively.

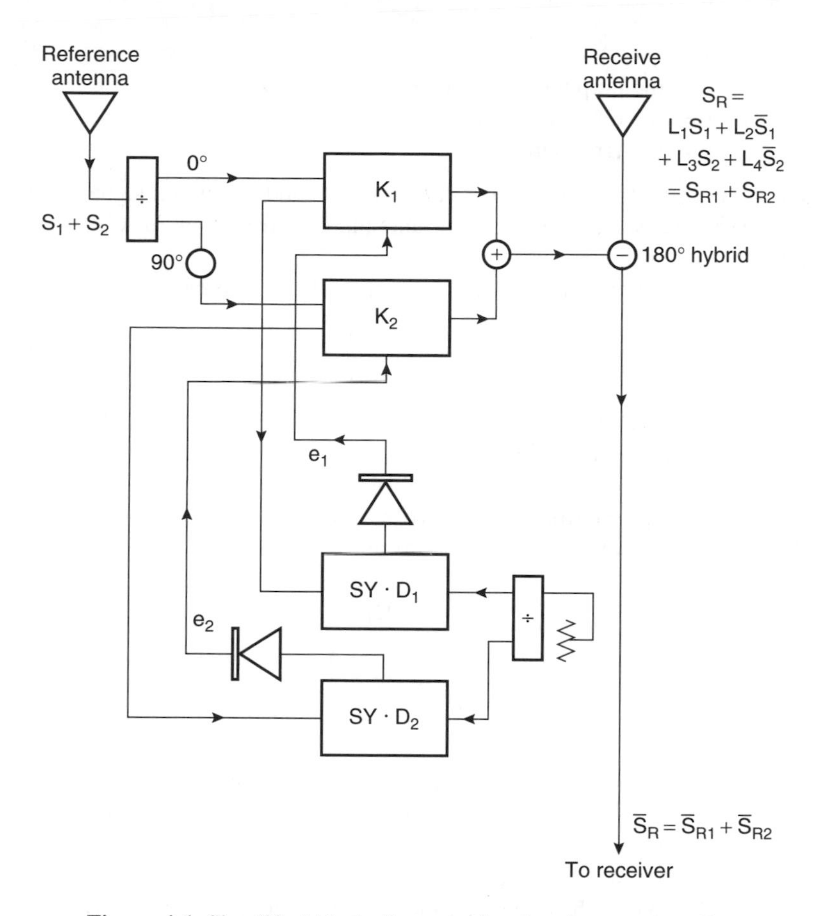

Figure 4-1 Simplified block diagram of an interference canceller.

If the signal controllers in the zero phase and quadrature phase lines introduce amplitude reduction factors K_1 and K_2, respectively, the signal at the receiver, as shown in Figure 4-1, can be written as

$$\bar{S}_R = L_1 S_1 - K_1 S_1 + L_3 S_2 - K_1 S_2 + L_2 \bar{S}_1 - K_2 \bar{S}_1 + L_4 \bar{S}_2 - K_2 \bar{S}_2. \tag{4.2}$$

To determine the effect of the control loop in the adaptive canceller, it is necessary to characterize the function of the synchronous detector. We have considered one such characterization in Eq. (3.35), given by

$$\begin{aligned} e(t) &= g(t) && \text{when } f(t) \geq 0 \\ &= -g(t) && \text{when } f(t) < 0 \end{aligned}$$

where

$e(t)$ = the output of the synchronous detector
$f(t)$ = the reference port input
$g(t)$ = the error signal.

In general, both $e(t)$ and $f(t)$ contain the desired and undesired signal or interference, particularly when a reference antenna or sensor is used to sample the undesired signal or interference.

The above relation among $e(t)$, $f(t)$, and $g(t)$ can also be written as

$$e(t) = g(t)\frac{f(t)}{|f(t)|}. \tag{4.3}$$

For the purpose of simplicity in analysis, we will assume that the synchronous detector in this case is a multiplier.[1] To determine the degree of cancellation, one needs to consider the dynamics of the control loops. Thus, when the canceller employs a velocity-type control, one may write the characteristic equation governing the dynamics of the signal-controller gains as

$$\begin{aligned} \frac{dK_1}{dt} &= M_1 e_1 \\ \frac{dK_2}{dt} &= M_2 e_2 \end{aligned} \tag{4.4}$$

where

K_1 and K_2 = the amplitude factors introduced by the 0° and 90° signal controllers, respectively, as shown in Figure 4-1
M_1 = loop gain of the 0° loop
M_2 = loop gain of the 90° loop
e_1 and e_2 = the outputs of the 0° and 90° loops, respectively.

Following the subtraction of the cancelling signal or the counterinterference from the desired signal and interference present at the receive line, one obtains the residual signal input $\bar{S}_R$ to the receiver as

$$\bar{S}_R = (L_1 - K_1)S_1 + (L_3 - K_1)S_2 + (L_2 - K_2)\bar{S}_1 + (L_4 - K_2)\bar{S}_2. \tag{4.5}$$

Based on the model of the synchronous detector given in Eq. (4.3), then, one may write the synchronous detector outputs as

$$\begin{aligned} e_1 &= \bar{S}_R \frac{(S_1 + S_2)}{|S_1 + S_2|} \\ e_2 &= \bar{S}_R \frac{(\bar{S}_1 + \bar{S}_2)}{|\bar{S}_1 + \bar{S}_2|}. \end{aligned} \tag{4.6}$$

[1] The difference in loop performance when the synchronous detector is a multiplier and when it is characterized as shown in Eq. (3.35) will be discussed later in this section.

Assuming that S_1 is uncorrelated with S_2 and $\bar{S}_2$ and is also uncorrelated with $\bar{S}_1$, and similarly S_2 is uncorrelated with S_1, $\bar{S}_1$, and $\bar{S}_2$, one obtains the average value of

$$\bar{S}_R(S_1 + S_2) = (L_1 - K_1)\langle S_1^2\rangle + (L_3 - K_1)\langle S_2^2\rangle. \tag{4.7}$$

Similarly, assuming that $\bar{S}_1$ is uncorrelated with S_1 and S_2, and $\bar{S}_2$ is uncorrelated with S_2, S_1, and $\bar{S}_1$, one may write the average value of

$$\bar{S}_R(\bar{S}_1 + \bar{S}_2) = (L_2 - K_2)\langle \bar{S}_1^2\rangle + (L_4 - K_2)\langle \bar{S}^2\rangle. \tag{4.8}$$

Assuming further that $|S_1 + S_2|$ and $|\bar{S}_1 + \bar{S}_2|$ can be replaced by their rms value, one may write

$$|S_1 + S_2| = [\langle S_1^2\rangle + \langle S_1^2\rangle]^{1/2}.$$

Also, since $(\bar{S}_1 + \bar{S}_2)$ differs from $(S_1 + S_2)$ only by a 90° phase shift as is seen from Figure 4-1

$$|\bar{S}_1 + \bar{S}_2| = |S_1 + S_2| = [\langle S_1^2\rangle + \langle S_2^2\rangle]^{1/2}. \tag{4.9}$$

Thus, from Eqs. (4.7), (4.8), and (4.9)

$$\begin{aligned} &\frac{dK_1}{dt} + M_1\left[\frac{(L_1 - K_1)\langle S_1^2\rangle + (L_3 - K_1)\langle S_2^2\rangle}{[\langle S_1^2\rangle + \langle S_2^2\rangle]^{1/2}}\right] \\ &\frac{dK_2}{dt} + M_2\left[\frac{(L_2 - K_2)\langle S_1^2\rangle + (L_4 - K_2)\langle S_2^2\rangle}{[\langle S_1^2\rangle + \langle S_2^2\rangle]^{1/2}}\right]. \end{aligned} \tag{4.10}$$

It is assumed in Eq. (4.10) that the averaging process used for the right-hand side of the equation is permissible since the outputs of the synchronous detectors are integrated before being used as control signals for the signal controllers. In other words, the averaging process is assumed to be effected as an integral part of the synchronous detector and the integrator operation that precedes the signal controller operation in the control loop.

The characteristic equations for the control loops can now be written as

$$\begin{aligned} \frac{dK_1}{dt} + K_1 M_1 C &= M_1 N_1 \\ \frac{dK_2}{dt} + K_2 M_2 C &= M_2 N_2 \end{aligned} \tag{4.11}$$

where
$C = [\langle S_1^2\rangle + \langle S_2^2\rangle]^{1/2}$

and

$$N_1 = \frac{L_1\langle S_1^2\rangle + L_3\langle S_2^2\rangle}{C}$$

$$N_2 = \frac{L_2\langle \bar{S}_1^2\rangle + L_4\langle \bar{S}_2^2\rangle}{C}. \tag{4.12}$$

The solutions of the differential equations in Eq. (4.11) are

$$K_1 = \frac{N_1}{C} + A_1 e^{-M_1 Ct}$$

$$K_2 = \frac{N_2}{C} + A_2 e^{-M_2 dt} \tag{4.13}$$

where
A_1 and A_2 = are arbitrary constants to be evaluated from the boundary conditions.

The steady-state solutions as obtained when t approaches infinity are

$$K_1 = \frac{N_1}{C} = \frac{L_1\langle S_1^2\rangle + L_3\langle S_2^2\rangle}{C^2}$$

$$K_2 = \frac{N_2}{C} = \frac{L_2\langle \bar{S}_1^2\rangle + L_4\langle \bar{S}_2^2\rangle}{C^2}. \tag{4.14}$$

From Eqs. (4.5) and (4.14), the steady-state solution of the signal at the receiver following the suppression of the undesired signal can be expressed as

$$\begin{aligned}\bar{S}_R &= (L_1 - K_1)S_1 + (L_3 - K_1)S_2 + (L_2 - K_2)\bar{S}_1 + (L_4 - K_2)\bar{S}_2 \\ &= L_1 S_1 + L_3 S_2 - \frac{L_1\langle S_1^2\rangle + L_3\langle S_2^2\rangle}{C^2}(S_1 + S_2) \\ &\quad + L_2\bar{S}_1 + L_4\bar{S}_2 - \frac{L_2\langle S_1^2\rangle + L_4\langle S_2^2\rangle}{C^2}(\bar{S}_1 + \bar{S}_2).\end{aligned} \tag{4.15}$$

It is evident from Eq. (4.15) that both the desired and undesired signals are affected to some extent by the cancellation process, depending on the parameters L_1, L_2, L_3, L_4, and C. Thus, the signal of the specie S_1, which appears at the receiving antenna as $(L_1 S_1 + L_2\bar{S}_1)$, becomes

$$\begin{aligned}\bar{S}_{IR} &= L_1 S_1 + L_2\bar{S}_1 - \frac{L_1\langle S_1^2\rangle + L_3\langle S_2^2\rangle}{C^2}S_1 - \frac{L_2\langle \bar{S}_1^2\rangle + L_4\langle \bar{S}_2^2\rangle}{C^2}\bar{S}_1 \\ &= \frac{b}{a+b}\{S_1(L_1 - L_3) + \bar{S}_1(L_2 - L_4)\}\end{aligned} \tag{4.16}$$

where
$\langle S_1^2\rangle = a$
$\langle S_2^2\rangle = b.$

Similarly, the signal of the specie S_2 which appears at the receiving antenna as $(L_3 S_2 + L_4 \bar{S}_2)$ becomes $\bar{S}_{2R}$ at the receiver following the cancellation process, where

$$\begin{aligned}
\bar{S}_{2R} &= L_3 S_2 + L_4 \bar{S}_2 - \frac{L_1 \langle S_1^2 \rangle + L_3 \langle S_2^2 \rangle}{C^2} S_2 - \frac{L_2 \langle \bar{S}_1^2 \rangle + L_4 \langle \bar{S}_2^2 \rangle}{C^2} \bar{S}_2 \\
&= \left[(L_3 - L_1) \frac{\langle S_1^2 \rangle}{C^2} \right] S_2 + \left[(L_4 - L_2) \frac{\langle \bar{S}_1^2 \rangle}{C^2} \right] \bar{S}_2 \\
&= \frac{a}{a+b} [S_2 (L_3 - L_1) + \bar{S}_2 (L_4 - L_2)]. \qquad (4.17)
\end{aligned}$$

Thus, if, for example, S_1 and S_2 denote, respectively, the undesired and desired signals at the reference antenna, it is seen from the simplified analysis outlined above that the suppression of the signal of the specie S_1 at the receiver becomes perfect when $\langle S_2^2 \rangle$ and $\langle \bar{S}_2^2 \rangle$ are zero. Such a situation implies that the reference ports of the synchronous detectors, shown in Figure 4-1, contain the uncontaminated undesired signal exclusively, as is likely to be expected when the reference line is coupled directly to the undesired signal-transmitter line.

Another performance parameter, which is often of more significance than the degree of cancellation in an interference environment, is the signal-to-noise ratio at the receiver or its improvement as a result of the interference cancellation. Based on the performance analysis outlined above, this is indicated by the ratio of $\bar{S}_{2R}$ to $\bar{S}_{1R}$ as given in Eqs. (4.16) and (4.17), respectively, since we assumed that S_2 and S_1 are, respectively, the desired and undesired signals. Thus, the signal-to-noise ratio S/N, defined as the ratio of the desired signal power to the undesired signal power following the cancellation, can be written as

$$\frac{S'}{N} = \frac{|\bar{S}_{2R}|^2}{|\bar{S}_{1R}|^2} = \left| \frac{1 + Q \dfrac{\langle \bar{S}_2^2 \rangle}{\langle S_2^2 \rangle}}{1 + Q \dfrac{\langle \bar{S}_1^2 \rangle}{\langle S_1^2 \rangle}} \right|^2 \qquad (4.18)$$

where
$Q = [(L_2 - L_4)]/[(L_1 - L_3)]$.

The parameters L_1, L_2, L_3, and L_4 are related to the propagation paths for the desired and undesired signals, and the signal parameters S_1, $\bar{S}_1$, S_2, and $\bar{S}_2$ are not always known or easy to measure. One may thus formulate the expression for S/N differently.

Let $\bar{M}_1$ and $\bar{M}_2$ be the transfer functions of the propagation paths for undesired and desired signals, respectively, between their respective sources and the receiving antenna. Thus, if $S_1'(\omega)$ and $S_2'(\omega)$ denote the transforms for the undesired and desired signals at their respective sources

$$\begin{aligned} \bar{M}_1(\omega)S'_1(\omega) &\rightarrow L_1S_1 + L_2\bar{S}_1 \\ \bar{M}_2(\omega)S'_2(\omega) &\rightarrow L_3S_2 + L_4\bar{S}_2 \end{aligned} \tag{4.19}$$

based on the notations used in Figure 4-1.

Similarly, let $\bar{N}_1$ and $\bar{N}_2$ be the corresponding transfer functions between the reference antenna and the respective sources for the undesired and desired signals. Thus

$$\begin{aligned} \bar{N}_1(\omega)S'_1(\omega) &\rightarrow S_1 \\ \bar{N}_2(\omega)S'_2(\omega) &\rightarrow S_2 \end{aligned} \tag{4.20}$$

again, based on the notations used in Figure 4-1. From Eqs. (4.18), (4.19), and (4.20)

$$\left(\frac{S}{N}\right)_{\text{Receive Line}} = \frac{|S'_1(\omega)|^2}{|S'_2(\omega)|^2}\frac{|\bar{N}_1(\omega)|^2}{|\bar{N}_2(\omega)|^2}. \tag{4.21}$$

In the previous chapter, we have seen the correspondence between the transfer function of a signal controller and that of the weighting function of a Wiener filter. We may extend this concept to examine the expression for *S/N* for the Wiener filter used for the suppression of an interference.

Figure 4-2 shows a schematic representation of a Wiener filter with an adjustable weighting function.

The transfer functions for various propagation paths are shown in the figure.

From Eq. (3.84), we have seen that the weighting function W is obtained from the ratio of the cross spectrum ϕ_c between the primary input and reference antenna signals to the power-density spectrum ϕ_a of the reference antenna signals. Thus

$$W = \phi_c/\phi_a.$$

Now, based on the notations used in Figure 4-2, the power-density spectrum of the reference antenna signals can be written as

$$\phi_a = \langle S'^2_1\rangle|\bar{N}_1(\omega)|^2 + \langle S'^2_2\rangle|\bar{N}_2(\omega)|^2 \tag{4.22}$$

where

$S'_1(\omega)$ and $S'_2(\omega)$ = the transforms of the undesired and desired signals, respectively, at their corresponding sources

$\bar{N}_1(\omega)$ and $\bar{N}_2(\omega)$ = the transfer functions of the propagation paths as defined earlier.

Similarly, the cross spectrum between the reference antenna signals and the primary input or receiving antenna signals can be written as

$$\begin{aligned} \phi_c = {} & [S'_1(\omega)\bar{M}_1(\omega) + S'_2(\omega)\bar{M}_2(\omega)] \\ & \cdot [S'^*_1(\omega)\bar{N}^*_1(\omega) + S'^*_2(\omega)\bar{N}^*_2(\omega)]. \end{aligned} \tag{4.23}$$

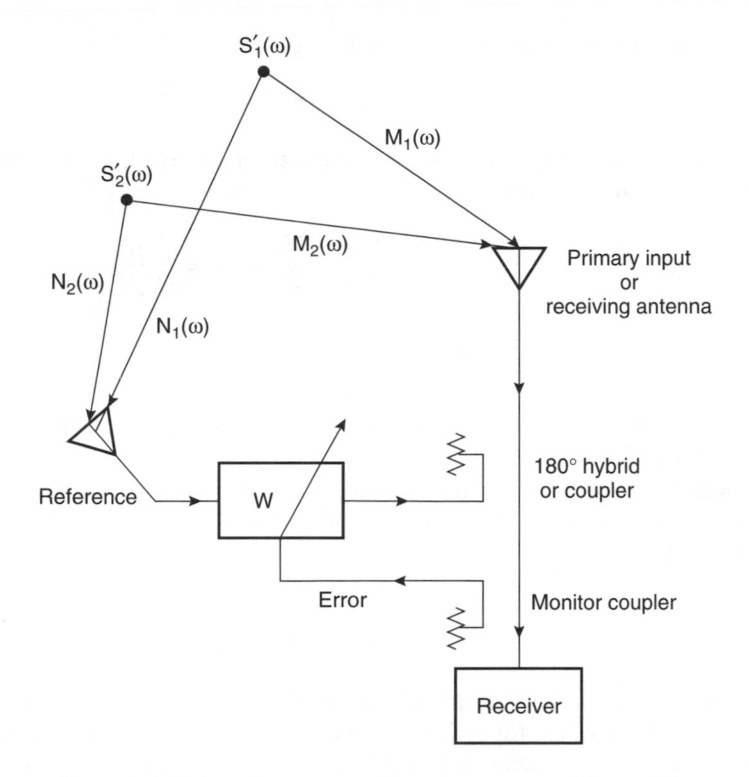

Figure 4-2 Schematic representation of an adaptive Wiener filter.

By definition, then

$$W = \frac{\langle S_1'^2 \rangle M_1 N_1^* + \langle S_2'^2 \rangle M_2 N_2^*}{\langle S_1'^2 \rangle |N_1^2| + \langle S_2'^2 \rangle |N_2^2|} \tag{4.24}$$

whereby the functional dependence on ω of the terms in the numerator and denominator are implied.

Now, from Figure 4-2, the transfer function of the propagation path for the undesired signal is

$$T_{S_1'} = \bar{M}_1 - \bar{N}_1 W. \tag{4.25}$$

Similarly, the transfer function of the propagation path for the desired signal is

$$T_{S_2'} = \bar{M}_2 - \bar{N}_2 W. \tag{4.26}$$

Then, the spectrum of the undesired signal following the cancellation is

$$\phi_{S_1'} = \langle S_1'^2 \rangle \{ |\bar{M}_1 - \bar{N}_1 W|^2 \}. \tag{4.27}$$

The corresponding spectrum of the desired signal is

$$\phi_{S'_2} = \langle S'^2_2 \rangle \{|\bar{M}_2 - \bar{N}_2 W|^2\}. \tag{4.28}$$

Finally, the signal-to-noise ratio of the signals at the receive line following the cancellation can be written as

$$\begin{aligned}\frac{\phi_{S'_2}}{\phi_{S'_1}} &= \frac{\langle S'^2_2 \rangle}{\langle S'^2_1 \rangle} \cdot \frac{\langle S'^2_1 \rangle^2 |\bar{N}_1|^2 \cdot |\bar{M}_2 \bar{N}_1 - \bar{M}_1 \bar{N}_2|^2}{\langle S'^2_2 \rangle^2 |\bar{N}_2|^2 \cdot |\bar{M}_1 \bar{N}_2 - \bar{M}_2 \bar{N}_1|^2} \\ &= \frac{\langle S'^2_1 \rangle}{\langle S'^2_2 \rangle} \frac{|\bar{N}_1|^2}{|\bar{N}_2|^2}\end{aligned} \tag{4.29}$$

which is identically the same as in Eq. (4.21), derived for the canceller using the signal controller as shown in Figure 4-1.

Another important relation between the *S/N* at the reference antenna and that at the receive line following the cancellation should be noted here. From Eq. (4.20), it is seen that

$$(S/N)_{\text{Reference}} = \frac{S_2^2}{S_1^2} = \frac{\langle S'^2_2 \rangle |N_2|^2}{\langle S'^2_1 \rangle |N_1|^2}. \tag{4.30}$$

From Eqs. (4.21) and (4.30), it is seen that the *S/N* at the output of the canceller, that is, at the receive line following the cancellation, is simply the reciprocal of that at the reference-line signal. The same result is obtained in the case of a Wiener filter also, as is evident from Eqs. (4.22) and (4.29). The significance of this result is important, particularly for the undesired signal canceller where the undesired signal source is collocated with the victim receiver. For such a case, the reference signal may be obtained by directly coupling into the undesired signal transmitter line, as noted earlier. Since no desired signal then can appear at the reference line, the $(S/N)_{\text{Reference}}$ must approach zero. Consequently, for such a case, the receive-line *S/N* following the cancellation approaches infinity, implying a perfect cancellation of the undesired signal. One may thus conclude that the degree of cancellation is the maximum realizable for a particular adaptive canceller when the reference-line signal contains the undesired signal to be cancelled exclusively.

4.3 DIRECT COUPLED REFERENCE FOR COUNTERINTERFERENCE SYNTHESIS

When a sample of the interference or undesired signal is obtained directly from its source as a reference as shown in Figure 4-3, there cannot be any desired signal at the signal controller or at the input of the synchronous detector. The absence of the desired signal suggests that, regardless of the signal-controller settings, or the values of K_1 and K_2, no desired signal can be injected into the receive line to modify thereby the desired signal at the receive line. Thus, for the direct coupled reference,

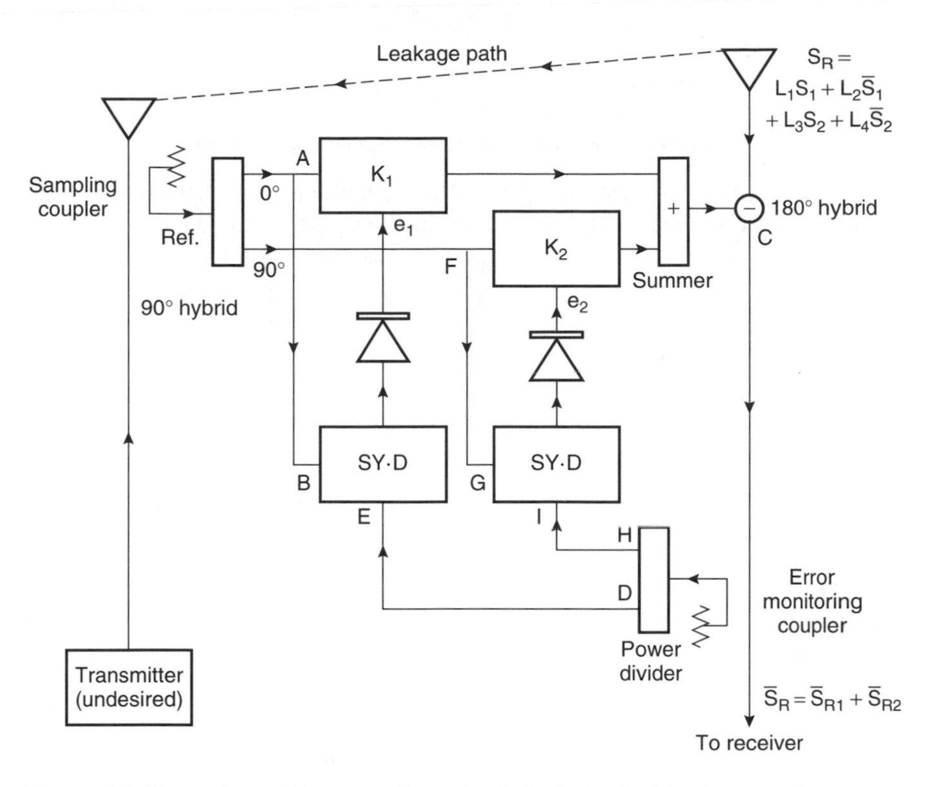

Figure 4-3 Illustration of direct coupling of undesired signal. Adaptive canceller simulates the leakage path for the undesired signal so that the undesired signals propagating through the leakage path and through the canceller cancel each other at the 180° hybrid.

one may not expect any change in the desired signal at the receive line, except for the insertion loss due to the 180° hybrid and the error-sampling coupler which are at the through receive line from the receiving antenna to the receiver. This, however, is contrary to the conclusion that may be reached from Eq. (4.17) as b approaches zero, corresponding to the case where there is no desired signal at the reference line. Whether an adaptive canceller that reduces the interference from the receive line can also modify the desired signal is a subject that requires further examination.

For such a case, the signal and the counterinterference at the receive line can be written as

$$\bar{S}_R = L_1 S_1 - K_1 S_1 + L_2 \bar{S}_1 - K_2 \bar{S}_1 + L_3 S_2 + L_4 \bar{S}_2. \tag{4.31}$$

To examine the effect of the control loop, we assume the characteristic of the synchronous detector to be the same as in Eq. (4.3). That is,

$$e(t) = g(t) \frac{f(t)}{|f(t)|},$$

where
e = the output
f = the reference port signal
g = the error or input signal of the synchronous detector.

Also, the synchronous detector is assumed to be a multiplier. Thus, the error signals e_1 and e_2, corresponding to the 0° and 90° loop, respectively, become

$$e_1 = \bar{S}_R \frac{S_1}{|S_1|} \quad \text{and} \quad e_2 = \bar{S}_R \frac{\bar{S}_1}{|\bar{S}_1|}. \tag{4.32}$$

From Eqs. (4.31) and (4.32)

$$\begin{aligned} e_1 &= \frac{(L_1 - K_1)\langle S_1^2 \rangle}{|S_1|} \\ e_2 &= \frac{(L_2 - K_2)\langle \bar{S}_1^2 \rangle}{|\bar{S}_1|}. \end{aligned} \tag{4.33}$$

The characteristic equations for this case corresponding to Eq. (4.10) can now be written as

$$\begin{aligned} \frac{dK_1}{dt} &= M_1(L_1 - K_1)\sqrt{a} \\ \frac{dK_2}{dt} &= M_2(L_2 - K_2)\sqrt{b} \end{aligned} \tag{4.34}$$

where
$a = \langle S_1^2 \rangle$
$b = \langle \bar{S}_1^2 \rangle$

the solutions for K_1 and K_2 being

$$\begin{aligned} K_1 &= L_1 + B_1 e^{-M_1\sqrt{a}\,t} \\ K_2 &= L_2 + B_2 e^{-M_2\sqrt{b}\,t} \end{aligned} \tag{4.35}$$

and B_1, B_2 are arbitrary constants.
The steady-state solutions, obtained as $t \to \infty$, then become

$$K_1 = L_1 \quad \text{and} \quad K_2 = L_2. \tag{4.36}$$

Substituting the results in Eq. (4.36) into Eq. (4.31), one obtains

$$\bar{S}_R = L_3 S_2 + L_4 \bar{S}_2 \tag{4.37}$$

which is the undistorted desired signal as received by the receiving antenna. The insertion loss for the desired signal due to the 180° hybrid and the error monitoring coupler as shown in Figure 4-3 is not taken into account in Eq. (4.37). Thus, it is possible to eliminate the undesired signal without affecting the desired signal when the reference signal for the canceller is obtained by coupling into the transmitting antenna line for the undesired signal, thereby avoiding any desired signal at the reference line and at the reference port of the synchronous detectors.

As noted earlier, the parameters L_1, L_2, L_3, and L_4 relate to the characteristics of the propagation paths for the undesired and desired signals from their respective sources to the receiving antenna, and these parameters are not always known or easy to measure. The same is true for the signal parameters S_1, $\bar{S}_1$, S_2, and $\bar{S}_2$.

To examine the performance of the same adaptive canceller shown in Figure 4-3, with a direct-coupled undesired signal reference with known or measurable parameters relating to propagation paths and signals, let $\bar{M}_1$ and $\bar{M}_2$ be the transfer functions of the propagation paths for the undesired and desired signals, respectively, between their respective sources and the receiving antenna. Also, let $S_1'(\omega)$ and $S_2'(\omega)$ denote, respectively, the transforms of the undesired and desired signals at their respective sources. With reference to Figure 4-3, the input of the signal controllers is $S_1'(\omega)$, which is directly coupled to the undesired signal-transmitter line.

The inputs of the receiving antenna line can now be written as

$$S_R(\omega) = \bar{M}_1(\omega)S_1'(\omega) + \bar{M}_2(\omega)S_2'(\omega). \tag{4.38}$$

The counterundesired signal as synthesized by the signal controllers is

$$S_c = K_1S_1'(\omega) + jK_2S_1'(\omega) = \sqrt{K_1^2 + K_2^2}\,S_1'(\omega)e^{j\theta(\omega)} \tag{4.39}$$

where

$$\theta = \tan -1\frac{K_2}{K_1}. \tag{4.40}$$

Thus, the signals at the receive line following the subtraction of the counter-undesired signal from the signals received by the receiving antenna can be written as

$$\bar{S}_R = |\bar{M}_1|S_1'(\omega)e^{j\phi} - KS_1'(\omega)e^{j\theta} + |\bar{M}_2|S_2'(\omega)e^{j\theta_D} \tag{4.41}$$

where

$S_1'(\omega)$ and $S_2'(\omega)$ are assumed to be real

$K^2 = K_1^2 + K_2^2$

$|\bar{M}_1|$ and $|\bar{M}_2|$ = magnitudes of $\bar{M}_1$ and $\bar{M}_2$, respectively

ϕ and θ_D = the phase angles associated with $\bar{M}_1$ and $\bar{M}_2$, respectively.

The signals $\bar{S}_R$ at the receive line will not only be the input to the receiver, but also will be the input to the two synchronous detectors. As in Eq. (4.3), we will assume the output of the synchronous detector $e(t)$ as

$$e(t) = g(t)\frac{f(t)}{|f(t)|}$$

where

$g(t)$ and $f(t)$ = the input and reference signals for the synchronous detector, respectively.

In the earlier part of this section, we assumed a multiplier model of the synchronous detector. Equation (4.3), however, has a special significance when the reference signal for synthesizing the counterundesired signal is obtained by directly coupling into the undesired signal-transmitter line. More specifically, the synchronous detector, according to Eq. (4.3), is a switch controlled by the zero crossings of the reference-port signal, the output of the detector being $(g \cdot 1)$ or $g \cdot (-1)$, depending on whether the value of $f(t)$ is positive or negative at any time t. Thus, for the direct coupled reference signal, the switch is controlled by the zero crossings of the undesired signal only. This is primarily because there is no desired signal or other signals present at the reference port. Also, when the level of the reference-port undesired signal is very high in comparison with the noise level at the reference port, the effect of this noise on the switching operation, with an amplitude of 1 or (-1), can also be ignored.

For the special case when the undesired signal at the reference and error ports of the synchronous detector are in phase, the output of the detector becomes the absolute value of the residual undesired signal at the error port. For the generalized case, where the undesired signal at the reference port and at the output of the signal controller are different from each other in amplitude and phase, as is usually the case, the performance of the synchronous detector, perhaps, can be better understood from Figure 4-4. For the purpose of illustration, a time-varying undesired signal, as it appears at the reference port of the 0° synchronous detector, is shown in Figure 4-4(a). The periods for zero crossings are also shown in the figure. The corresponding counterundesired signal as synthesized by the 0° signal controller (Figure 4-3), denoted as C, and the 0° phase component of the same undesired signal as it appears at the receiving antenna, denoted as B, are shown in Figure 4-4(b). The difference between the undesired signal components B and C constitute the error signal for the control loop (Figure 4-3) and appears at the error port of the 0° synchronous detector. Now, the switching based on zero crossings of the reference-port undesired signal will increase the average value of the synchronous detector output by changing the sign of the difference-undesired signal $(B - C)$ from the negative to positive during the negative half-cycle period of the reference-port undesired signal. This effect at the output of the 0° synchronous detector is shown in Figure 4-4(c).

When the reference undesired signal is obtained through direct coupling from its transmitter or the transmitting antenna line, the reference-port undesired signal will be $S_1'(\omega)$, and from Eq. (4.41), the undesired signal component at the error port, corresponding to $(B-C)$ in Figure 4-4, will be $(|\bar{M}_1| \cos\phi - K \cos\theta) S_1'(\omega)$. When $S_1'(\omega)$ is a periodic signal, as is most often the case of practical interest, the average value of $S_1'(\omega)$ will be zero over a period of time equal to or much longer than the period of one cycle of $S_1'(\omega)$. But because of the switching action

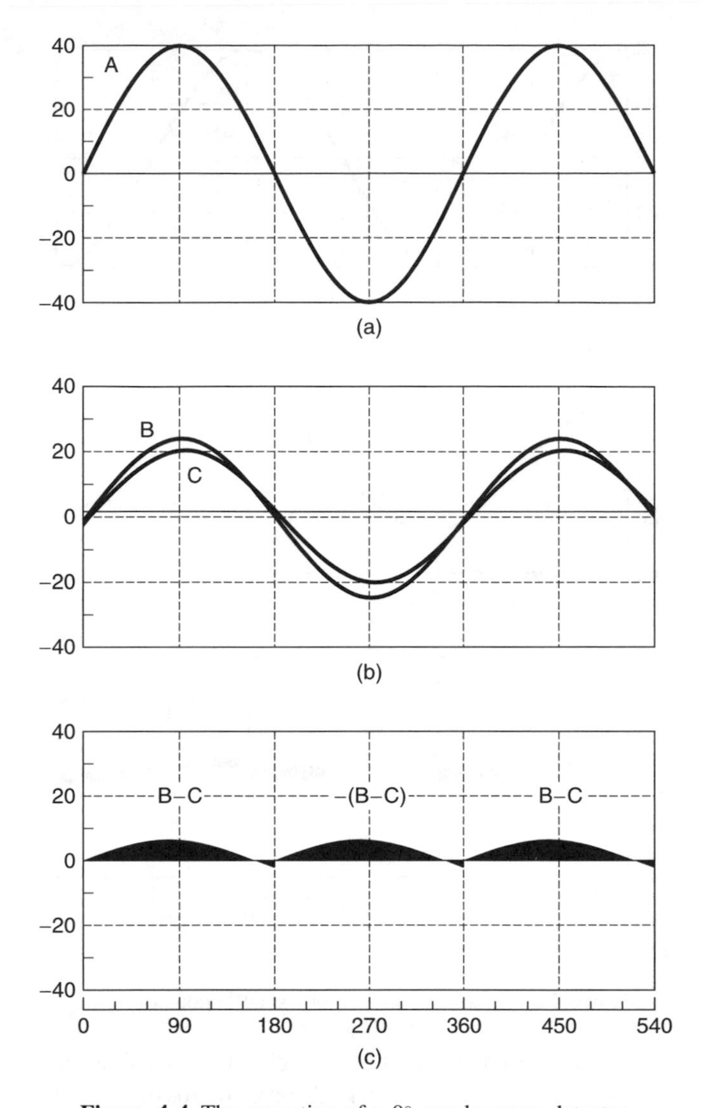

Figure 4-4 The operation of a 0° synchronous detector.

that makes $(B - C)$ become $-(B - C)$ during the negative half of the reference undesired signal cycle, the output of the synchronous detector will not be zero unless $(|\bar{M}_1| \cos \phi - K \cos \theta)$ is zero. And, since the output of the 0° synchronous detector is the driving force or controlling signal for the 0° signal controller as shown in Figure 4-3, the attenuation $K_1 = K \cos \theta$ will change continuously until $(|\bar{M}_1| \cos \phi - K \cos \theta)$ becomes zero.

The operation of the 90° synchronous detector is shown schematically in Figure 4-5. For the same time-varying undesired signal as assumed in Figure 4-4,

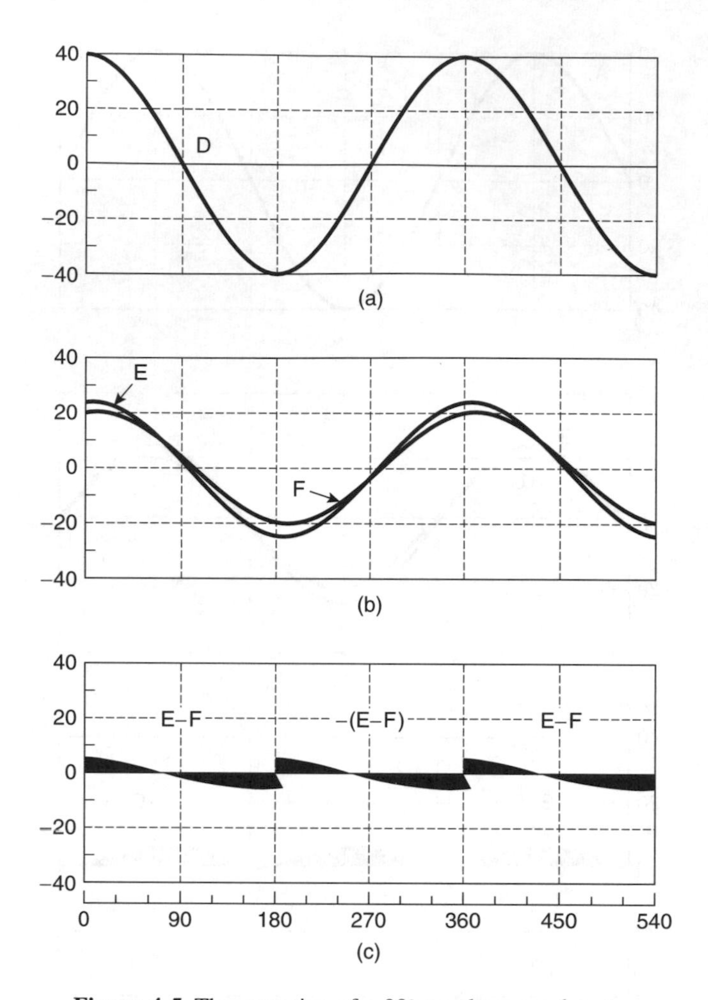

Figure 4-5 The operation of a 90° synchronous detector.

the reference-port undesired signal at 90° synchronous detector, denoted as D, is shown in Figure 4-5(a). Again, the periods of zero crossing are also shown in the figure. In Figure 4-5(b), the corresponding counterundesired signal as synthesized by the 90° signal controller (Figure 4-3) is denoted as F, and the 90° phase component of the undesired signal as it appears at the receiving antenna is denoted as E. Again, in the absence of switching action that makes $(E - F)$ become $-(E - F)$ during the negative half of the reference-port undesired signal cycle, the average value of the synchronous detector output would be zero over a period of time equal to or much longer than the period of one cycle of the reference-port undesired signal. As a result of switching, this output will not be zero unless $(|\bar{M}_1| \sin\phi - K \sin\theta)$ is zero. For the nonzero value of the 90° synchronous detector output, the 90° signal controller (Figure 4-3) will cause the value of $K_2 = K \sin\theta$ to change continuously until $(|\bar{M}| \sin\phi - K \sin\theta)$ becomes zero.

It should be noted that when the phase angles ϕ and θ and the phase of the reference-port undesired signal are the same with their amplitudes being different, the output of the synchronous detectors will be maximum. Consequently, it may appear that when these phases are unequal, the effectiveness of the adaptive canceller may be less than what is desired. This, however, is not the case because of the closed-loop operation, since if the output of either synchronous detectors is not adequate, the value of K_1 and K_2 will not be what is required for cancellation. There will be a nonzero error signal as a result, causing a nonzero synchronous detector output and a change in the value of K_1 or K_2 until the new value makes the error signal zero, and an equilibrium is reached in the closed-loop operation.

When the interference has a reasonable bandwidth, as is almost always the case in practice, a time-delay matching for the reference and input ports of the synchronous detector is essential insofar as the reference signal is concerned. This means that, with reference to Figure 4-3, the time delay for the path *AB* at the reference line of the 0° synchronous detector must be equal to *ACDE*, involving the 0° signal controller, the summer for the 0° and 90° controllers, the 180° hybrid, the monitoring coupler, the error-signal power divider, and the line connecting the path *DE* as shown in the figure. Similarly, for the 90° synchronous detector, the time delay of the path *FG* must be equal to the time delay of the path *FCHI*, involving the 90° signal controller, the summers for the 0° and 90° controllers, the 180° hybrid where the countersignal is subtracted from the receive-line signals, the monitoring coupler, the error-line power divider, and the line connecting the path *HI*. It is this time-delay matching that makes the reference signal synchronous at the reference and error ports of the synchronous detector. It should also be noted that a similar time-delay matching at both sides of the 180° hybrid making the time delays for the interference at the leakage and signal-controller paths shown in Figure 4-3 the same is also essential for cancellation of an interference having any reasonable bandwidth. Thus, the interference to be cancelled must be synchronous at both sides of the 180° hybrid such that the cancellation should be ideal at any frequency component of the interference since the time delay is the same for all frequency components of the interference. Figure 4-6 shows the effect of time-delay matching on the instantaneous bandwidth of interference cancellation where a better match leads to a better canceller performance. Thus, for example, a time mismatch of 1 ns provides not only a higher degree of cancellation, but also over a wider frequency range than what could be expected for a 10 ns mismatch.

It should also be noted that the instantaneous bandwidth of cancellation, as shown in Figure 4-6, is not the same as the operating frequency bandwidth over which an adaptive canceller is effective. Figure 4-7 shows an example of the operating frequency bandwidth and the degree of realizable cancellation at each

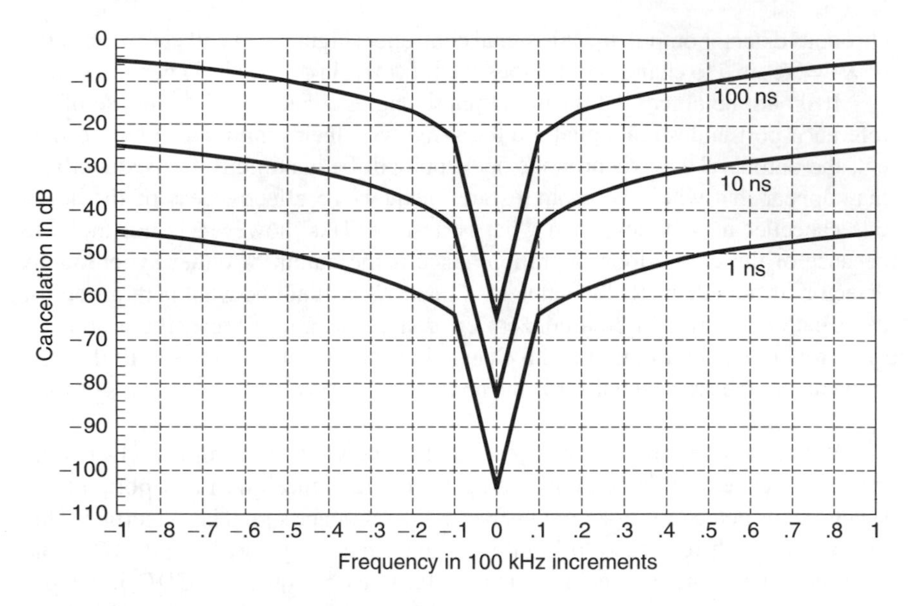

Figure 4-6 Cancellation degradation due to time delay mismatch at or near the frequency of maximum cancellation.

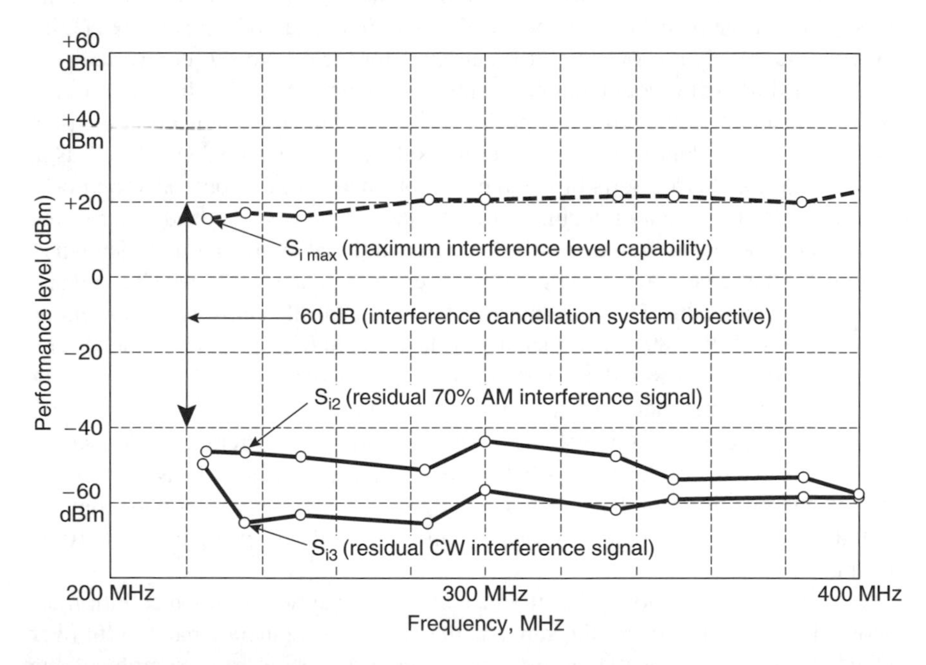

Figure 4-7 Adaptive interference cancellation over a wide operating frequency band.

frequency within this operating bandwidth. This canceller characteristic does not depend on time-delay mismatch. Instead, it shows the frequency band over which the signal controller can provide the amplitude and phase variations as needed to cancel the interference at the receive line.

Some of the practical consequences of the use of an adaptive canceller on the receive line and the basic requirement for achieving a cancellation goal may, perhaps, be further clarified by an example. Let us suppose that we are concerned with an interference caused at the receiver of a high-frequency radio (30–80 MHz), due to a radiating transmitter of a collocated, second radio. Both radios transmit signals at 50 W or 47 dBm level, and the propagation path attenuation from the transmitting antenna of one radio to the receiving antenna of the other is 40 dB. We further assume that the radios are located physically close to each other to permit a direct coupling from the interference-transmitter line for the interference sampling, as shown in Figure 4-3. Figure 4-8 illustrates in block-diagram form the leakage and signal controller paths of the adaptive canceller and the receiving antenna line of the victim receiver. In this case, the interference level at the receive-line side of the 180° hybrid is 7 dBm, and hence, for a cancellation, the counterinterference power level as generated at the output of the signal controller also must be 7 dBm at the 180° hybrid. From the viewpoint of the requirement, the signal controller

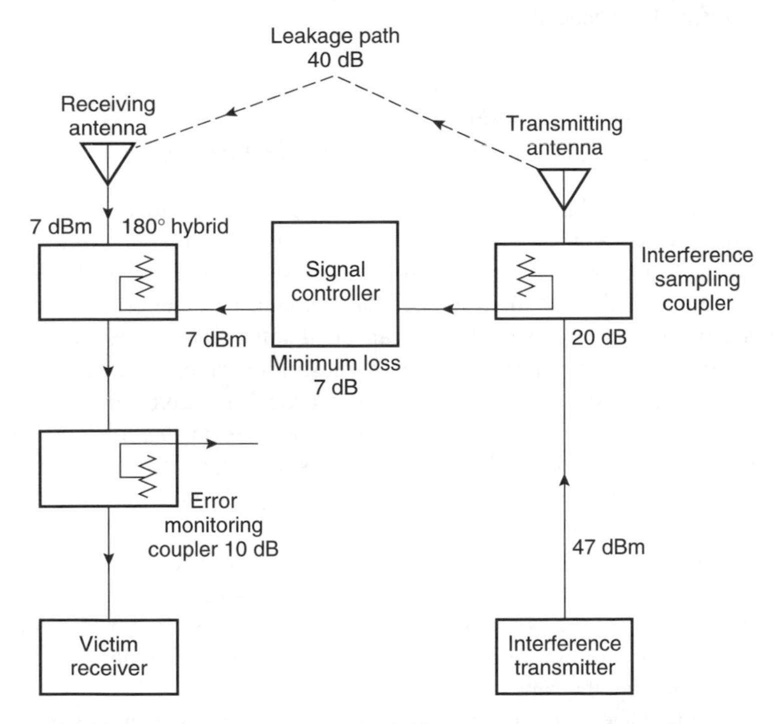

Figure 4-8 Leakage path, signal controller, and receiving line.

must handle an output power of 7 dBm, and must have an input power level equal to or greater than $(7 + L)$ dBm, L being the minimum insertion loss at the signal controller. For the p-i-n diode type signal controllers as discussed in Chapter 3, an output power-handling ability of approximately 20 dBm is not unusual, and the insertion loss is on the order of 7 dB. If now one selects a 20 dB directional coupler for sampling the interference at its transmitting antenna line, the input power of the signal controller will be $47 - 20 = 27$ dBm. This will be an adequate power input for the controller since the output power requirement is only 7 dBm. A 20 dB coupler at the interference transmitter line will, in turn, introduce a theoretically minimum insertion loss of about 0.044 dB, which is only a modest transmitter power loss for most applications, although the actual power loss may be slightly higher because of the connectors, etc. The receiving antenna line will experience an insertion loss due to the 180° hybrid and the coupler needed for monitoring the error signal. Since the insertion loss due to the hybrid will be slightly more than 3 dB and a 10 dB error-sampling coupler will introduce an insertion loss of about 0.46 dB, the total receive-line insertion loss due to the canceller will be less than 5 dB which, again, is acceptable in most cases. There are, of course, considerable design flexibilities to reduce the receiving antenna line loss, transmitting antenna line loss, and the inherent loss at the signal controller, depending on the circumstances and the particular use of the canceller.

4.4 REFERENCE SENSORS FOR COUNTERINTERFERENCE SYNTHESIS

From the mathematical model of the adaptive canceller as outlined in Section 4.2, and from the discussion on direct-coupled reference for the counterinterference synthesis in Section 4.3, it is evident that the most effective use of the adaptive cancellers is realized when the reference line contains no other signal but the interference or the undesired signal to be cancelled at the receive line. Also, such a situation arises only when the transmitter or transmitting antenna line is collocated with the receiver and the receive line. In some practical situations, however, it may be desirable to use a reference sensor or antenna, even for the collocated interference source. More specifically, when the undesired signal-transmitting antenna and the receiving antenna are so located that nonnegligible propagation multipaths between those antennas exist due to local scatterers, the undesired signal as received by the receiving antenna through multipaths may differ significantly from that coupled directly from its transmitting antenna line. This is particularly the case when the undesired signal is a broadband signal, as multipath distortions for such cases will make the reference undesired signal different from that at the receive line, thereby making the cancellation less perfect.

To consider the effect of propagation multipaths between the undesired signal antenna and the receiving antenna, let $S(\omega)$ denote the transform of the undesired signal at its source, that is, at its transmitting antenna, and let there be N number of propagation multipaths, as shown schematically in Figure 4-9(a). Let the ith

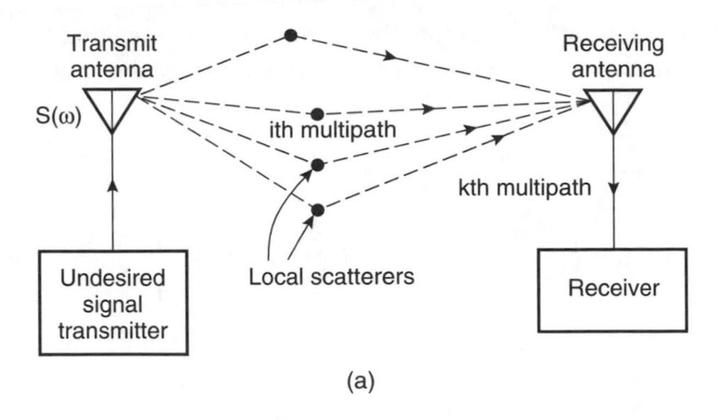

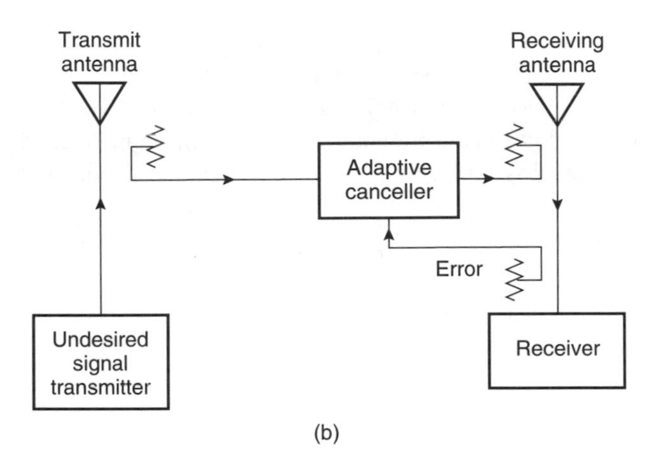

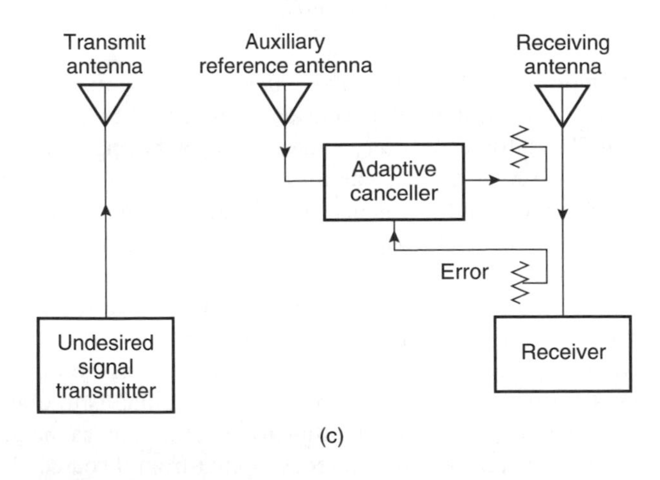

Figure 4-9 (a) Illustration of propagation multipaths caused by local scatterers. (b) Adaptive canceller with direct-coupled reference from the undesired signal transmitter line. (c) Adaptive canceller with auxiliary reference antenna even for a collocated undesired signal source.

propagation path introduce an attenuation K_i and a phase ωT_i such that the total signal received by the receiving antenna from the undesired signal source is

$$\begin{aligned} S_R &= \sum_{i=1}^{N} K_i S(\omega) e^{-j\omega T_i} \\ &= S(\omega) \left\{ \sum_{i=1}^{N} K_i \cos \omega T_i - j \sum_{i+1}^{N} K_i \sin \omega T_i \right\}. \end{aligned} \tag{4.42}$$

It may be noted that the expression S_R for the undesired signal in Eq. (4.42) may be considered as a generalized expression of the multipath-received signal since it represents any form of time-varying undesired signal with any arbitrary amplitude, waveform, and frequency.

To cancel S_R at the receiving antenna, one must synthesize a counterundesired signal identical to that of S_R so that at the output of the 180° hybrid at the receive line, there will be no undesired signal. For the direct-coupled reference of undesired signal as considered in the previous section and shown in Figure 4-9(b), the counterundesired signal synthesized by the two quadratured signal controllers can be written as

$$S_C = S(\omega)[K_1 + jK_2]. \tag{4.43}$$

Thus, for the cancellation of s_R as given in Eq. (4.42), one needs

$$\begin{aligned} K_1 &= \sum_{i=1}^{N} K_i \cos \omega T_i \\ K_2 &= \sum_{i=1}^{N} K_i \sin \omega T_i. \end{aligned} \tag{4.44}$$

Equation (4.44) implies that K_1 and K_2 must, in general, be a function of ω. But from the discussion of signal controllers considered in Section 3.2, K_1 and K_2 are not functions of ω. Thus, for an arbitrary undesired signal having a transform $S(\omega)$, Eq. (4.44) could be satisfied only when it is a CW signal. Consequently, in the presence of propagation multipaths and for a wideband undesired signal, a direct-coupled reference obtained through a direct coupling from the undesired signal-transmission line may yield a cancellation much less than what is desirable or acceptable, regardless of whether the undesired interference source or its transmitting antenna is collocated with the receiving antenna.

One way to avoid the effect of a lack of correlation between the reference undesired signal and the same undesired signal at the receive line is to obtain a reference which is affected by propagation multipaths the same way as the undesired signal at the receive line, instead of a direct coupling from it source. Thus, when the reference undesired signal is obtained by a sensor or antenna located very close to the receive antenna, the reference undesired signal becomes more correlated with the undesired signal at the receive line than what is feasible through a direct coupling. Obviously, when the reference and receive antennas are exactly at the same place, the correlation will be perfect, regardless of the number and characteristics of propagation multipaths. Since such a situation is not possible, particularly when the reference and receive antennas must be different,[2] for the purpose of cancellation, it is necessary to locate the reference antenna as close as is practicable. Although an independent reference antenna lessens the effect of propagation multipaths, it is not an ideal remedy since other signals, including the desired signal, may also be received by the reference antenna. Such signals also affect the cancellation, as we will see in the next chapter. In addition to affecting the cancellation of undesired signal, the reference antenna introduces some desired signal into the receive line through the signal controller, thereby creating a multipath propagation for the desired signal which may not have existed in the absence of the reference antenna.

4.5 CANCELLATION AT HIGH-POWER LEVELS

The necessary criterion for the cancellation of an arbitrary undesired signal at the receive line by a subtraction-type cancellation process is that a counterundesired signal be delivered at the 180° hybrid (Figure 4-1) at an amplitude and phase which are identical to those of the undesired signal at the receive line. Thus, the power level of the undesired signal at the receive line and that of the counterundesired signal must be the same at the 180° hybrid. In many practical situations where the source of interference or the undesired signal is collocated with the receive antenna, the power level of the undesired signal at the receive line is very high because of the proximity of a high-power transmitting antenna. Consequently, the power output of the signal controller that delivers the counterundesired signal at the 180° hybrid must handle the necessary high-power level. Further, a signal controller has inherent losses, making the power-handling requirement for the signal controller even more than the power level of the undesired signal at the receive line. A signal controller, using p-i-n diodes as discussed in Chapter 3, have severe limitations on the power-handling capability, and when the undesired signal power level at the receive line far exceeds 1 W or 30 dBm, p-i-n diode signal controllers, as discussed earlier, may not be practical.

There are several means available to increase the power-handling capability of an adaptive canceller. Noteworthy among them are:

- high-power signal controllers
- attenuation amplification in receive line
- amplification of signal-controller output prior to delivering the counterundesired signal to the 180° hybrid
- switchable attenuators and delay lines capable of handling high powers.

[2]The practical feasibility of an adaptive cancellation when the reference and receive antennas are the same is considered later in the text.

High-Power Signal Controllers

A very simple means for increasing the power-handling capability of p-i-n diode signal controllers is to replace each single p-i-n diode by a number of p-i-n diodes in parallel. The sharing of the load by a number of p-i-n diodes permits the passage of a higher power counterundesired signal through the signal controller without distortion. It is assumed in this case that each p-i-n diode in the parallel network assembly will be chosen to carry the highest power level that an individual p-i-n diode can carry and as it is available commercially.

Alternate forms of signal controllers which do not employ p-i-n diodes, such as those shown in Figures 3-5 and 3-6, may also be considered for high-power adaptive cancellers. As shown in Figure 3-5, the variable resistance can be designed to accommodate a power level much higher than what is usually permissible in p-i-n diode signal controllers, particularly where the rate of change of the signal-control parameters, K or ϕ, is high. The electromechanical control, however, is often impractical at high radio and microwave frequencies. The associated system reliability problem with such controls is also not desirable in many applications. A signal controller employing variable capacitors, as shown in Figure 3-5, can also be designed to accommodate a higher power level than what is usual for a p-i-n diode signal controller. The variability of the capacitance can be achieved by electromechanical controls or through ferroelectric devices where the dielectric constant for the capacitor is varied by a variable dc bias voltage [4] placed across the capacitor.

Attenuation-Amplification in Receive Line

In some situations, an overwhelmingly large interference or undesired signal at the receive line, received through the receive antenna or coupled into the line by other means, can be accommodated by an adaptive canceller by attenuating the signal at the receive line prior to cancelling, as shown in Figure 4-10. Following the cancellation of the undesired signal, the receive-line signals can be amplified to the desired level as needed for the receiver. Such a means for signal controlling is permissible when the primary noise of concern that affects the signal-to-noise ratio (S/N) for the desired signal is external to the receiver, that is, that noise is received through the receive antenna or coupled into the receive line by other means. For such cases, the noise of concern is also attenuated the same way as the desired signal by the attenuator without materially affecting the S/N. Since any attenuation prior to cancellation at the receive line necessarily degrades the equivalent noise figure for the receiver, the attenuation amplification of the undesired signal at the receive line is often not permissible. There is also the problem of added noise in the receive line due to the amplifier, if and when such an amplifier is used.

Amplification of Signal-Controller Output

A corollary of the same approach is to amplify the output of the signal controller so that the amplifier output of the undesired signal to be cancelled is at the same power level as the same undesired signal at the receive line. Such an approach does

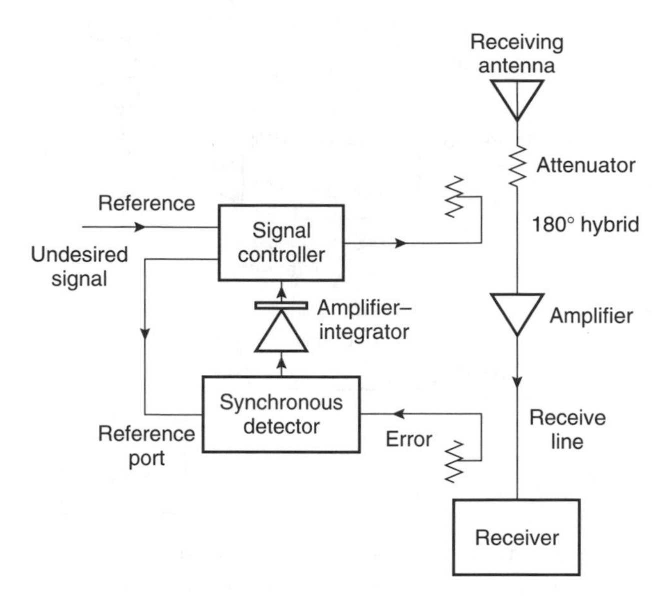

Figure 4-10 Illustration of amplifier attenuator approach. Although accommodating a high-power undesired signal, the approach introduces noise in the receive line due to the amplifier.

not affect the effective noise figure of the receiver, although it is still not desirable in many situations since a power amplification of the undesired signal in the signal controller line, as shown in Figure 4-11, often introduces into the receive line higher harmonics of the undesired signal and intermodulation products arising from the amplifier nonlinearity, very common in high-power amplifiers. In addition, the amplifier noise is also injected into the receive line, affecting the signal-to-noise ratio for the desired signal. The amplification-attenuation approach, however, is still a viable option, particularly when, in balance, the added noise and harmonics, etc., introduced into the receive line are less objectionable than the overwhelming undesired signal level at the same line.

Phase-Varying Signal Controller

Another form of a signal controller, particularly suitable for a high-power microwave undesired signal, makes use of varying phases instead of attenuations to control the amplitude and phase of the signal flowing through the signal controller. To illustrate the concept of the phase-varying signal controller, one may consider the network with two branches, similar to the attenuation-type signal controller, except for two phase-varying elements replacing the attenuators, as shown in Figure 4-12.

Let the input to the signal controller be an arbitrary amplitude-modulated signal of the form

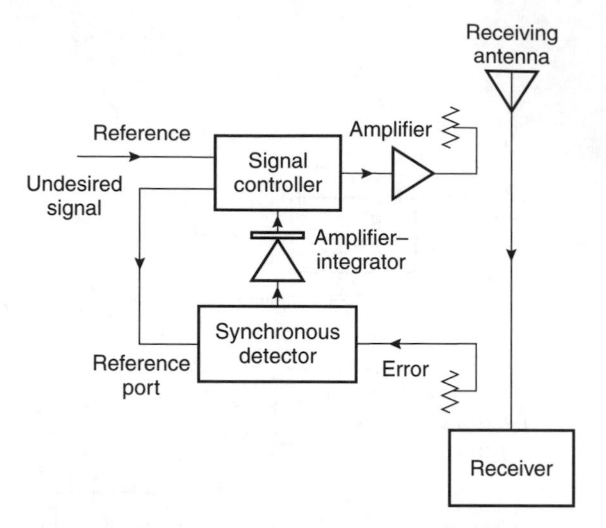

Figure 4-11 Amplification of signal controller output. Although accommodating a high-power undesired signal, the approach introduces noise and harmonics and intermodulation products in the receive line due to the amplifer.

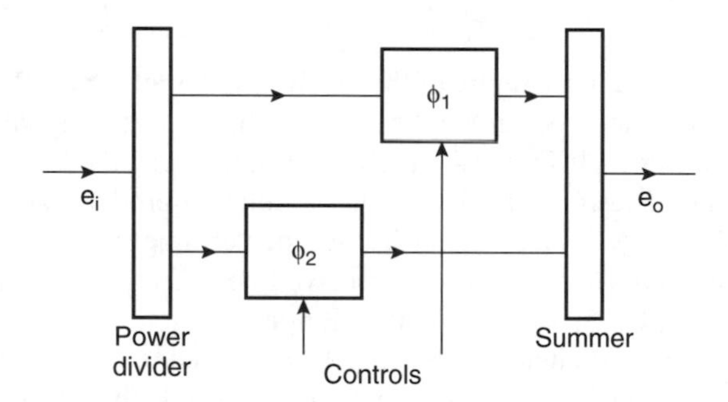

Figure 4-12 Signal controller employing two phase-varying elements in parallel.

$$e_i = A(t)\sin(\omega t - \phi_0) \tag{4.45}$$

where

$A(t)$ = the time-varying amplitude

ω = some angular frequency

ϕ_0 = some arbitrary phase.

The output of the signal controller, e_0, can then be written as

$$\begin{aligned} e_0 &= A(t)\sin(\omega t - \phi_1 - \phi_0) + A(t)\sin(\omega t - \phi_2 - \phi_0) \\ &= A(t)K\sin(\omega t - \theta - \phi_0) \end{aligned} \tag{4.46}$$

where

K and θ = the gain and phase shift introduced by the signal controller, respectively, and

$$\begin{aligned} K &= [2 + 2\cos(\phi_1 - \phi_2)]^{1/2} = 2\left|\cos\left(\frac{\phi_1 - \phi_2}{2}\right)\right| \\ \theta &= \tan^{-1}\frac{\sin\phi_1 + \sin\phi_2}{\cos\phi_1 + \cos\phi_2}. \end{aligned} \tag{4.47}$$

It is evident from Eq. (4.47) that one can obtain the required signal-controller functions, that is, the ability to change the amplitude and phase of the signal flowing through the signal controller arbitrarily by varying ϕ_1 and ϕ_2.

To examine whether there is a unique set of values of K_1 and θ for a specific set of values of ϕ_1 and ϕ_2, one may first plot the values of K as a function of $(\phi_1 - \phi_2)$ based on the relation shown in Eq. (4.47), as shown in Figure 4-13(a). Now, from Eq. (4.47)

$$\theta = \tan^{-1}\frac{\sin\phi_1 + \sin\phi_2}{\cos\phi_1 + \cos\phi_2}. \tag{4.48}$$

If one denotes $(\phi_1 + \phi_2)/2 = A$ and $(\phi_1 - \phi_2/2) = B$, then

$$\theta = \tan^{-1}\tan A = A = \frac{\phi_1 + \phi_2}{2}. \tag{4.49}$$

A plot of θ as a function of $\phi_1 + \phi_2$ is shown in Figure 4-13(b). Thus, when the required signal-controller functions K and θ are known, one may obtain two sets of values of $\phi_1 - \phi_2$ from Figure 4-13(a) and one value of $\phi_1 + \phi_2$ from Figure 4-13(b). From this, one may obtain a pair of values for ϕ_1 and ϕ_2 which will be able to provide the required signal-controller parameters. It should be noted that if ϕ_1 and ϕ_2 are positive and may vary between 0 and π, the maximum value of θ obtainable from this process is π. If, however, the range of possible values of ϕ_1 and ϕ_2 is 0–2π, one may realize any value of K and θ by this kind of signal controller.

From Eq. (4.47), it is evident that when $|K| \leq 1$, $|\phi_1 - \phi_2|$ is between 120° and 180° if both ϕ_1 and ϕ_2 are positive. Thus, when the phases ϕ_1 and ϕ_2 are positive and each can assume a value ranging from 0 to π, one may obtain a value of $|K| \leq 1$ and $\theta \leq \pi$.

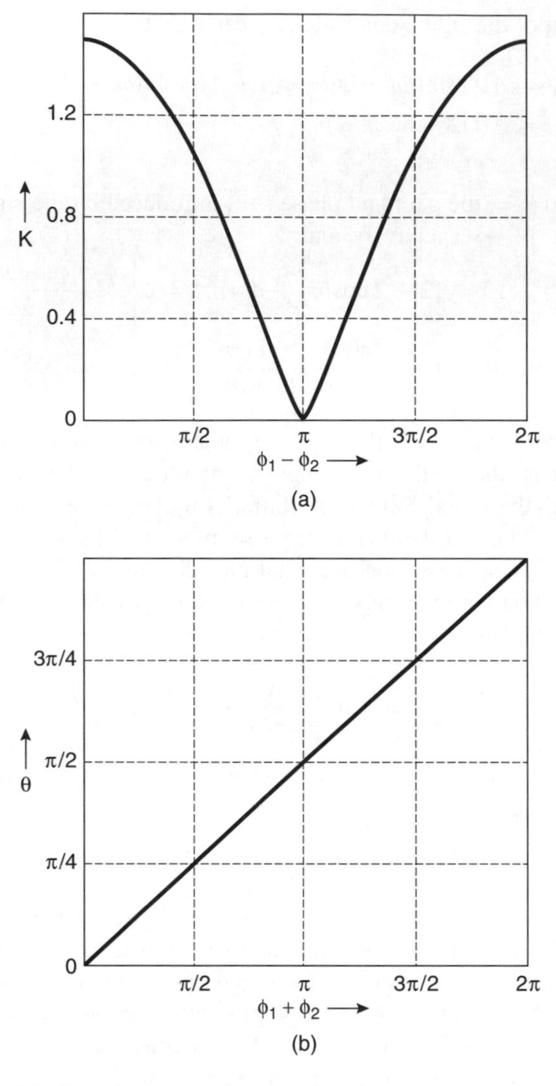

Figure 4-13 (a) Amplitude (K) variation in signal controller. (b) Phase (θ) variation in signal controller.

The signal-controller arrangement shown in Figure 4-12, then, is inadequate when θ needs to vary from 0 to 2π for the requirement of any arbitrary phase angle for the signal controller. If, however, ϕ_1 and ϕ_2 can vary from 0 to 2π, one can obtain an unrestricted signal-controller function with two phase-varying elements as shown in Figure 4-12, provided, of course, that there are practical means to control the values of ϕ_1 and ϕ_2.

The concept of a phase-varying signal controller is attractive when a waveguide-type phase shifter, using ferrites, can be employed to realize variables ϕ_1 and ϕ_2. The control in such a case is provided by the dc bias that changes the value of the phase. Since waveguides can be designed to handle a large power, particularly at microwave frequencies, the phase-varying signal controller becomes an alternative, specifically where a high-power signal controller is needed at microwave frequencies. Also, the through-line losses in a waveguide-type signal controller will be less than that expected in a p-i-n diode attenuator-type signal controller.

The physical realizability of a microwave phase-varying element has been suggested by [5], [6], [8], and others, all using ferrites.

Electronically variable phase shifters, as is needed for the signal-controller operation, can also be obtained without waveguides and ferrites. Figure 4-14(a) shows, for example, a lumped element phase shifter using a quadrature hybrid with a matched pair of *L–C* networks. The variable *L–C* network, linked to the output ports 2 and 3 of the 90° hybrid, perform as sliding short circuits. These "shorts," placed at the output ports of the hybrid, will reflect energy incident to them. The reflected energy will appear at the isolated port 4 of the hybrid undiminished in amplitude. When the "sliding short" is varied by an electrical distance ϕ in effect, the phase of the reflected signal will vary by 2ϕ. By properly selecting the *L–C*

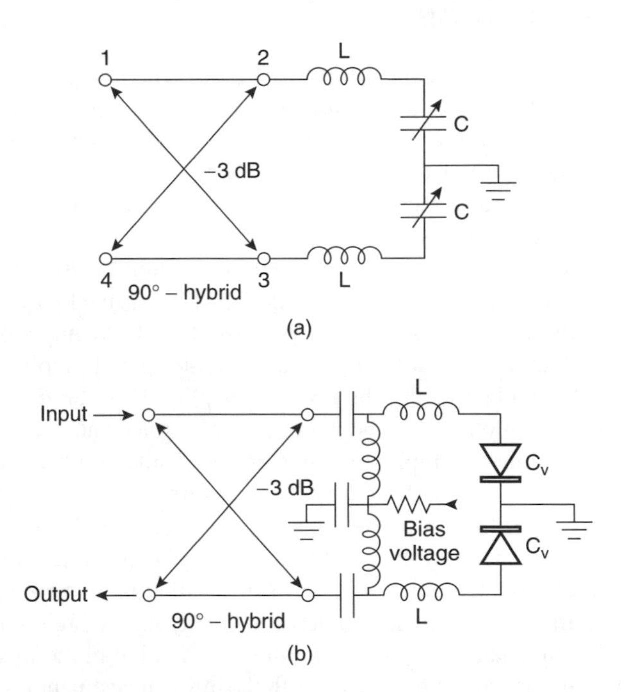

Figure 4-14 (a) Illustration of a lumped element phase shifter. (b) A lumped element electronically variable phase shifter.

elements, one may obtain a one-way phase variation of 90°. This will result in an overall phase shift of 180°, or π rad. To obtain a phase shift of 2π rad suitable for a signal controller operation, one needs to connect two 0–π type continuously variable phase shifters in series. Also, for the signal controller operation, one needs to vary the "sliding shorts" electronically. Figure 4-14(b) illustrates one such means. Here, two voltage-variable capacitors or varactors are used to change the value of the capacitors, and hence the effective electrical distance ϕ, the variable voltage being obtained from the control signals of the adaptive canceller for a closed-loop operation.

The power-handling capability of a varactor-type signal controller will be limited by the physically realizable capacitance of the varactors. Alternatively, one may replace the varactors by capacitors which may be switched into the phase-shifter networks by an electronically controlled switch. It should be noted that the control of the phase-varying signal controllers in a closed-loop operation will be different from what has been discussed earlier employing two attenuators in the signal controller.

4.6 SAME-FREQUENCY REPEATER AMPLIFICATION

An important practical application of the cancellation of an undesired signal originating from a collocated source is found in a repeater where a weak signal is received and amplified prior to its retransmission at the same frequency. A same-frequency repeater is needed in intracontinental microwave relay links and in airborne or satellite transponders to facilitate communications well beyond the horizon. Often, the degree of possible amplification, however, becomes limited because of the external coupling of the amplifier output to its input. Thus, for example, when in a microwave link a weak signal is received by one antenna at the left, as shown in Figure 4-15, and is retransmitted, following a power amplification, by another antenna at the right, the amplifier gain has to be restricted to avoid self-oscillation because of the finite decoupling between the right and left antennas. The same problem arises in a microwave transponder when a signal is received by an antenna, is amplified at the "receive" side of a "transmit–receive" hybrid, and is then transmitted through the "transmit" side of the hybrid and the same antenna. Any reflection of the transmitted signal from the antenna due to a possible mismatch appears at the amplifier input, and depending on the degree of mismatch, there is usually a probability of self-oscillation, particularly for a very high-gain amplifier. For a terrestrial microwave relay link, a high-gain amplification at each intermediate relay station permits a reduction in the total number of intermediate stations, thereby reducing both the initial investment for the link and recurring maintenance cost for the intermediate stations saved due to the high-gain amplification at the others. For a transponder, a high-gain amplification permits

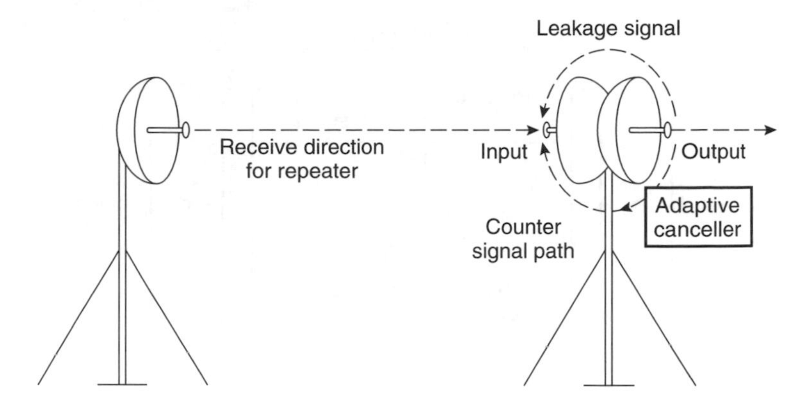

Figure 4-15 Illustration of same-frequency repeater using an adaptive canceller.

a longer communication range. Thus, there is a strong economic motivation to design and implement a high-gain microwave amplification system that ensures the avoidance of a self-oscillation risk.

If the self-oscillation that limits the high-gain amplification is due to the finite decoupling between the amplifier output to its input, its avoidance may be effected by further reducing the coupling. Since passive means available, if any, in most situations to enhance decoupling are usually restricted, one seeks active approaches by adaptively cancelling the "leakage" signal at the amplifier input resulting from finite input–output decoupling. One such cancellation means involves the synthesis of a counterleakage signal equal in amplitude and 180° out of phase with respect to the leakage signal, as is usually obtained by a signal controller, such that when the counterleakage signal is added with the leakage signal at the amplifier input, the effect of the leakage signal is nulled. Since for a positive feedback situation, similar to the case under consideration, a self-oscillation may be initiated by noise and may occur at any frequency, and since the amplitude and phase of the leakage signal are different at different frequencies and are seldom *a priori* known, any cancellation to effect output–input decoupling has to be adaptive. Also, to be effective, any cancellation means must be operable over a wide frequency band.

The problem of high-gain amplification at a relay station is illustrated in Figure 4-16. Here, the wideband microwave signal is received by Antenna 1 at the left from a communication base station or another relay station at the left. This signal, flowing through the "Receive" side of the Transmit–Receive Hybrid 1, is amplified and transmitted through the "Transmit" side of the Hybrid 2 and Antenna 2 at the right. Similarly, another signal is received by Antenna 2 and the "Receive" side of Hybrid 2, and following amplification, is transmitted through the "Transmit" side of Hybrid 1 and Antenna 1. Notwithstanding the high-directivity gains of the antennas, a fraction of the signal transmitted by one antenna is coupled

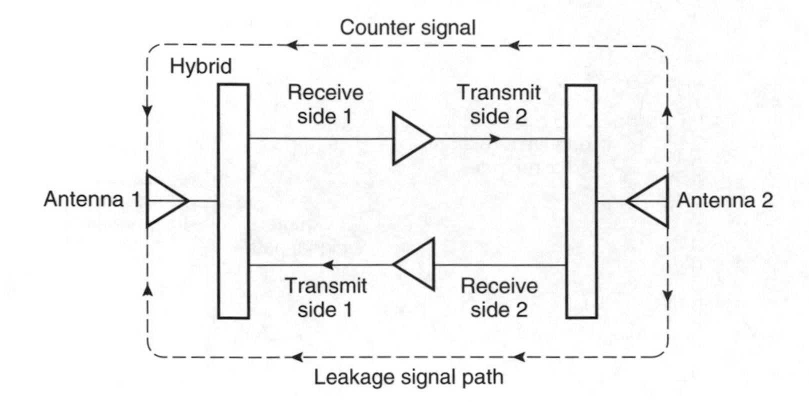

Figure 4-16 Schematic illustration of high-gain amplification in a relay station of a microwave link.

into the other antenna as a leakage signal, and for the same frequency amplification, this leakage signal is indistinguishable in spectral characteristics from the signal that the latter antenna receives. A self-oscillation negating the function of the relay station usually occurs when the amplifier gain exceeds the decoupling factor between Antennas 1 and 2 because of the oscillation causing positive feedback. Since the decoupling factor between the antennas could be on the order of 20–30 dB, the system power gain through amplification at the relay station, in any direction, is limited by that amount or less.

A similar problem arises in a same-frequency transponder or repeater, as shown in Figure 4-17. Here, a microwave signal is received by the antenna and the "Receive" side of the "Transmit–Receive" hybrid, and following amplification, is transmitted through the "Transmit" side of the same hybrid and the same antenna. When the antenna is not perfectly matched, as is often the case, the reflected signal from the antenna is coupled into the "Receive" side of the hybrid. Again, this coupled signal is indistinguishable from the signal that the antenna receives in spectral characteristics. In this case, a self-oscillation also occurs when the amplification

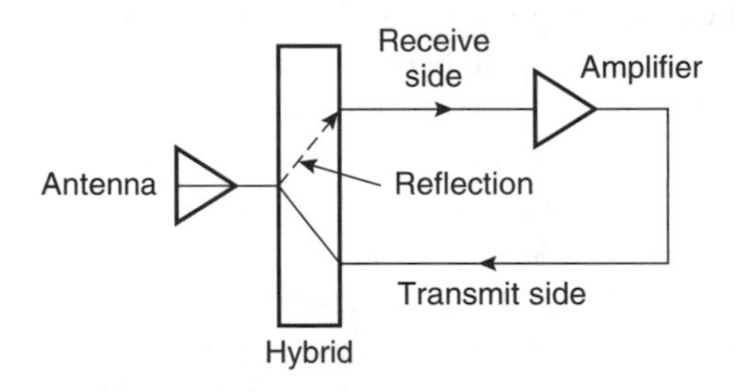

Figure 4-17 Schematic illustration of high-gain amplification in a transponder.

gain exceeds the decoupling factor for the reflected or leakage signal. For a wide-band signal, this decoupling factor could be on the order of 20 dB or less. Hence, the power gain of the repeater or transponder becomes limited to about 20 dB.

Where a higher amplifier gain is required or desired, means must be devised to reduce the net positive feedback in the presence of increased amplification gain to avoid a self-oscillation. One such means is an adaptive canceller that selectively cancels the oscillation-causing leakage signal. Conceptually, the cancellation of the leakage signal involves the injection of a countersignal, similar to that required for an adaptive canceller, into the input circuit of the amplifier, the countersignal being equal in amplitude but 180° out of phase with respect to the leakage signal that causes the oscillation. The countersignal in this case is synthesized from a sample of the amplifier output, and since the leakage signal as it appears at the input of the amplifier differs from the output of the amplifier by an amplitude factor and a time delay, the means for synthesis of the countersignal becomes a signal controller that can effect an amplitude and time-delay change with respect to the same signal at the controller input. However, the exact amplitude factor and the time delay necessary to null the leakage signal can seldom be *a priori* known or determined analytically, or from past measurement, particularly in a field environment. The synthesis of the countersignal therefore must be adaptive. This, as we have noted earlier, is precisely the function of an adaptive canceller.

The concept of an amplifier gain enhancement, without risking a self-oscillation, is shown in Figure 4-18. Here, the coupling of the amplifier output to its input, shown by a dotted line, may be regarded as a leakage signal path. As the amplifier gain exceeds the path loss for the leakage path, a self-oscillaton occurs, making the amplifier ineffective. If, now, a countersignal which is equal in amplitude and 180° out of phase with respect to the leakage signal is synthesized and is injected at the amplifier input, the net coupled signal from the amplifier output to its input is nulled, thereby eliminating the risk of self-oscillation. Since the coupling is linear in most cases, the leakage signal, coupled into the amplifier input, differs from the amplifier output signal by an amplitude factor K and a time delay T. The countersignal needed to null the leakage signal at the amplifier input can then be synthesized from a sample of the amplifier output signal by appropriately controlling its amplitude and phase or time delay until such a null occurs.

To further illustrate the operation of a same-frequency, high-gain amplifier, let the leakage signal as it appears at the amplifier input from the same amplifier output be

$$e_1 = KS(t - T) \tag{4.50}$$

where

K = the amplitude reduction factor
T = the time delay introduced by the leakage path
$S(t)$ = the amplifier output signal.

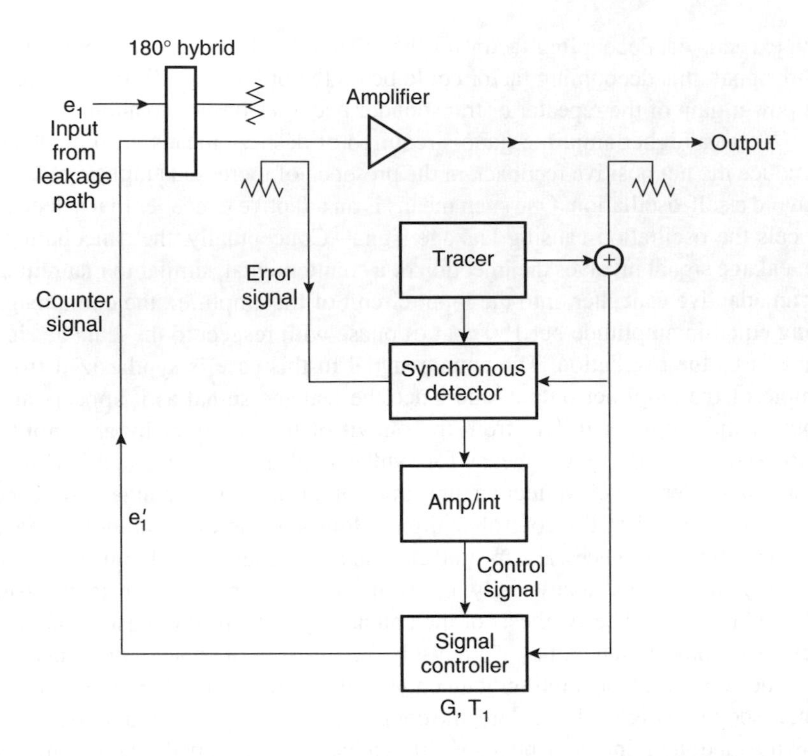

Figure 4-18 Illustration of amplifier gain enhancement by cancelling feedback.

If the signal controller is capable of effecting a change in gain G and a time delay T_1 from its input to output, then the output of the signal controller shown in the figure will be

$$e_1' = GS(t - T_1). \tag{4.51}$$

If, now, $K = G$ and $T = T_1$ at the difference port of the 180° hybrid, and hence at the amplifier input, there will be no leakage signal of the type $S(t)$, since e_1' will cancel e_1 at this port. If, however, the values of G and T_1 as set by the signal controller are not exact to make $G = K$ and $T_1 = T$, as will be the case at the start, the difference port of the 180° hybrid will contain a nonzero leakage signal, and hence a nonzero error signal. This nonzero error signal will result in dc signals at the output of the synchronous detectors. These dc signals can then be amplified and integrated to create control signals for the signal controller such that G and T_1 may change only when an error signal is present. As the nonzero error signal causes the control signals to change, the values of G and T_1 will change until such values become what are exactly needed to drive the error signal to zero.

This condition, then, constitutes simultaneously the equilibrium condition for the closed-loop control and the suppression of the oscillation causing leakage signal at the difference port of the 180° hybrid, and hence at the amplifier input. Here, the closed loop comprises the coupler for the error signal, the synchronous detector, the amplifier–integrator, the signal controller, and the 180° hybrid, as shown in Figure 4-18.

As is the case for all adaptive cancellers considered earlier, the only parameters involved in the signal controller that synthesizes the counterleakage signal are G and T_1, both of which are independent of $S(t)$ such as its frequency spectrum, amplitude, and modulation. Also, if the signal controller can truly synthesize the required amplitude gain and timc delay, a wideband leakage signal reduction is possible since the gain and time delay required for the cancellation of one frequency component of the leakage signal are, in general, the same as needed for any other frequency component of the leakage signal spectrum. Additionally, if the characteristics of the principal amplifier, primarily its gain, or the characteristics of the leakage path change with time, the closed-loop control automatically accommodates the changes by tracking, and always suppressing, the oscillation causing feedback.

Like most adaptive cancellers, the degree of suppression of the leakage signal obtainable by an adaptive control depends primarily on the loop gain and noise. When the loop gain is on the order of 100 dB, the cancellation of the leakage signal ranging from 60 to 70 dB is usually realized, depending on the effective noise bandwidth of the loop and the imperfections of the system components. For such cases, the amplifier gain can be enhanced almost by that amount, without risking a self-oscillation.

The discussion of the adaptive control to cancel the net leakage signal at the amplifier input will not be complete without reference to the need of a tracer signal, as shown in Figure 4-18. It is seen that whatever signal appears as an error signal also appears as the amplifier input signal, except, perhaps, for a different gain factor associated with the error signal coupler. Thus, when the error signal is driven to zero, and the synchronous detector and the closed-loop control operate on the intended signal for the amplifier input, there will not remain any amplifier input signal including the intended signal since the intended signal at the amplifier input will also be driven to zero by the closed-loop control. Since such a situation defeats the purpose of the amplifier, means must exist to operate the control without nulling the intended input signal of the amplifier. A low-level tracer signal, fed simultaneously at the output of the amplifier and the signal controller, resolves this problem. For such a case, the leakage signal path, shown in Figure 4-18, carries the tracer signal to the amplifier input, along with the fraction of the amplifier output signal. The tracer signal also travels through the signal controller and appears at the 180° hybrid, along with the countersignal. The error signal now contains the tracer signal, and if the synchronous detector and the closed-loop control operate on the tracer signal exclusively, there will be no tracer signal at the amplifier input, although the intended input signal for the amplifier will not be driven to zero. Now, if the low-level tracer signal is chosen so that the characteristics of the leakage path for the tracer signal and the intended signal for the amplifier are the same, one can cancel the oscillation causing feedback at the amplifier input, without nulling the intended input signal for the amplifier. This condition is satisfied when the tracer signal frequency is close to the intended signal frequency. Alternatively, the tracer signal can be modulated on the intended signal carrier frequency and the synchronous detector, and the control loop can be made to operate on the tracer signal alone, as derived by a synchronous demodulation of the error signal.[3] In either case, the synchronous detector in Figure 4-18 must be referenced to the tracer signal to ensure the control of the loop exclusively by the tracer signal.

One of the distinct advantages of the same frequency broadband microwave amplification is that the amplifier can be used for a number of communication channels for "repeating" purposes, and no restriction is imposed on the random access of these communication channels to the repeater. Figure 4-19 illustrates this concept of a repeater for multiple communication channels [7]. Only two transmitting stations and two receiver stations are shown in the figure as examples, the repeater station containing primarily a wideband amplifier with a gain enhancement means without risking a self-oscillation. Figure 4-20 shows test data on the same frequency amplification, solid and dotted lines being, respectively, the maximum distortionless amplifier output without and with an adaptive canceller.

The following possibilities resulting from the use of a multichannel repeater may be noteworthy:

(a) The amplification process can accommodate a large and a varying number of communication channels at any time,

(b) Random access for "repeating" purposes is available to all channels in the specified band,

(c) There is no restriction on the type of modulation of the signals that may be amplified through the process, thus permitting even different types of modulations for different channels,

(d) Since no frequency conversion is involved, the conversion noise typically experienced in a repeater system is avoided, and

(e) No logic circuits within the repeater system is necessary to hold the repeater signal within its specified channel since the reception and transmission frequencies are the same.

[3] R. Ghose, "Microwave repeater at 2–4 GHz band," Technical Note, Technology Research International, Calabasas, CA, 1985.

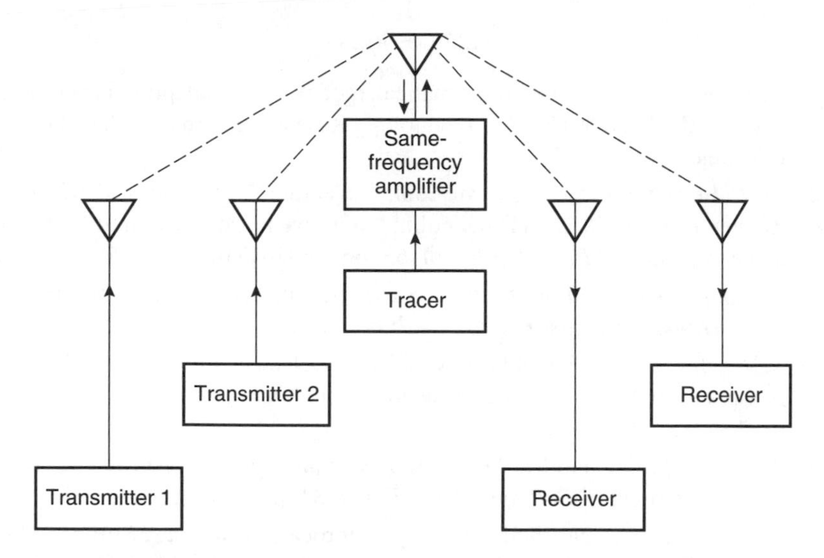

Figure 4-19 Schematic illustration of same-frequency amplification linking multiple communication channels.

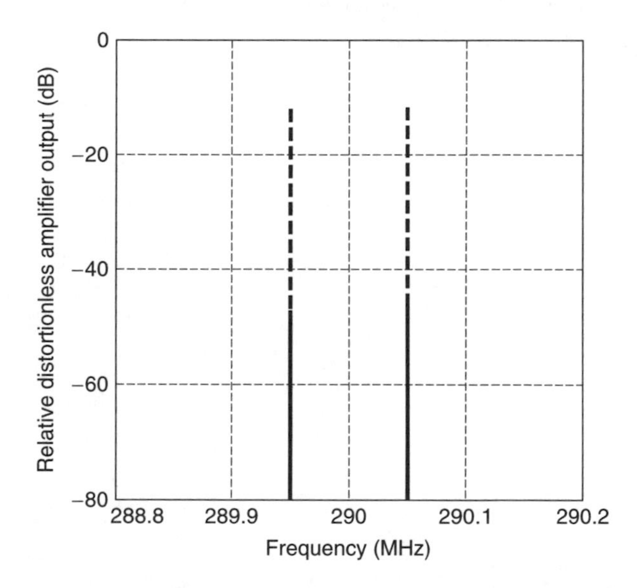

Figure 4-20 Same-frequency amplification without and with adaptive canceller. Solid and dotted lines show, respectively, the maximum distortionless amplifier output without and with an adaptive canceller.

REFERENCES

[1] R. Ghose, "Theory and mathematical model of the adaptive cancellation system," Tech. Note 160064, Technology Research International, Calabasas, CA, June 1986.

[2] B. Widrow, J. Glover, Jr., J. McCool, J. Kaunitz, C. Williams, R. Hearn, J. Zeidler, E. Dong, Jr., and R. Goodlin, "Adaptive noise cancelling: Principles and applications," *Proc. IEEE*, vol. 63, pp. 1692–1716, Dec. 1975.

[3] R. Ghose, "Collocation of receivers and high-power broadcast transmitter," *IEEE Trans. Broadcasting*, vol. 34, June, 1988.

[4] V. Varadan, D. Ghodgaonkar, V. Varadan, J. Kelly, and P. Glikerdas, "Ceramic phase shifter for electronically steerable antenna systems," *Microwave J.*, pp. 116–127, Jan. 1992.

[5] A. Fox, S. Miller, and M. Weiss, "Behavior and application of ferrites in the microwave region," *Bell Syst. Tech. J.*, vol. 34, pp. 5–103, Jan. 1955.

[6] C. Boyd, Jr., "A dual mode latching reciprocal ferrite phase shifter," *IEEE Trans. Microwave Theory Tech.*, vol. MTT-18, pp. 1119–1124, Dec. 1970.

[7] R. Ghose, "Same frequency repeater amplification," *J. Inst. Electron. Telecommun. Eng.*, vol. 37, no. 4, pp. 357–362, 1991.

[8] H. Scharfman, "Three new ferrite phase-shifters," *Proc. IRE*, vol. 44, pp. 1456–1459, Oct. 1956.

5

CANCELLATION OF INTERFERENCE FROM REMOTE SOURCES

By definition, a direct sample of the interference, which needs to be suppressed at the receive line, from the interference source or its transmitter as a reference, is not possible for interferences of remote origin and, hence, the reference interference has to be received by an antenna, which may receive other signals as well, including the desired signal and noise, along with the interference. The consequential performance of an adaptive canceller will be affected by the contamination of the reference signal. Quantitative effects of such a contamination on the performance of the canceller are addressed in this chapter. Additionally, the use of an adaptive array with appropriate nulls along the directions of the interferences to mitigate interference is also discussed.

5.1 BASIC CONCEPT AND CANCELLER REQUIREMENTS

Practical problem areas where a cancellation of an undesired signal or interference of remote origin is needed include radar operations or communications from a satellite where the undesired signal is received through a sidelobe of a high-gain antenna, communications subject to an intentional or unintentional interference or jamming, etc. For such cases, a reference antenna is necessary to sample the interference or the undesired signal which needs to be removed from the receive line. A reference antenna[1] on the other hand, also receives the desired signal and other undesired signals which may or may not have been received by the receiving antenna. The desired and undesired signals thus received by the reference antenna enter the receive line through the signal-controller path, although they are modified by the transfer function of the signal controller which attempts to cancel a particular interference for which the canceller is designed. Ideally, then, when a reference antenna

[1] An adaptive canceller can be implemented to rid an interference where the provision of a reference antenna is not practicable or economically feasible. Such a case is discussed later in this chapter.

must be used for an adaptive canceller, it will be highly desirable to design the reference antenna such that it exclusively receives the particular undesired signal which needs to be cancelled from the receive line. This, however, may not be possible in most cases. The effect of the desired signal, received by the reference antenna and injected into the receive line through the 180° hybrid (Figure 4-1), will be to modify and even reduce or distort the desired signal at the receive line and at the receiver, although an ideal adaptive canceller is presumed not to affect the desired signal, except for the through-line insertion loss introduced by the 180° hybrid.

We have already seen in Section 4.1 that when the reference antenna receives the desired signal in addition to an undesired signal which needs to be cancelled at the receive line, the adaptive canceller modifies both the desired and undesired signals. As seen in Eq. (4.16), for example, the undesired signal which is received as S_1 at the reference antenna and $(L_1 S_1 + L_2 \bar{S}_1)$ at the receiving antenna becomes, following cancellation, $\bar{S}_{1R}$, where

$$\begin{aligned}\bar{S}_{1R} &= \left[(L_1 - L_3)\frac{\langle S_2^2\rangle}{C^2}\right] S_1 + \left[(L_2 - L_4)\frac{\langle \bar{S}_2^2\rangle}{C^2}\right]\bar{S}_1 \\ &= \frac{b}{a+b}\{S_1(L_1 - L_3) + \bar{S}_1(L_2 - L_4)\}\end{aligned}$$

where
S_1 = undesired signal at the reference antenna
S_2 = desired signal at the reference antenna
$\bar{S}_1$ = quadrature component of S_1 at the reference antenna
$\bar{S}_2$ = quadrature component of S_2 at the reference antenna.

$\langle S_1^2\rangle = a$, $\langle S_2^2\rangle = b$, $C^2 = a + b$, and L_1, L_2, L_3, and L_4 are constants relating to different propagation paths for the desired and undesired signals, between the reference antenna and the sources of both signals, and between the receiving antenna and the same sources of signals, as shown in Figure 4-1.

Similarly, the desired signal which appears at the receiving antenna as

$$S_{2R} = L_3 S_2 + L_4 \bar{S}_2$$

becomes, following the cancellation,

$$\begin{aligned}\bar{S}_{2R} &= \left[(L_3 - L_1)\frac{\langle S_1^2\rangle}{C^2}\right] S_2 \\ &\quad + \left[(L_4 - L_2)\frac{\langle \bar{S}_1^2\rangle}{C^2}\right]\bar{S}_2 \\ &= \frac{a}{a+b}[L_3 S_2 + L_4 \bar{S}_2 - (L_1 S_2 + L_2 \bar{S}_2)]. \end{aligned} \tag{5.1}$$

It may be surprising at first glance that the deviation of the desired signal from what is received at the receiving antenna depends on the propagation path parameters L_1 and L_2 for the undesired signal among others. This dependence is due to the fact that the desired signal as received by the reference antenna must necessarily flow through the signal controller along with the undesired signal, and the signal-controller parameters K and ϕ are directly related to L_1 and L_2 associated with the undesired signals S_1 and $\bar{S}_1$.

The relations shown in Eq. (5.1) can also be expressed in terms of more familiar functions involving the transforms of the desired and undesired signals and the transfer functions of various propagation paths. Thus, for example, if $S_1'(\omega)$ and $S_2'(\omega)$ denote, respectively, the transforms of the undesired and desired signals, the input of the receiving antenna may be expressed as

$$S_R(\omega) = \bar{M}_1(\omega)S_1'(\omega) + \bar{M}_2(\omega)S_2'(\omega) \tag{5.2}$$

where

$\bar{M}_1$ and $\bar{M}_2$ = the transfer functions of the propagation paths for the undesired and desired signals, respectively, between their respective sources and the receiving antenna, as shown in Figure 5-1.

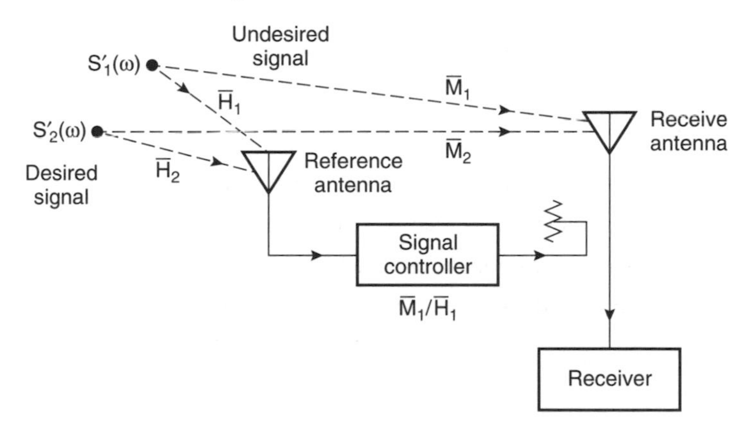

Figure 5-1 Adaptive canceller using a reference antenna, thereby introducing a desired signal distortion.

The same signals as received by the reference antenna can be written as

$$S_{\text{Ref}}(\omega) = \bar{H}_1(\omega)S_1'(\omega) + \bar{H}_2(\omega)S_2'(\omega) \tag{5.3}$$

where

$\bar{H}_1$ and $\bar{H}_2$ = the transfer functions of the propagation paths for the undesired and desired signals, respectively, between their sources and the reference antenna.

From Figure 4-1 and Eqs. (5.2) and (5.3)

$$
\begin{aligned}
\bar{M}_1 S_1' &= L_1 S_1 + L_2 \bar{S}_1 \\
\bar{M}_2 S_2' &= L_2 S_2 + L_4 \bar{S}_2 \\
\bar{H}_1 S_2' &= S_1 \\
H_2 S_2' &= S_2
\end{aligned}
\tag{5.4}
$$

where
the parameters L_1, L_2, L_3, and L_4 are related to the more familiar propagation-path transfer functions, $\bar{M}_1$, $\bar{M}_2$, $\bar{H}_1$, and $\bar{H}_2$.

Equation (5.1) shows the extent of the modification of the desired signal even though the cancellation process addresses only the undesired signal. Quantitatively, one may define this modification or distortion, D, of the desired signal at the receiving antenna following the cancellation as the ratio of the spectrum of the unwanted contribution ΔS_2 of the desired signal at the receiver (Figure 4-1) to the spectrum of the desired signal component received by the receiving antenna. From Eq. (4.17)

$$\Delta S_2 = \frac{a}{a+b}[L_1 S_2 + L_2 \bar{S}_2]. \tag{5.5}$$

Now, if one denotes

S_{2R} = desired signal as received by the receiving antenna

S_{2A} = desired signal as received by the reference or auxiliary antenna

S_{1A} = undesired signal as received by the reference antenna

S_{1R} = undesired signal as received by the receiving antenna

then

$$\Delta S_2 = \frac{a}{a+b} S_{1R} \frac{S_{2A}}{S_{1A}} \tag{5.6}$$

and

$$D = \frac{a}{a+b} \frac{S_{1R}}{S_{2R}} \frac{S_{2A}}{S_{1A}}. \tag{5.7}$$

Also, the factor (S_{2A}/S_{1A}) may be regarded as the signal-to-noise ratio (SNR) at the reference antenna, S_{2A} being the desired signal and S_{1A} being the undesired signal or noise. Similarly, the factor (S_{2R}/S_{1R}) may be regarded as the signal-to-noise ratio at the receiving antenna. Thus

$$\begin{aligned} D &= \frac{a}{a+b} \frac{(\text{SNR})_{\text{Reference antenna}}}{(\text{SNR})_{\text{Receiving antenna}}} \\ &\simeq \frac{(\text{SNR})_{\text{Reference antenna}}}{(\text{SNR})_{\text{Receiving antenna}}} \end{aligned} \tag{5.8}$$

when $a \gg b$, as is usually the case. Equation (5.8) suggests that a low distortion of the desired signal results from a low (SNR) at the reference antenna and a high (SNR) at the receiving antenna.

In many practical cases of concern where a directional reference antenna is used to selectively obtain a sample of an interfering or jamming signal, one may express the reduction of the jamming and desired signals in terms of the ratio of the jamming signal to desired signal field intensities, and the receiving and reference antenna gains for the desired and jamming signals. Thus, for example, if one defines

$$R = S_J/S_D \qquad M = G_{AJ}/G_{AR} \qquad N = G_{RJ}/G_{RD} \tag{5.9}$$

where

S_J and S_D = the field intensities for the jamming and desired signals, respectively, measured, for example, in volts/meter

G_{AJ} = effective gain of the reference antenna for the jamming signal

G_{AR} = effective gain of the reference antenna for the desired signal

G_{RJ} = effective gain of the receiving antenna for the jamming signal

G_{RD} = effective gain of the receiving antenna for the desired signal.

one may then write

$$\begin{aligned} \frac{\bar{S}_{1R}}{S_{1R}} &= \frac{1}{R^2M^2+1}\left(1-\frac{M}{N}\right) \\ \frac{\bar{S}_{2R}}{S_{2R}} &= \frac{R^2M^2}{R^2M^2+1}\left(1-\frac{M}{N}\right) \end{aligned} \tag{5.10}$$

where

S_{1R} = the jamming signal at the receive line before cancellation

$\bar{S}_{1R}$ = the jamming signal at the receive line after cancellation

S_{2R} = the desired signal at the receive line before cancellation

$\bar{S}_{2R}$ = the desired signal at the receive line after cancellation.

It appears from Eq. (5.10) that, for a jamming signal suppression system, one needs to increase the value of M and decrease the value of N for any given R, which in most cases cannot be altered easily at the receiving station. The degree of cancellation of the jamming signal will be significantly improved for higher values of R.

To illustrate the significance of Eq. (5.10), one may consider an example of protecting a receiver which employs an omnidirectional antenna by a jamming signal suppression process through an adaptive canceller. If the field intensity of the jamming signal at the receiver antenna is 20 dB higher than that of the desired signal, and if a directional reference antenna is used to receive the sample of the jamming signal in preference to the desired signal with a gain ratio $(G_{AJ}/G_{AR}) =$ 20 dB, the maximum degree of cancellation achievable by the canceller will be

$$\frac{\bar{S}_{1R}}{S_{1R}} = \frac{1}{10{,}000 + 1}|1 - 10| \simeq 60\,\text{dB}. \tag{5.11}$$

It should be noted that this general result given in Eq. (5.11) is equally applicable to estimate the degree of suppression of the jamming signal when the difference in polarizations of the desired and undesired signals is utilized for discrimination, instead of the antenna directivity gain. Thus, for example, if an adaptive canceller is used to protect a vertically polarized jamming signal, one can introduce within the antenna feed system a horizontal probe which receives a component of the jamming signal, but no desired signal. For such a case, the value of M may vary from 6 to 20 dB. The corresponding value of N should be on the order of -6 to -20 dB. For a value of R corresponding to 15 dB, then the degree of suppression of the jamming signal will be approximately 30 dB. The corresponding reduction of the desired signal will be negligible, particularly when M/N is 20 dB or more. The implementation of an interference canceller for a radar where the difference of polarizations of the target-return signal and the interference is utilized for discrimination and control is further discussed later in this chapter.

5.2 CANCELLATION OF A JAMMING SIGNAL BY AN ADAPTIVE ARRAY

As noted early in Section 5.1, an adaptive cancellation is often highly desirable to rid an undesired or jamming signal received through a sidelobe of a high-gain antenna being used for a radar or a satellite communication. This is primarily because the characteristics of such an undesired signal and its direction of arrival at the antenna are seldom *a priori* known and could be different at different times. A strong unwanted signal incident on the sidelobe of an antenna can severely interfere with the reception of a weaker signal at the main beam.

The conventional method for reducing a sidelobe interference is adaptive beamforming [2]–[4] by an adaptive array.

To examine the adaptive beamforming concept, let us consider first an adaptive array consisting of N number of array elements, where each array element is connected to a weighting network W_1, W_2, etc., as shown in Figure 5-2. The object of the weighting networks is to maximize, in general, the signal-to-noise (SNR) of the array output which, in this case, is the sum of the outputs of the weighting networks. For an adaptive array, the weighting networks adapt to the signal and noise environments to maximize the SNR.

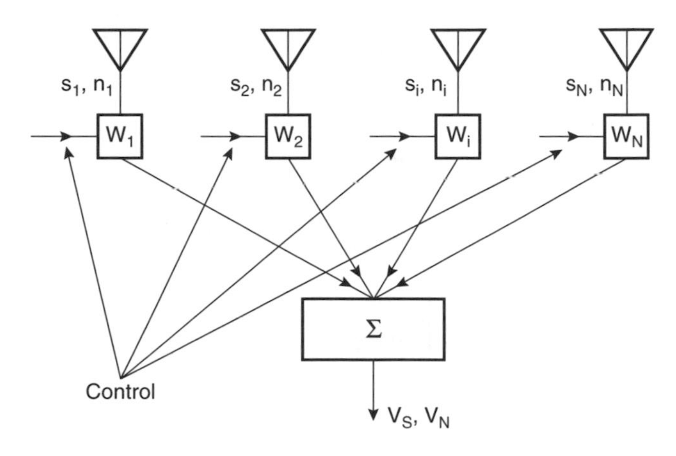

Figure 5-2 Functional representation of an adaptive array. The weighting networks are adaptively adjusted to optimize performance.

Let the desired signal received by the Kth array element be $\alpha S_K \exp(-j\theta_{SK})$, where S_K is the amplitude, θ_{SK} is the phase angle and α defines the level and time variation of the desired signal. If the weighting network for the Kth array element is $W_K \exp(-j\phi_{WK})$, then the desired signal at the output of the array, as shown in Figure 5-2, can be written as

$$V_S = \alpha \sum_{K=1}^{N} S_K W_K \exp(-j\theta_{SK} - j\phi_{WK}). \tag{5.12}$$

Similarly, if n_K denotes the noise received by the Kth array element, then the noise output of the array, V_n, can be written as

$$V_n = \sum_{K=1}^{N} n_K W_K \exp(-j\phi_{WK}). \tag{5.13}$$

In matrix notation

$$\begin{aligned} V_S &= \alpha \tilde{W} S = \alpha \tilde{S} W \\ V_n &= \tilde{W} n = \tilde{n} W \end{aligned} \tag{5.14}$$

where
W = the column vector representing, the weighting networks
S = the column vector representing the desired signal
n = the column vector representing the noise.

That is,

$$W = \begin{bmatrix} W_1 \exp(-j\phi_{W1}) \\ W_2 \exp(-j\phi_{W2}) \\ \vdots \\ W_N \exp(-j\phi_{WN}) \end{bmatrix} \tag{5.15}$$

$$S = \begin{bmatrix} S_1 \exp(-j\theta_{S1}) \\ S_2 \exp(-j\theta_{S2}) \\ \vdots \\ S_N \exp(-j\theta_{SN}) \end{bmatrix} \tag{5.16}$$

and

$$n = \begin{bmatrix} n_1 \\ n_2 \\ \vdots \\ n_N \end{bmatrix} \tag{5.17}$$

where
$\tilde{W}$, $\tilde{S}$, and $\tilde{n}$ = the transposed matrices W, S, and n, respectively.

Since the square of the absolute values of V_S and V_n are, respectively, proportional to the signal and noise power, a measure of SNR is obtained from the relation

$$\text{SNR} = \frac{|V_S|^2}{|V_n|^2}. \tag{5.18}$$

Now, the noise power P_n can be obtained from the expected value of $|V_n|^2$. Thus, from Eq. (5.13)

$$\begin{aligned} P_n &= \{|V_n|^2\} \\ &= E\left\{\sum_{i=1}^{N}\sum_{k=1}^{N} n_i n_k^* W_i W_k^* (-j\phi_{Wj} + j\phi_{Wk})\right\} \end{aligned} \tag{5.19}$$

where
E = the expectation operator
$*$ denotes the complex conjugate operation.

Since the weighting network parameters are discrete functions, the expectation operator E will only affect the noise terms.

Let us assume[2] that in the absence of an interference or a jamming signal, the noise power received by each array element is the same and the noise n_j is uncorrelated with n_K, where $j \neq k$ such that

$$\begin{aligned} E(n_j n_k^*) &= P_0 \qquad j = k \\ &= o \qquad j \neq k. \end{aligned} \tag{5.20}$$

From Eqs. (5.19) and (5.20), then

$$P_n = P_o \sum_{k=1}^{N} |W_K^2| \tag{5.21}$$

where

W_k = a real number, as assumed in Eqs. (5.12) and (5.13).

Now, from Eq. (5.12)

$$|V_S|^2 = \alpha^2 \sum_{i=1}^{N} \sum_{k=1}^{N} S_i S_k^* W_i W_k^* \exp(-j\theta_{si} - j\phi_{Wi} + j\theta_{sk} + j\phi_{Wk}) \tag{5.22}$$

and from the Schwarz inequality

$$|V_S|^2 \leq \alpha^2 \sum_{k=1}^{N} |S_k^2| \cdot \sum_{k=1}^{N} |W_K^2|.$$

Thus

$$\frac{|V_S|^2}{|V_n|^2} \leq \frac{\alpha^2}{P_o} \sum_{k=1}^{N} |S_k^2| \tag{5.23}$$

S_K is assumed to be a real number, as in Eq. (5.12). Equation (5.23) sets an upper bound of the SNR.

Now, if we assume

$$[W_K \exp(-j\theta_{WK})] = \mu[S_K \exp(-j\phi_{SK})]^* \tag{5.24}$$

where

μ = a constant,

[2]This assumption is not a restriction on the array theory that follows as Applebaum [4] has shown that this apparent restriction can be eliminated by a suitable transformation of the matrix representing the weighting networks. Such a transformation is also discussed in Section 6.2.

and substitute this relation in Eqs. (5.19) and (5.22), we obtain

$$P_n = \mu^2 P_o \sum_{k=1}^{N} |S_k^2|$$

$$|V_S|^2 = \alpha^2 \mu^2 \left[\sum_{k=1}^{N} |S_K^2| \right]^2 \tag{5.25}$$

and

$$\mathrm{SNR} = \frac{\alpha^2}{P_o} \sum_{k=1}^{N} |S_K^2|. \tag{5.26}$$

From Eqs. (5.23) and (5.26), it is seen that the maximum SNR that is permissible under Eq. (5.23) is obtained when Eq. (5.24) is satisfied. Thus, Eq. (5.24) may be assumed as the control law for optimizing the array operation when the noise power received by each array element is the same and the array element noise is uncorrelated.

Using the matrix notation introduced in Eqs. (5.15)–(5.17), one may write

$$P_n = p_o |\tilde{W}^* W| \tag{5.27}$$

corresponding to Eq. (5.21) and

$$\frac{|V_S|^2}{|V_n|^2} \leq \frac{\alpha^2}{p_o} |\tilde{W}^* W| \cdot |\tilde{S}^* S| \tag{5.28}$$

corresponding to Eq. (5.23).

The corresponding control law for maximizing the SNR as given in Eq. (5.24) now becomes

$$W_{\mathrm{opt}} = \mu S^* \tag{5.29}$$

where

W_{opt} = the optimum weighting network parameters in matrix form.

In Eq. (5.20), the noise received by each array element is assumed to be uncorrelated with others, and the noise power received by each element was assumed to be the same. This, however, may not always be the case, particularly when a jamming signal or an interference is involved. For such a case, one may denote

$$E\{n_j^*, n_k\} = \mu_{jk} \tag{5.30}$$

where

μ_{jk} = the covariance of the noise components j and k.

From Eq. (5.19), the corresponding equation for P_n becomes

$$P_n = \sum_{i=1}^{N} \sum_{k=1}^{N} \mu_{ik} W_i W_k^* (-j\phi_{Wi} + j\phi_{Wk}) \tag{5.31}$$

and in matrix notation

$$P_n = \tilde{W}^* M W \tag{5.32}$$

where
$M = E\{n^* \tilde{n}\} = [\mu_{jk}]$,

M being the covariance matrix of the noise components.

The control law[3] for optimizing the array becomes

$$M W_{\text{opt}} = \mu S^*. \tag{5.33}$$

So far, we have considered the general theory of an adaptive array and the control law for the optimization of the weighting networks without regard to its operation or performance in the presence of an interference. We will now see that if an adaptive array is designed for a given desired signal and a noise environment as defined in Eq. (5.20), that is, one without a jamming signal, and if the same control law, given in Eq. (5.33), is followed to change the weighting network parameters to accommodate the jamming signal, the resulting weighting parameters will automatically reduce the jamming signal at the output of the array. To illustrate this concept and to determine quantitatively the degree of a jamming signal suppression due to the adaptive process, let us consider a linear, uniformly spaced array designed for a given desired signal and a set of weighting parameters for the noise environment defined in Eq. (5.20). Let M_0 and W_0 denote, respectively, the covariance and weighting parameter matrices for this case. Clearly, M_0 in matrix form for this case is

$$M_0 = \begin{bmatrix} p_0 & 0 & 0 & 0 \\ 0 & p_0 & 0 & 0 \\ 0 & 0 & p_0 & 0 \\ 0 & 0 & 0 & p_0 \end{bmatrix}$$

$$= p_0 I_N \tag{5.34}$$

where
p_0 = the noise power output of each element
N = the number of array elements
I_N = an identity matrix of order N.

[3]For a proof of the relation among the signal components, covariance, and weighting parameters as given in Eq. (5.33), one may refer to Applebaum [4].

For a desired signal being received from a direction θ_0 from the physical boresight, one may write

$$W_0 = \begin{bmatrix} W_1 \\ W_2 \exp(-j\beta_0) \\ W_3 \exp(-j2\beta_0) \\ \vdots \\ W_N \exp(-j(N-1)\beta_0) \end{bmatrix} \tag{5.35}$$

where
W_1, W_2, etc., are real numbers,

$$\beta_0 = \frac{2\pi d}{\lambda} \sin\theta_0$$

and
$d =$ the spacing between two adjacent array elements
$\lambda =$ the wavelength of the desired signal.

The array pattern as a function of an angle θ, also measured from the same physical boresight, can be written as

$$G_0(\theta) = \sum_{k=1}^{N} W_K \exp(j[K-1](\beta - \beta_0)) \tag{5.36}$$

where

$$\beta = \frac{2\pi d}{\lambda} \sin\theta.$$

In matrix notation, the array pattern $G_0(\theta)$ as given in Eq. (5.36) can be written as

$$G_0(\theta) = \tilde{B} W_0 \tag{5.37}$$

where

$$B = \begin{bmatrix} 1 \\ \exp(j\beta) \\ \exp 1(j2\beta) \\ \vdots \\ \exp(j[N-1]\beta) \end{bmatrix}.$$

Assuming that the control law for the adaptive array is obeyed for the noise environment defined in Eq. (5.20), we may write from Eqs. (5.33) and (5.34)

$$M_0 W_0 = \mu S^* = p_0 W_0. \tag{5.38}$$

Now, suppose that a jamming signal or an unwanted interference appears at the array on top of the noise environment defined in Eq. (5.20), and the same control law as given in Eq. (5.33) is followed by the adaptive array to accommodate the jamming signal. Since the desired signal remains unchanged, one may write from Eq. (5.38)

$$MW = M_0 W_0 = p_0 W_0 \qquad \text{or} \qquad W = p_0 M^{-1} W_0 \tag{5.39}$$

M is now the covariance matrix for the combined noise and jamming environment; W denotes, in matrix form, the new weighting network parameters, presumably changed due to the presence of the jamming signal; and M^{-1} is the inverse of the matrix M.

If the jamming signal arrives at the array from an angle θ_i, measured from the boresight, and if it is denoted by $J(t)$ at the first array element, then at the kth array element, the jamming signal can be expressed as

$$j_k = J(t)\exp([k-1]j\beta_i) \tag{5.40}$$

where
$\beta_i = (2\pi d/\lambda)\sin\theta_i$, and a narrow jamming signal bandwidth is assumed.

Assuming, further, a linear medium where the jamming signal is superimposed on the noise environment, the covariance matrix M of the total undesired signal as noted in Eq. (5.20) will be the sum of those of the original noise and the jamming signal. Thus

$$\begin{aligned} M &= M_0 + M_i \\ &= P_0 I_N + M_i \end{aligned} \tag{5.41}$$

where
I_N = an identity matrix of order N.

But M_i from Eq. (5.40) is

$$M_{i(k,\ell)} = p_i \exp(-j[k-\ell]\beta_i) \tag{5.42}$$

where
p_i = the envelope jamming power in each array element.

In matrix form, M_i is seen to be a Hermitian matrix of order N, in which all terms on the same diagonal are equal and $\tilde{M}_i = M_i^*$. Here, M_i is given by

$$M_i = p_i \begin{bmatrix} 1 & \exp(j\beta_i) & \exp(j2\beta_i) & \cdot & \exp[(N-1)j\beta_i] \\ \exp(-j\beta_i) & 1 & \exp(j\beta_i) & \cdot & \exp[(N-2)j\beta_i] \\ \cdot & \cdot & 1 & & \\ \cdot & & & & \\ \cdot & & & & \\ \exp[-(N-1)j\beta_i] & \cdot & \cdot & \cdot & 1 \end{bmatrix} \tag{5.43}$$

One may also write

$$M_i = P_i H^* U H \tag{5.44}$$

where

H = a diagonal matrix given by

$$H = \begin{bmatrix} 1 & \cdot & \cdot & \cdot \\ \cdot & \exp(j\beta_i) & \cdot & \cdot \\ \cdot & \cdot & \cdot & \exp[(N-1)j\beta_i] \end{bmatrix}$$

U = an $N \times N$ matrix of ones.

From Eqs. (5.41) and (5.44)

$$M = P_0 I_N + P_i H^* U H \tag{5.45}$$

and

$$M^{-1} = \frac{1}{P_0}\left\{ I_N - \left(\frac{P_i}{P_0 + N p_i}\right) H^* U H \right\}. \tag{5.46}$$

The proof of the validity of Eq. (5.46) is not provided here, although it can be verified as follows:

$$\begin{aligned} M^{-1}M &= I_N \\ &= \left[I_N - \frac{p_i}{p_0 + N p_i} H^* U H \right] \cdot \left[I_N + \frac{p_i}{p_0} H^* U H \right] \\ &= I_N + \frac{p_i}{p_0} H^* U H - \frac{p_i}{(p_0 + N p_i)} H^* U H \\ &\quad - \frac{p_i^2}{(p_0 + N p_i) p_0} H^* U H \cdot H^* U H \\ &= I_N + \frac{p_i}{p_0} \frac{N p_i}{(p_0 + N p_i)} H^* U H \\ &\quad - \frac{p_j}{p_0} \frac{p_i}{(p_0 + N p_i)} H^* U H \cdot H^* U H. \end{aligned} \tag{5.47}$$

Now, considering the matrix product $H^*UH \cdot H^*UH$, one may write

$$H^*UH \cdot H^*UH = H^*U(HH^*)UH = H^*UUH \tag{5.48}$$

since $HH^* = I_N$.

Also, since U is an $N \times N$ matrix of ones

$$U \cdot U = NU. \text{ Thus } H^*UUH = NH^*UH. \tag{5.49}$$

Substituting the result of Eq. (5.49) into Eq. (5.47), we note that

$$\left[\frac{I_N}{p_0} - \frac{p_i}{p_0(p_0 + Np_i)} H^* U H\right] \cdot M = I_N;$$

hence

$$M^{-1} = \left[\frac{I_N}{p_0} - \frac{p_i}{p_0} \frac{1}{(p_0 + Np_i)} H^* U H\right]. \tag{5.50}$$

Now, from Eqs. (5.39) and (5.50)

$$\begin{aligned} W &= p_0 M^{-1} W_0 \\ &= W_0 - \frac{p_i}{(p_0 + Np_i)} H^* U H W_0. \end{aligned} \tag{5.51}$$

But from the definition of W_0 as given in Eq. (5.35) and H as given in Eq. (5.44)

$$H W_0 = \begin{bmatrix} W_1 \\ W_2 \exp(j[\beta_i - \beta_0]) \\ W_3 \exp(j2[\beta_i - \beta_0]) \\ \vdots \\ W_N \exp(j(N-1)[\beta_i - \beta_0]) \end{bmatrix}. \tag{5.52}$$

Now

$$U H W_0 = \begin{bmatrix} 1 & 1 & 1 & \cdot & 1 \\ 1 & \cdot & \cdot & \cdot & \cdot \\ \cdot & \cdot & \cdot & \cdot & 1 \\ \cdot & & & & \\ 1 & 1 & & \cdot & 1 \end{bmatrix} \begin{bmatrix} W_1 \\ W_2 \exp(jx) \\ W_3 \exp(j2x) \\ \vdots \\ W_N \exp(j[N-]x) \end{bmatrix}$$

$$= \sum_{k=1}^{N} W_k \exp(j[k-1]x) \begin{bmatrix} 1 \\ 1 \\ \vdots \\ 1 \end{bmatrix} \tag{5.53}$$

where
$x = \beta_i - \beta_0$.

From Eqs. (5.36) and (5.53)

$$UHW_0 = G(\theta_i)\begin{bmatrix} 1 \\ 1 \\ 1 \\ \vdots \\ 1 \end{bmatrix} \tag{5.54}$$

where
$\beta i = (2\pi d/\lambda)\theta_i$.

Finally

$$H^*UHW_0 = G(\theta_i)\begin{bmatrix} 1 \\ 0\exp(j\beta_i) \\ 0 \\ 0\cdots\exp(j[N-1]\beta_i) \end{bmatrix} \tag{5.55}$$

$$= G(\theta_i)B_i^*$$

where

$$B_i = \begin{bmatrix} 1 \\ \exp(j\beta_i) \\ \exp(j2\beta_i) \\ \vdots \\ \exp(j[N-1]\beta_i) \end{bmatrix}.$$

A substitution of the result of Eq. (5.55) into Eq. (5.51) yields

$$W = W_0 - \frac{p_i}{(p_0 + Np_i)}G(\theta_i)B_i^*. \tag{5.56}$$

From Eqs. (5.37) and (5.56), the array pattern $G(\theta)$ as a function of θ, measured from the boresight, can be written as

$$\begin{aligned} G(\theta) &= \tilde{B}W \\ &= \tilde{B}W_0 - \left(\frac{p_i}{p_0 + Np_i}\right)G(\theta_i)\tilde{B}B_i^*. \end{aligned} \tag{5.57}$$

Again, from Eq. (5.37)

$$\tilde{B}W_0 = G_0(\theta).$$

Now, from the definitions of $\tilde{B}$ and B_i, we have

$$\tilde{B}B_i^* = [1\ \exp(j\beta)\cdots\exp(j[N-1]\beta)]\cdot\begin{bmatrix}1\\ \exp(-j\beta_i)\\ \exp(-2j\beta_i)\\ \vdots\\ \exp(-j[N-1]\beta_i)\end{bmatrix}$$

$$= 1+\exp(-jx)+\exp(-j2x)\cdots\exp(-j[N-1]x)$$

$$= \bar{S}(x) \tag{5.58}$$

where
$x = \beta_i - \beta_0$.

Multiplying $\bar{S}$ by exp $(-jx)$ and subtracting $\bar{S}$ from the result yields

$$\bar{S}(\exp(-jx)-1) = 1-\exp(-jNx).$$

Therefore

$$\bar{S}(x) = e^{-j[N-1]x/2}\frac{\sin(Nx/2)}{\sin(x/2)}. \tag{5.59}$$

From Eqs. (5.57) and (5.59), then

$$G(\theta) = G_0(\theta) - \left(\frac{p_i}{p_0+Np_i}\right)G_0(\theta_i)\bar{S}(x). \tag{5.60}$$

Equation (5.60) shows that the array pattern indicating the array gain at any angle θ in the presence of a jammer when the adaptive control law is followed consists of two parts. The first part is $G_0(\theta)$, that is, the pattern that would have existed in the absence of the jammer while optimizing the desired signal reception. The second part which is subtracted from the first part is a function of $x = (\beta_i - \beta)$ and is centered on the jammer. The result is shown in Figure 5-3.

To show the effect of the adaptive array or the advantage of the array toward reducing the array gain along the direction of the jamming signal, we note from Eq. (5.60) that

$$G(\theta_i) = G_0(\theta_i) - \frac{p_i}{p_0+Np_i}G_o(\theta)\bar{S}(0) \tag{5.61}$$

corresponding to $x = 0$.

But from Eq. (5.59)

$$\bar{S}(0) = N. \tag{5.62}$$

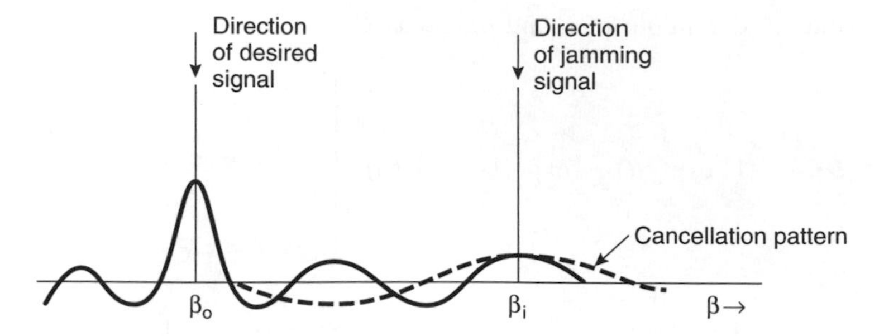

Figure 5-3 Illustration of jamming signal reduction by adaptive array. Here, the cancelling pattern reduces the array gain along the jamming signal direction.

Hence

$$G(\theta) = G_0(\theta_i)\frac{p_0}{p_0 + Np_i}. \tag{5.63}$$

If the array weights did not change when the jamming signal appeared, from their settings when there was no jamming signal, the array gain along the direction of the jamming signal would have been $G_0(\theta_i)$. Thus, the adaptive array has reduced the array gain along the jamming-signal direction by a factor

$$G(\theta_i)/G_0(\theta_i) = \frac{1}{1 + (Np_i/p_0)} \tag{5.64}$$

where
(Np_i/p_0) = the jammer-to-noise ratio in the cancellation beam.

Since $(Np_i/p_0) >> 1$, it is seen that the adaptive array gain is $1/(Np_i/p_0)$, that is, the higher the jammer-to-noise ratio in the cancellation beam, the more will be the advantage of the adaptive array toward reducing the jamming signal. The adaptive array advantage shown in Eq. (5.64) is a ratio of the two voltage gains.

Although a significant reduction of an unwanted or jamming signal is achieved, particularly for a large array, that is, for large N, automatically when the control law given in Eq. (5.33) is followed, the implementation of such a control law needs to be considered for practical applications. Applebaum [4] has given an illustration of one such implementation.

The basic idea to achieve the control law $MW_{\text{opt}} = \mu S^*$ is shown in Figure 5-4. Here, an N-element adaptive array is assumed, and the signal e_k, as received by the Kth element, is multiplied by the weight W_k and the weighted signals for all channels are summed, as is the case for any controllable array. Usually, the multiplication and summation, shown in Figure 5-4 occur on carrier frequencies at an intermediate frequency (IF) much lower than the carrier frequency of the desired signal. The weight W_k for the channel K is derived by correlating e_k with

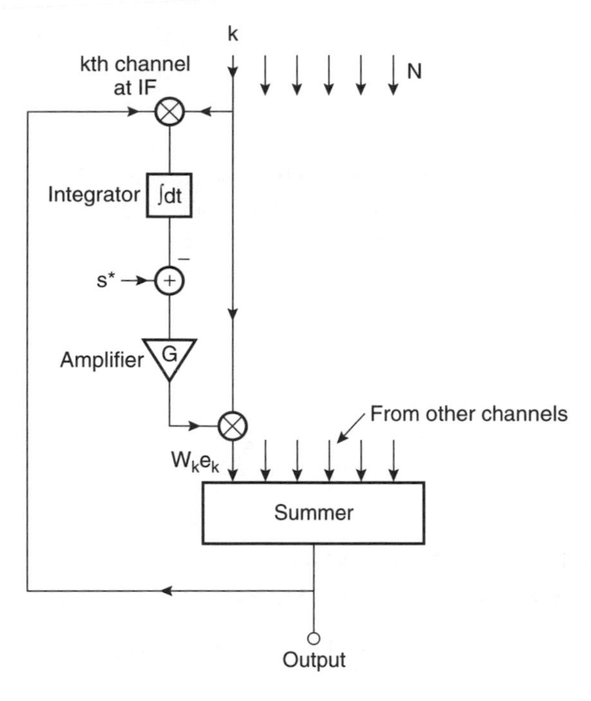

Figure 5-4 Illustration of a control loop for kth channel. Implementation of adaptive array control law is illustrated in the figure.

the sum signal, subtracting the correlation from the desired signal component S_k^*, and then using a high-gain amplifier. The process, as can be described from Figure 5-4, is

$$W_K = G\left[S_K^* - e_k^* \sum_{i=1}^{N} W_i e_i\right] \tag{5.65}$$

or

$$\sum_{i=1}^{N} W_i \left[e_k^* e_i + \frac{\delta_{ik}}{G}\right] = S_K^* \tag{5.66}$$

where

$K = 1, 2, \ldots, N$

$$\delta_{ik} = \begin{cases} 1 & \text{when } K = i \\ 0 & \text{when } K \neq i. \end{cases}$$

Recalling that [4] $e_k^* e_i$ is an element of the covariance matrix M, the N equations, one for each i ranging from 1 to N, as given in Eq. (5.66), can be written in rm as

$$\left(M + \frac{I_N}{G}\right) W = S^*. \tag{5.67}$$

The equation (5.67) differs from the control law equation (5.33) by the addition of the second term (within the parentheses), which is inversely proportional to the gain of the amplifier, as shown in Figure 5-4. Since μ in Eq. (5.33) is a constant, it can be set equal to 1. Also, since this second term approaches zero as G approaches infinity, its effect can be made negligible for a sufficiently large value of G.

It should be noted here that the control loop shown in Figure 5-4 has to be repeated for each channel, and when N is large to offer a real advantage toward the adaptive reduction of an arbitrary jamming or unwanted signal, the complexity of the implementation means cannot be ignored.

5.3 SIDELOBE CANCELLATION BY AN ADAPTIVE CANCELLER

We have seen in the previous section how an adaptive receiving array can be used to reduce or eliminate a directional interference by adaptive beamforming, thereby improving the signal-to-noise ratio of the array output. The problem where there is an existing, satisfactory, high-gain antenna, such as a large dish, which is designed for a specific communication or radar operation corresponding to a desired $G_0(\theta)$ as discussed in the previous section, but which needs to be protected from a jamming signal that may appear anywhere at the sidelobes of the main antenna pattern, is not uncommon. Since the characteristics or look angle of the jamming signal are usually not known, one needs an adaptive solution to rid the jamming signal from the receive line or receiver. An auxiliary antenna array can be used to work with the existing antenna in an adaptive manner to cancel an arbitrary jamming signal at the antenna sidelobe without affecting the operation of the main antenna. The schematic arrangement for such a sidelobe canceller is shown in Figure 5-5.

Let it be assumed that the gains of the auxiliary antennas are designed to approximate the average sidelobe level of the main receiving antenna gain pattern. The amount of desired signal received by the auxiliary antennas is negligible compared to the same desired signal received at the main beam of the main receiving antenna. For the sidelobe cancelling arrangement, the auxiliary antennas provide independent replicas of the jamming signals in the sidelobes of the main receiving antenna pattern. Since the desired signal collected by the auxiliary antennas is negligible, and the main antenna has a carefully designed pattern, we may choose

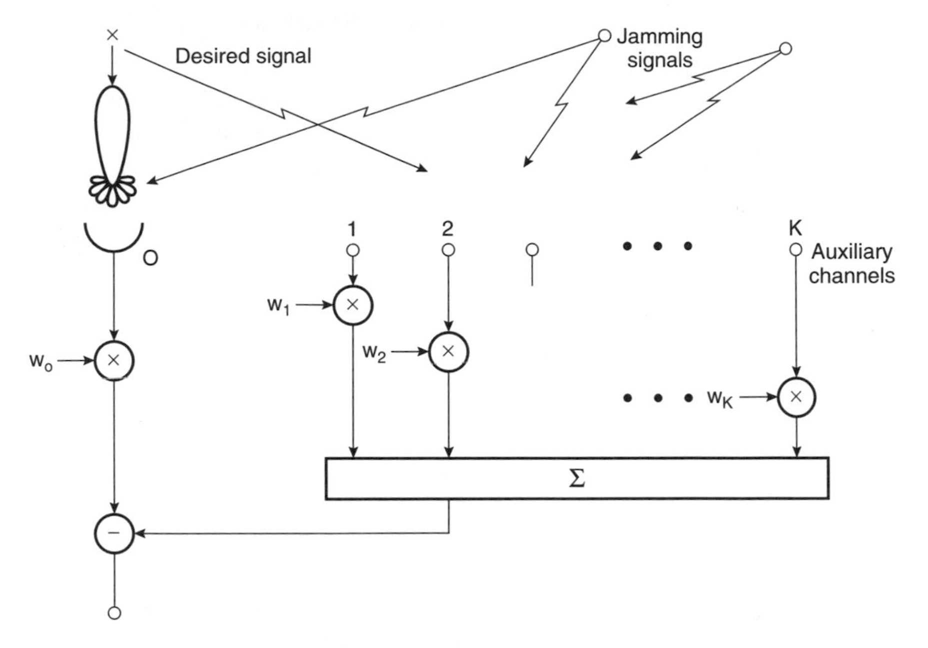

Figure 5-5 Sidelobe cancellation utilizing adaptive beamforming concept.

the signal vector S as an $(N+1)$ column vector given by

$$S = \begin{bmatrix} 1 \\ 0 \\ 0 \\ 0 \\ \vdots \\ 0 \end{bmatrix}. \tag{5.68}$$

It should be noted that here the signal vector S is similar to the signal vector in Eq. (5.16), which does not include any information on the level and time variation of the desired signal.

From Eq. (5.33), it is seen that the optimum control law for an adaptive array is

$$M'W' = \mu S^* \tag{5.69}$$

where
M' = the $(N+1) \times (N+1)$ covariance matrix of all the channels or array elements

W' = the $(N+1)$ column vector of all the weights
μ = a constant.

Now, let M be the $N \times N$ covariance matrix of the auxiliary channels only, and let W be the N column vector of the auxiliary channel weights. Equation (5.69) can be partitioned as follows:

$$\begin{bmatrix} p_0 & \tilde{A}^* \\ A & M \end{bmatrix} \begin{bmatrix} W_0 \\ W \end{bmatrix} = \begin{bmatrix} \mu \\ 0 \end{bmatrix} \tag{5.70}$$

where
$p_0 = \mu'_{00}$ of M' = the noise power output of the main channel

and

$$A = \begin{bmatrix} \mu'_{10} \\ \mu'_{20} \\ \vdots \\ \mu'_{NO} \end{bmatrix}. \tag{5.71}$$

It should be noted that μ'_{K0}, according to the above definition, is the cross correlation of the noise output of the kth auxiliary antenna, with the corresponding noise output at the main or 0th channel. Equation (5.70) may further be written as two separate equations such that

$$p_o W_0 + \tilde{A}^* W = \mu \tag{5.72}$$

and

$$MW = -W_0 A, \tag{5.73}$$

Eq. (5.72) being a scalar equation.

The control law expressed in Eq. (5.69) can be implemented using $N+1$ control loops, each similar to that shown in Figure 5-4. However, because of the unique form of the S vector as given in Eq. (5.68), it is possible to achieve an optimum control with only N control loops. To show this, we note that if the weight vector W' is optimum for a given M', then any multiple of W' will also be optimum since, from Eq. (5.69), the value of μ' will be different, with no consequence to the control law. Thus, instead of fixing μ and trying to control W_0 to satisfy Eq. (5.72), we can fix W_0 to satisfy Eq. (5.72) provided $W_0 \neq 0$. Now, with W_0 fixed at $\bar{W}_0$ where $\bar{W}_0$ satisfies Eq. (5.72), we may write Eq. (5.73) as

$$MW = -W_0 A. \tag{5.74}$$

Equation (5.74) then satisfies both Eqs. (5.72) and (5.73), and hence the overall control law as given in Eq. (5.69) or Eq. (5.70). Equation (5.74), therefore, is the sidelobe cancellation control law for minimizing output jamming signal power.

To illustrate the effectiveness of interference mitigation realizable by an antenna array by an example, Figure 5-6(a) and (b) show, respectively, the array gain pattern $G_0(\theta)$ in the absence of any jammer and the array pattern following the interference rejection. The array gain pattern $G_0(\theta)$, as shown in Figure 5-6(a), is a computer-simulated [4] pattern for a 21-element linear array with one-half wavelength spacing between antenna elements. This pattern is designed to have the main beam along the direction of the signal of interest and a sidelobe level not to exceed -40 dB with respect to the main beam directivity. Figure 5-6(b) shows the array pattern based on $G_0(\theta)$ and the nulls along the directions of eight randomly located jammers as shown by dots. The null depths along the directions of the jammers are approximately -60 dB or less.

Although a sidelobe canceller using an adaptive array, as shown in Figure 5-5, is an effective means for reducing an arbitrary interference or jamming signal at the main antenna, even when the exact characteristics and direction of arrival of the interference are not known, it has been observed that the presence of the desired signal in the Σ port of the auxiliary channels could be harmful. Such a problem, however, is readily remedied by eliminating the desired signal from the Σ port in Figure 5-5. An adaptive canceller, referenced to the main antenna receive line that contains the desired signal predominantly, can be used for this purpose. The presence of some jamming signal at the main antenna receive line, and hence at the reference line of the adaptive canceller, would be acceptable as long as the desired signal is the dominant signal at this line.

In the arrangement shown in Figure 5-5, one still needs multiple controls, corresponding to the number of antenna elements or channels in the auxiliary array. An adaptive interference canceller, as discussed earlier, can also be used for the same purpose, that is, for sidelobe cancellation, thus providing in effect a spatial selectivity in a signal conditioning process, instead of its usual role in signal processing in the time and frequency domain. The principal difference between adaptive beamforming and the adaptive cancellation means may be regarded as the number of independent controls necessary for the adaptive operation. More specifically, an adaptive beamforming needs multiple control circuits, one for each array element as shown in Figure 5-4, for one unwanted or jamming signal, while a single control is used for the adaptive canceller of the type shown in Figure 4-1 to achieve the same objective. Thus, for an adaptive canceller attempting to rid an interference appearing at the sidelobe of a high-gain antenna, a sample of the interference is used as a reference, and the signal controllers synthesize a counterinterference from the reference or sampled signal. This counterinterference, generated from the reference, when coupled into the receive-antenna line cancels the interference as received in that line through the receiving antenna sidelobe.

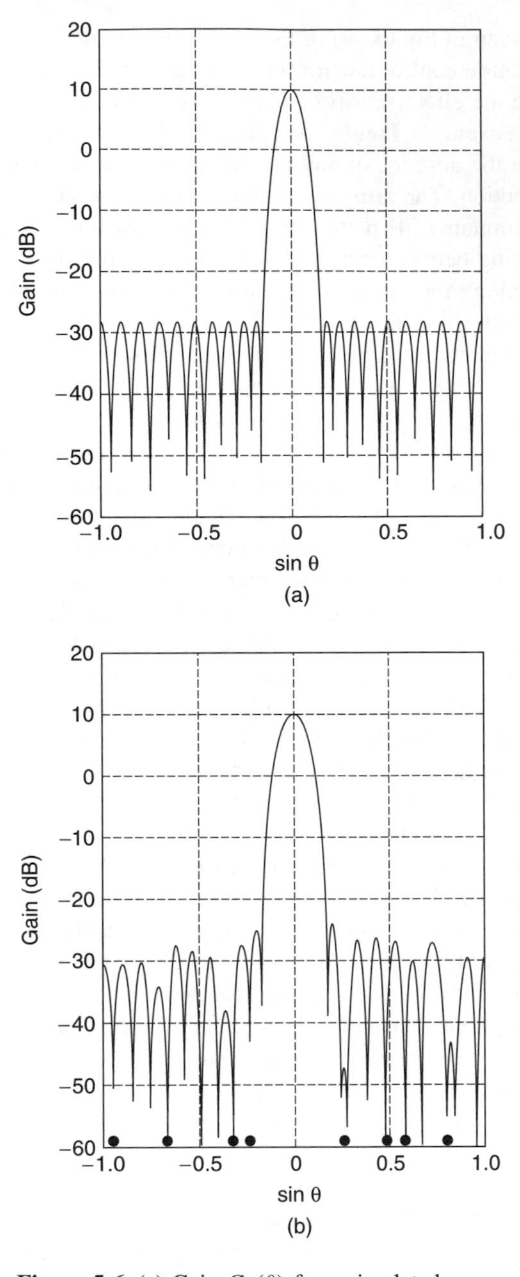

Figure 5-6 (a) Gain $G_0(\theta)$ for a simulated array in the absence of jammers. (The angle θ is the angle from boresight.) Main beam is along the desired direction, and sidelobe levels are at or below −40 db. (b) Modified array gain pattern. The array accommodates eight randomly-located jammers, as shown by dots.

The concept of sidelobe cancellation with a single adaptive control for each jamming signal is illustrated in Figure 5-7. Here, the receiving antenna is seen to receive the desired signal at its main beam, and an undesired or jamming signal is presumed to be received through its sidelobe. A reference antenna that receives the jamming signal exclusively or predominantly is used to synthesize a counter-jamming signal for the adaptive canceller such that this counterjamming signal and the jamming signal at the receive antenna create a jamming signal null. It is desirable, in this case, that the reference antenna receive little or no desired signal, as otherwise the desired signal at the receive line could be modified as a result.

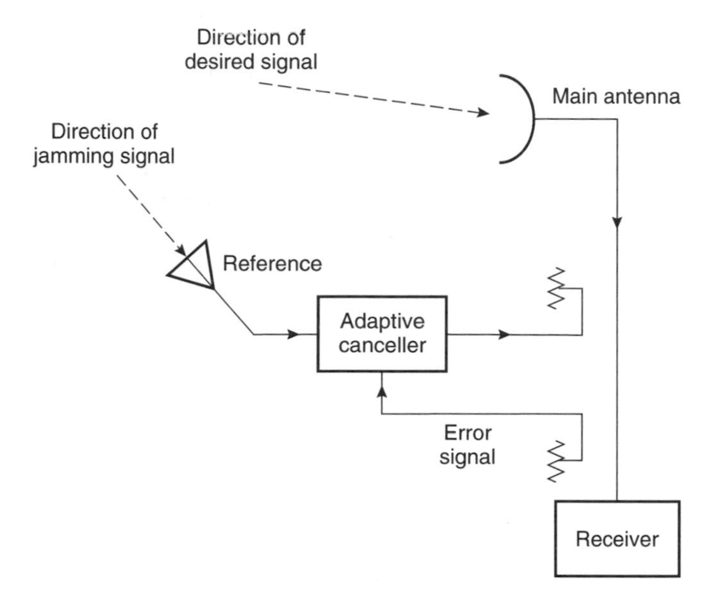

Figure 5-7 Illustration of basic concept of sidelobe cancellation.

The simplest form of sidelobe cancelling originally devised by Howells [2] and Applebaum [4] is illustrated in Figure 5-8. Here, two omnidirectional antennas, each with a uniform directivity in all directions, are used, one being used as a receiving antenna, while the other is a reference antenna as in a typical adaptive canceller system. When there is one desired signal and one undesired or jamming signal present, both the receiving and reference antennas will receive the desired and undesired signals. If the interference is much stronger than the desired signal at the reference antenna, the operation of the adaptive canceller, shown in Figure 5-8, will be dominated by the interference. Thus, when the closed-loop control of the adaptive canceller reaches an equilibrium, the interference or undesired signal at the receive line will be cancelled by the adaptive canceller, presumably leaving only the desired signal at the receive line. The desired signal will not be cancelled since the signal-controller parameters (K, ϕ) or (K, τ) that cause a cancellation of interference at the receive line are almost always different from those needed

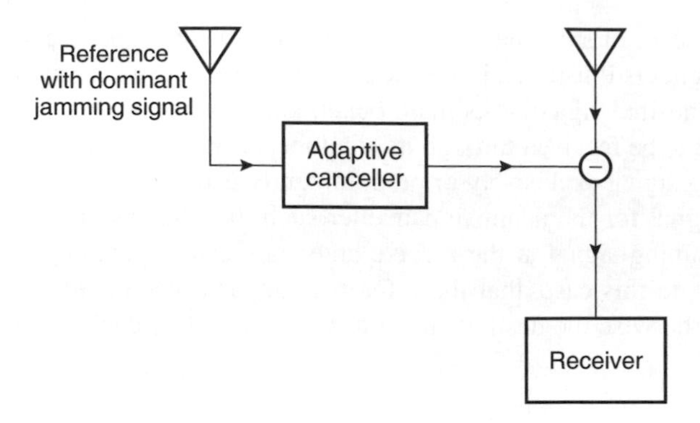

Figure 5-8 Sidelobe jamming signal cancellation employing two omnidirectional monopoles.

to cancel the desired signal using the same reference antenna. Thus, when the interference is cancelled, the desired signal should remain nonzero. The contribution of the desired signal coupled into the receive line from the signal controller of the adaptive canceller shown in Figure 5-8, however, could be very damaging at times. In fact, the canceller may operate on the desired signal instead of the jamming signal, as assumed before, when the desired signal at the reference antenna is the dominant signal, and cancel the desired signal entirely. Widrow [6] has shown that the sidelobe cancellation scheme in its simplest form, as shown in Figure 5-8, can be effective only when noise other than the jamming signal, such as the receiver noise, is present at the omnidirectional antenna. This problem, however, is easily remedied by injecting an artificial noise as suggested by Jablon [7].

Referring back to the sidelobe cancellation concept shown in Figure 5-7, one may use an interference canceller as shown in Figure 4-1 or an adaptive filter shown conceptually in Figures 2-8 and 2-9 for the adaptive canceller. To further illustrate this adaptive filter approach using a tapped delay line to cancel the jamming signal appearing at the sidelobe of a high-gain antenna, one may consider Figure 5-9. Here, the antenna output signal following the cancellation can be written as

$$y(t) = x(t) - \sum_{K=0}^{N-1} W_k a(t - K\Delta) \tag{5.75}$$

where

Δ = the tap spacing

$T = (N-1)\Delta$ = the total length of the delay line

$x(t)$ and $a(t)$ = the outputs of the main and reference antennas, respectively

W_K = the adaptive weight in amplitude and phase for the Kth channel, where $K = 1, 2, 3$, etc.

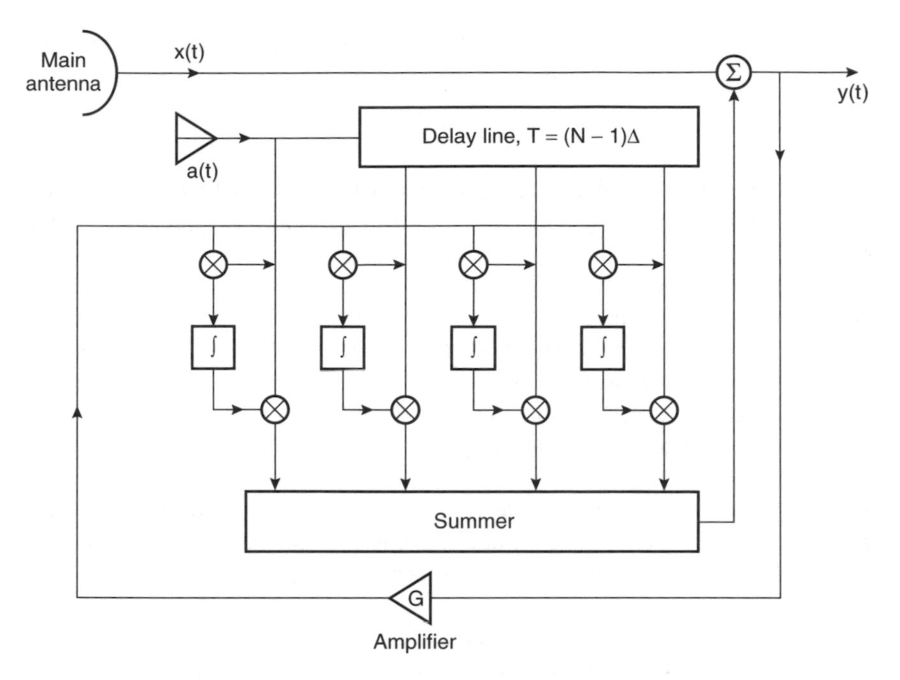

Figure 5-9 Adaptive sidelobe canceller with a single auxiliary reference antenna and a delay line.

As we have noted earlier, an adaptive filter based on the Wiener solution attempts to adjust the weights so as to minimize the power output $y(t)$. Such a condition is obtained when the expression

$$E\{y(t)a^*(t - K\Delta)\} = 0 \qquad K = 0, 1, \cdots, N - 1. \tag{5.76}$$

This process is equivalent to applying the orthogonality principle which minimizes root-mean-square (rms) residue power. Other adaptive algorithms such as sequential regression [6] or direct matrix inversion [8] may also be used to achieve a faster convergence. However, the steady-state performance will remain the same.

The optimum steady-state weights that minimize the residue power based on the Wiener solution, expressed in matrix form, can be written as

$$RW_0 = r \tag{5.77}$$

where

R = an $N \times N$ matrix with

$$R_{K\ell} = E\{a(t - \ell\Delta)a^*(t - k\Delta)\} \tag{5.78}$$

r = an $N \times 1$ column vector where

$$r_K = E\{x(t)a^*(t - K\Delta)\} \tag{5.79}$$

W_0 = the column vector of steady-state weights.

The residue power is minimized when the weights $W_K = W_{0k}$ corresponding to the steady-state solution, and for such a case [9]

$$P_r \text{ min} = P_m - \tilde{r}^* R^{-1} r. \tag{5.80}$$

As noted earlier, an adaptive canceller shown schematically in Figure 4-1 can also be used instead of the adaptive filter for the cancellation of sidelobe jamming signal.

When a reference antenna is used to obtain a sample of the interference or jamming signal to be cancelled from the sidelobe of the main antenna as shown in Figure 5-7, the gain and coverage characteristics for the reference antenna are selected to follow the envelope of the sidelobes of the main antenna. Also, an adaptive cancellation needs to be performed for all angular regions of concern of the main antenna, except that of the main beam. When the direction of the interference is known, a directional antenna can be used for the reference antenna, pointing generally toward the source of the interference. When, on the other hand, the direction of the interference is not fixed or is not known, the reference antenna must be designed and oriented so as to receive the interference regardless of its look angle.

Since most adaptive cancellers, as discussed in Chapters 3 and 4, have some losses in the signal-controller path and amplifiers are not used to avoid any non-linearity, however small, in the signal-controller path, the gain or aperture of the reference antenna must be designed so that the level of the interference is at least the same as that of the interference at the main antenna receive line. Thus, if $\vec{E}_m$ and $\vec{E}_R$ denote the voltage output at the main and reference antennas, for a given interference direction, then

$$|\vec{E}_r| = |\vec{E}_m|(1\text{–}10^{\mathrm{SLR}/20}) \tag{5.81}$$

where
SLR = the desired degree of sidelobe rejection.

Equation (5.81) assumes that the sidelobe rejection is effected with the reference antenna output of the interference being 180° out of phase with respect to the corresponding output of the main antenna.

Thus, for example, when SLR = 10 dB

$$|\vec{E}_R| = 0.683|\vec{E}_m| \tag{5.82}$$

or the voltage output of the reference antenna must be 3.3 dB less than that of the main antenna in the direction of concern. Similarly, when two identical antennas are used as reference antennas, as shown in Figure 5-10, the amplitude gain and phase of both antennas being the same, the voltage output of each reference antenna must be 9.32 dB less than that of the main antenna, the desired degree of sidelobe rejection being the same, that is, 10 dB. The gain or aperture of the reference antenna can be determined from Eq. (5.81) and the field strength of the interference in the direction of concern. If there is an insertion loss in the signal controller path, as is usually the case, the gain of the reference antenna has to be increased to accommodate the loss factor, assuming that the incident field strength of the interference remains the same.

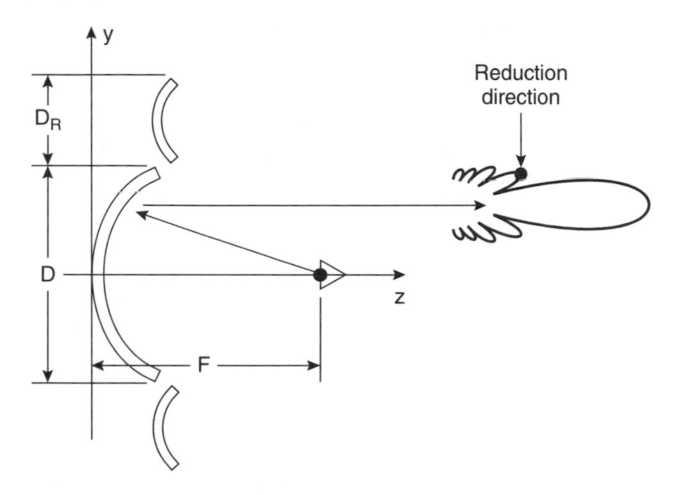

Figure 5-10 Illustration of two auxiliary antennas providing reference signal. D = diameter of main antenna. D_R = diameter of reference antenna. F = focal length of main reflector.

When the reference signal for sidelobe cancellation is obtained by an adaptive array, such as shown in Figure 5-5, one need not know the direction of arrival of the jamming signal as the array adapts to the jamming signal look angle. However, we have seen for such a case that one needs multiple control loops, the number of such loops being the same as the number of channels or array elements. If, instead, an adaptive canceller, such as shown in Figure 4-1, is used for the sidelobe cancellation approach as shown in Figure 5-7, there is only one control loop involved, but the approximate direction of arrival of the jamming signal needs to be known in that case to approximately point the reference antenna along that direction. It is possible in some cases to employ an omnidirectional reference for the sidelobe cancellation arrangement shown in Figure 5-7. Very often, however, such will not be the case, particularly when one needs to assure that there is little or no desired signal present at the reference antenna, and the dominant signal output of the reference antenna is the jamming signal which needs to be rejected from the main antenna. Also, a directivity gain of the reference antenna, with the maximum directivity along the direction of the jammer, is also helpful to enhance the degree of cancellation of the jamming signal as it usually provides a higher jamming-to-desired-signal ratio at the reference port of the synchronous detector (Figure 4-1). In some cases, it may be possible to create a two-element array directed toward the jamming signal source to provide such a directivity. One may employ an adaptive canceller, shown in Figure 4-1, again to achieve this objective. Such an arrangement is shown in Figure 5-11. Here, the adaptive canceller is used to eliminate the jamming signal in Antenna A by using the second antenna, B, as the reference antenna. As usual, for an adaptive canceller, the difference port of the 180° hybrid is used to close the control loop. The sum port of the 180° hybrid, however, provides the sample of the jamming-signal reference for the sidelobe canceller, as shown in Figure 5-10. Here, two adaptive cancellers are used for two distinct and different functions, one for adaptive beamforming and the other for adaptive sidelobe cancellation.

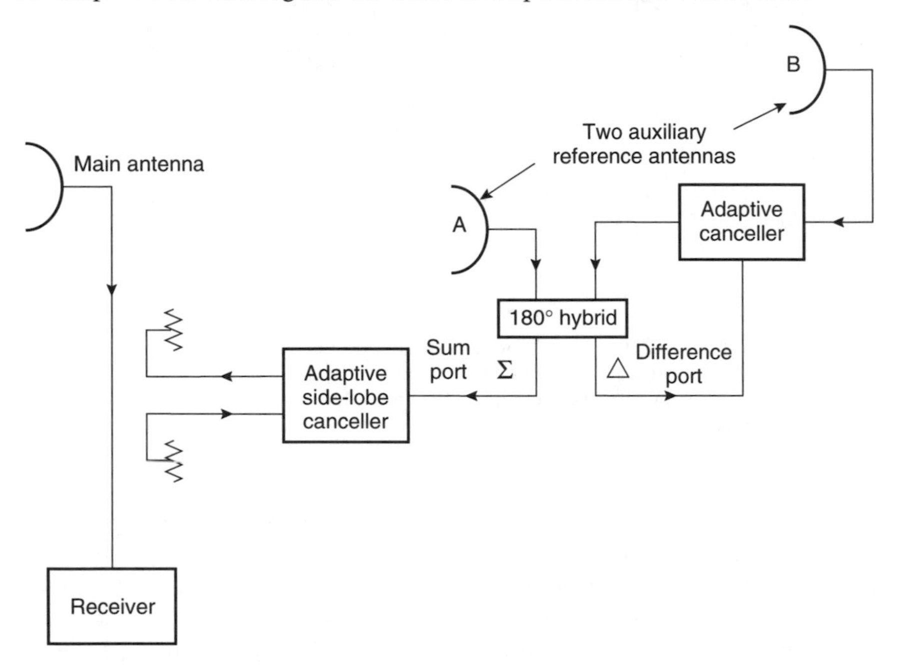

Figure 5-11 Formation of a two-element array directed automatically toward the jamming signal. Here, the sum port signal of the 180° hybrid provides the jamming signal reference for the sidelobe canceller.

In some antenna arrangements, such as the one found in amplitude-comparison monopulse radar operations, the sum and difference signal outputs of two adjacent antenna elements are often accessible. For such a case, the sum signal having a high directivity gain along boresight is used primarily for radar operation when the difference and sum signals are used for antenna tracking purposes. When a jamming signal is received at the sidelobe of the sum signal pattern, it also appears at the difference signal pattern. When the jamming signal is close to the main beam but not on the main beam, it becomes the dominant signal at the difference-port signal. Such a situation can be utilized to obtain a sample of the jamming signal from the difference port, and use such a sample as a reference for an adaptive canceller that cancels the jamming signal present at the sum-port line. For such an operation, it is not necessary to have a directional reference antenna pointing toward the jamming signal.

Another form of sidelobe cancellation employing a constrained adaptive beamforming has been suggested by Widrow *et al.* [10], Griffiths [11], and Frost [12]. According to these approaches, a pilot signal, simulating the desired signal from the desired look direction, is injected into the adaptive beamforming system such that the main beam of the array is along the specified look direction of the desired signal, while notches in the antenna pattern are formed along the directions of the interferences or jamming signals. Figure 5-12 illustrates the constrained beamforming concept. Here, the pilot signal that sets the main beam along the specified direction and the received signals from array elements enter into the adaptive signal processor No. 1. For this processor, the desired response is the pilot signal. The slaved adaptive processor No. 2 generates the actual array output signals, but it performs no adaptation and its input signals contain no pilot signal. The second processor is slaved to the first adaptive processor in such a way that its weights are exact copies of the corresponding weights of the adaptive system, so that it never needs to receive the pilot signal.

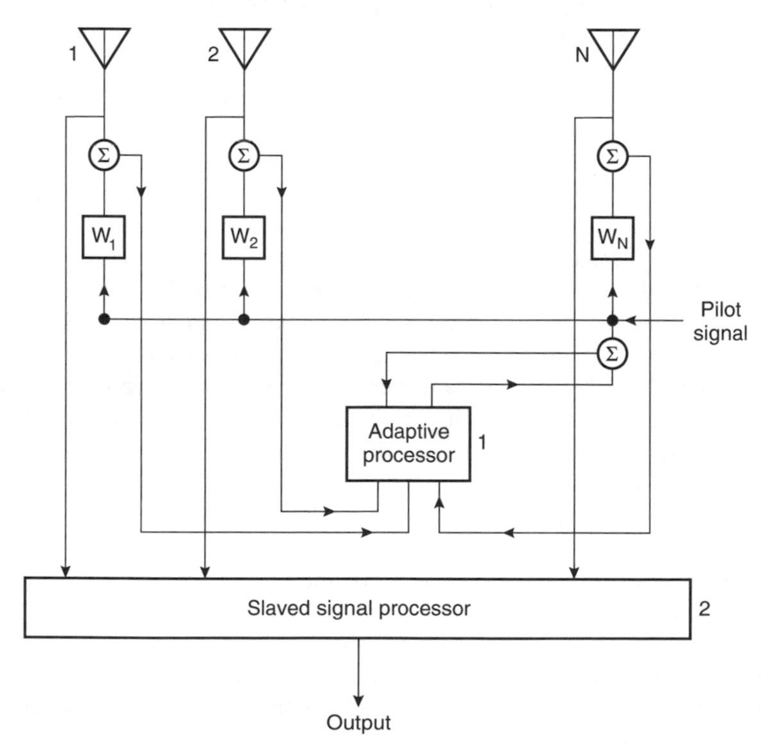

Figure 5-12 Pilot signal controlled adaptive processor.

In Figure 5-12, the pilot signal is on continuously. The adaptive processor is forced to reproduce the pilot signal as closely as possible and, at the same time, reject as well as possible all signals received by the antenna elements which are uncorrelated with the pilot signal. Thus, the adaptive processor forces a directivity pattern having the main-lobe sensitivity in the desired signal look direction and in the pilot signal passband, and at the same time forces nulls in the directions of interferences or jamming signals in their frequency bands.

It may be remarked at this point that any adaptive beamformer to eliminate a deliberate jamming signal from the receive line, including Howell–Applebaum's sidelobe canceller and those advanced by Griffith, Frost, Compton, and others, could be susceptible to attack even by simple jammers that radiate a bandpass noise, a sinusoid, or a sum of sinusoids suitably spaced in frequency. The interaction of such jamming signals with the desired signal in those adaptive algorithms may cause a cancellation of the desired signal components, even when the adaptive beamformers work perfectly. Such interactions, however, seldom exist for an adaptive canceller where a counterinterference is subtracted from the interference to be cancelled at the receive line.

5.4 CANCELLATION OF JAMMING SIGNAL ON MAIN BEAM

It is evident from the discussions on the adaptive canceller as presented earlier in this and other chapters that the signal to be cancelled at the receive line in the case of a canceller or the signal to be enhanced in the case of beamforming (Figure 5-11) must be the dominant signal at the reference port of the adaptive canceller. For a canceller, we have already seen that the higher the ratio of the undesired signal-to-desired-signal at the reference port, the higher will be the degree of cancellation of the undesired signal at the difference port of the 180° hybrid at the receive line. Therefore, when one needs to eliminate a jamming signal entering the receiving antenna through a sidelobe of the receiving antenna pattern, one needs a reference signal for the adaptive canceller such that the reference signal contains the jamming signal of concern as its dominant signal component. In the examples of the adaptive cancellers considered earlier, the discrimination of the jamming signal from the desired or other signals to establish its dominance has been based on the respective magnitudes of the signals. When the jamming signal appears at the mainlobe of the main receiving antenna, one can no longer obtain a sample of the jamming signal as a reference, such that the jamming signal is the dominant signal, particularly when the discrimination relies upon the respective magnitudes of the desired and jamming signals. In many cases, the requirement of the adaptive canceller involving the dominant jamming signal at the reference port can be met by utilizing a discrimination based on the different polarizations of the jamming and desired signals.

To further explain this approach, let us consider an antenna with the provision of receiving either a vertically or a horizontally polarized signal through the vertically or horizontally polarized feed point, respectively, as shown in Figure 5-13. Let the normal mode of operation for the receiving antenna be to receive a vertically polarized desired signal. If, now, a circularly polarized jamming signal appears at the receiving antenna and the horizontally polarized feed point of the antenna is used as a reference for an adaptive canceller to synthesize a counterjamming signal, then the reference-port signal will contain the jamming signal as its dominant component since no or little desired signal could be present at the horizontal feed point of the antenna. Even if there is some cross polarization caused by the propagation medium, the dominant signal at the horizontal feed point will still be the jamming signal. There will, of course, be a jamming signal at the vertically polarized feed point because of the circular polarization of the jamming signal. The adaptive canceller, however, will effectively cancel this jamming signal received through the vertical feed point, and this will be the case whether the jamming signal is received through the main beam or the sidelobe of the receiving antenna pattern.

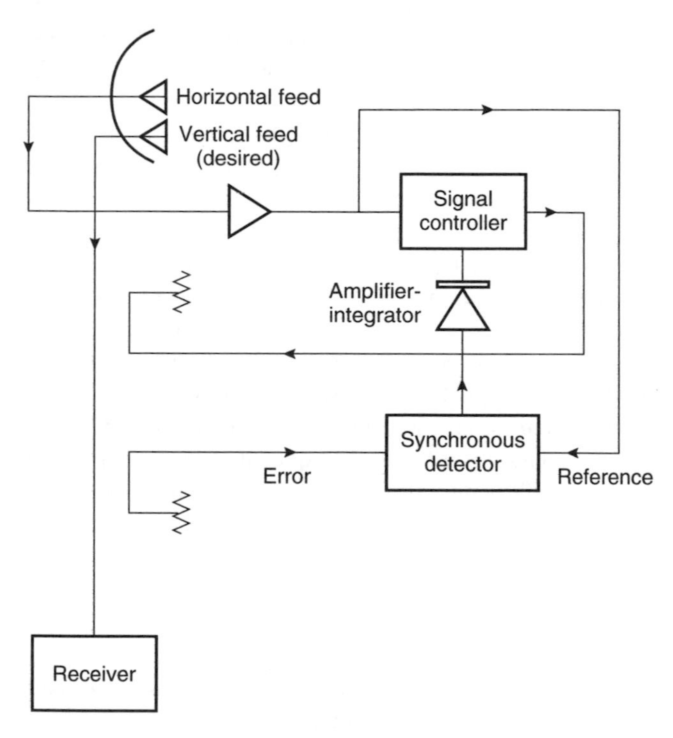

Figure 5-13 Jamming signal rejection at receiver line employing different polarizations as discriminating means.

Now, suppose that the jamming signal is horizontally polarized. There should be little jamming signal at the vertical feed point of the antenna, and hence at the receiver, because of the different polarizations, the desired signal being vertically polarized. Even if there is a little jamming signal present at the receive line, it will be cancelled by the same adaptive canceller since the reference port of the adaptive canceller, as noted earlier, will contain a dominant jamming signal.

If, instead, the polarization of the jamming signal is vertical and it appears at the main beam of the receiving antenna pattern, the adaptive canceller with its reference port connected to the horizontal feed point of the antenna will be of little value. This is true unless the radar operating mode is switched from the vertical to horizontal polarization and the adaptive canceller is used to operate with the jamming signal sampled at the vertical feed point as a reference to cancel any jamming signal present at the horizontal feed point where the desired radar signal is present.

Figure 5-13 shows a practical example of the main beam or sidelobe cancellation of a jamming signal for an airborne radar. The radar antenna shown at the left operates normally for vertically polarized signals. However, there is a nutator in front of the antenna that can convert the vertically polarized signal at the feed point of the antenna to a circularly polarized radiated field. Similarly, the same nutator converts the circularly polarized radar-return signal to a vertically polarized signal that can be received by the vertical feed point of the antenna. The polarization converter is switched on or off for providing protection against a jamming signal.

Although the antenna is designed to normally receive a vertically polarized signal, a horizontal probe is provided near the feed point so as to receive a horizontally polarized signal. As shown in the figure, the horizontally polarized RF signal received from the antenna probe is amplified and fed to the signal controller of an adaptive canceller. The output of the signal controller is further amplified and coupled back into the receive line through a 180° coupler.

A sample of the residual received signal at the radar antenna line, following the subtraction of the countersignal as coupled back into the receive line from the signal controller, is monitored through a sampling coupler, as shown in the figure. This sample is the error signal for the closed-loop control that minimizes the jamming signal. The error signal is heterodyned with a signal from the local oscillator to an IF frequency. Thus, the error signal at the IF constitutes the input port signal for the signal processor, which is essentially a synchronous detector. The reference-port signal of the synchronous detector is obtained by down-converting a sample of the signal at the input of the signal controller, as shown in the figure. The output of the synchronous detector provides the control signals for the signal controller, thus completing the loop. In this case, the synchronous detection is effected at the IF, instead of at the RF, as shown in Figure 4-1.

To illustrate the operation of this jamming signal suppression system, first let us assume that the radar operates on the vertically polarized mode such that both the radiating and target-return signals are vertically polarized. The horizontal sensor or feed of the antenna may contain a very small desired signal due to the depolarization of the signal along the propagation path or at the target.

Now, suppose that there is a jamming signal present at the radar antenna entering the antenna pattern either through the main lobe or the sidelobe of the antenna, the jamming signal being a left- or right-handed circularly polarized signal. The vertical component of the jamming signal will appear at the radar receiver along with the target-return signal, and will affect or disrupt the radar operation. The horizontal component of the circularly polarized jamming signal will be sensed by the horizontal sensor of the antenna. This signal, used as a reference for the adaptive canceller, will cancel the jamming signal component at the radar receive line. It should be noted here that the horizontal component of the jamming signal will be the dominant signal at the input of the signal controller, and hence the cancellation of the jamming signal at the radar receive line will be very effective.

If, now, the jamming signal as it appears at the radar antenna is horizontally polarized, the effect of this jamming signal will be small, not only because only a small part of the jamming signal, if any, will be coupled into the radar receive line, but also because the horizontal component will be sensed by the horizontal sensor at the antenna feed point, and the adaptive canceller will cancel any jamming signal present at the radar receive line.

Finally, let the jamming signal at the radar antenna be vertically polarized. This jamming signal will undoubtedly affect and disrupt the radar operation, particularly if the radar operation remains vertically polarized. At this point, the polarization converter can be switched on such that the radar will now illuminate the target with a circular polarization. The received radar-return signal will also be circularly polarized in this case, but the nutator referred to earlier will convert the circularly polarized received signal to the vertically polarized signal suitable for reception by the vertical feed of the antenna. Again, there will be little or no desired signal present at the horizontal feed point of the antenna. The jamming signal, however, being vertically polarized, will be converted to a circular polarization by the polarization-converting nutator. Consequently, the horizontal component of the jamming signal will be sensed by the horizontal feed point. The adaptive canceller now will create a counterjamming signal from the jamming signal sampled at the horizontal feed point, and cancel the jamming signal at the receive line. In this case, also, the sample of the jamming signal used as a reference will be the dominant signal at the input of the signal controller, there being little or no desired signal at the horizontal feed point of the antenna.

Thus, a strategy involving the polarization change switching, whenever the radar is being jammed, either at the main beam or at the sidelobe of the radar antenna, can be effective to reject the jamming signal from the radar receiver, regardless of the jammer strategy to disrupt the radar operation. Although the arrangement shown in Figure 5-14 may not be usable for many radars, the provision of sensing both horizontally and vertically polarized target-return signals can very often facilitate an effective use of an adaptive canceller.

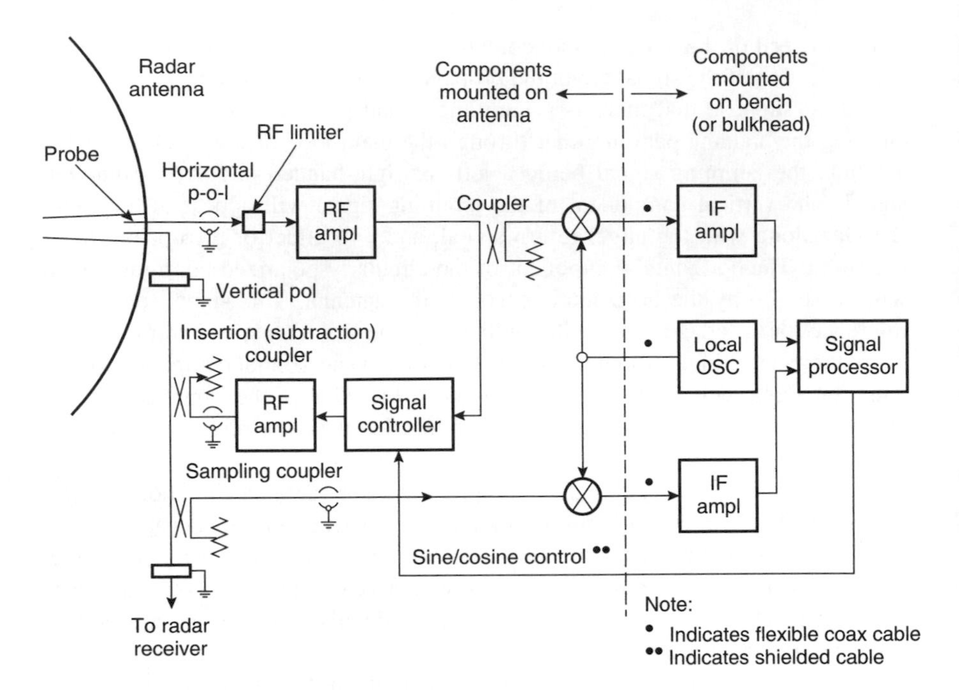

Figure 5-14 Block diagram of a main beam jamming signal suppressor using adaptive canceller.

5.5 ADAPTIVE SIGNAL SEPARATION

So far, we have considered cases where an adaptive interference canceller can be used to rid an interference from the receive line, provided one can obtain an independent sample of the interference as a reference. Also, for the canceller to be effective, the reference-port signals of the adaptive canceller must contain the interference or jamming signal as the dominant signal for this port. We have also seen that the higher the ratio of interference-to-desired-signal at the reference port of the canceller, the higher will be the degree of cancellation. Even when the difference in polarization of the desired and undesired signals is used for the discrimination needed for the control of the adaptive canceller, the reference port of the canceller is arranged to contain a higher level of the interference than the desired signal. Fortunately, for most cases of the interference cancellation requirements, this criterion for the adaptive canceller is easily met. However, there are cases when the interference level is not constant, and it is sometimes higher and sometimes lower than the desired signal level without *a priori* knowledge when it is lower or higher in level with respect to the desired signal. For example, consider the case of a communication between a satellite and an aircraft which is being interfered with by an interference source or jammer on the ground. The desired signal received from the satellite at the aircraft will be higher or lower than the jamming signal at times, depending on the aspect angle and orientation of aircraft antenna with respect to the satellite and ground jammer. The question that now arises is whether an adaptive canceller can be used for such a case effectively. Undoubtedly, the canceller, if feasible at all, must be adaptive since the relative levels of the interference and the desired signal are not fixed. It turns out that a pair of adaptive cancellers operating concurrently can be used to accommodate the situations involving variable desired and undesired signal levels. For such cases, the interference is cancelled at the output of one canceller, while the desired signal is cancelled at the output of the other. This signal-separation means [13] can then be used to receive the desired signal exclusively, regardless of the relative levels of the signals at any time.

Figure 5-15 shows schematically the general signal separation means when two arbitrary signals in the same frequency band, and time varying, and at unknown signal levels need to be separated. Two adaptive cancellers are used in a cross-connected mode, as shown in the figure. Two antennas receive two signals S_1 and S_2 originating from two sources, where S_2 is a desired signal and S_1 is an undesired signal or interference or another desired signal which is interfering with S_2. With no loss of generality, let the signal received by Antennas 1 and 2, as shown in

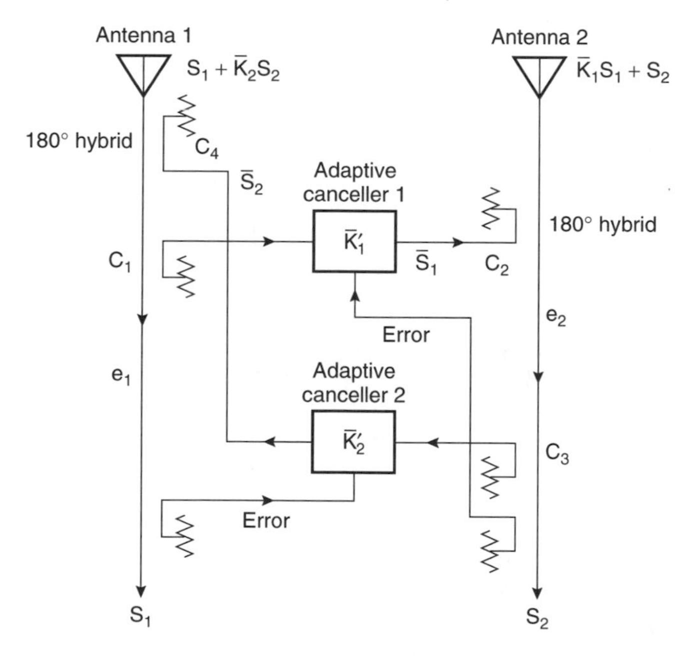

Figure 5-15 Signal separation system capable of separating two uncorrelated signals.

Figure 5-15, be $S_1 + \bar{K}_2 S_2$ and $\bar{K}_1 S_1 + S_2$, respectively, where $\bar{K}_1$ is the transfer function of the propagation path between Antenna 2 and the source of S_1 and $\bar{K}_2$ is the transfer function of the propagation path between Antenna 1 and the source of S_2.

If one assumes that S_1 is greater than S_2 at the instant of turn on of the signal separation system, Adaptive Canceller 1, with a transfer function $\bar{K}_1$, will tend to suppress S_1 at the Antenna 2 line. Following such a suppression, the Antenna 2 line will contain more desired signal, S_2, than before. This enhanced signal is used as a reference for the Adaptive Canceller 2 to synthesize a cancelling signal corresponding to $\bar{K}_2 S_2$ which is subtracted from the signals at Antenna line 1. Thus, following the cancellations effected by Adaptive Cancellers 1 and 2, the output of Antenna line 2 will contain S_2 and the output of Antenna line 1 will contain S_1 almost exclusively at the equilibrium condition of the two Adaptive Cancellers.

To illustrate the effect of simultaneous operations of the two cancellers, one may write the signal e_2 at the Antenna 2 line following the subtraction of the cancellation signal delivered by the Adaptive Canceller 1 as

$$e_2 = \bar{K}_1 S_1 + S_2 - [S_1 + \bar{K}_2 S_2 - \bar{S}_2]\bar{K}'_1 C_1 C_2 \tag{5.83}$$

where

$\bar{S}_2 = e_2 \bar{K}'_2 C_3 C_4$

$C_1, C_2, C_3, C_4 =$ the transfer functions for the couplers, as shown in Figure 5-15.

Similarly, the signal e_1 at the Antenna 1 line, following the subtraction of the cancellation signal delivered by the Adaptive Canceller 2, can be written as

$$e_1 = S_1 + \bar{K}_2 S_2 - \bar{S}_2. \tag{5.84}$$

From Eqs. (5.83) and (5.84)

$$e_2 = S_1 \left[\frac{\bar{K}_1 - \bar{K}'_1 C_1 C_2}{D}\right] + S_2 \left[\frac{1 - \bar{K}'_1 \bar{K}_2 C_1 C_2}{D}\right]$$

$$\begin{aligned} e_1 &= S_1 \left[1 - \frac{(\bar{K}_1 - \bar{K}'_1 C_1 C_2)}{D} \bar{K}'_2 C_3 C_4\right] \\ &\quad + S_2 \left[\bar{K}_2 - \frac{(1 - \bar{K}'_1 \bar{K}_2 C_1 C_2)}{D} \bar{K}'_2 C_3 C_4\right] \end{aligned} \tag{5.85}$$

where

$D = 1 - \bar{K}'_1 \bar{K}'_2 C_1 C_2 C_3 C_4.$

For the equilibrium conditions of the closed-loop controls for the Adaptive Cancellers 1 and 2

$$\bar{K}_1 = \bar{K}'_1 C_1 C_2 \tag{5.86a}$$

and

$$\bar{K}_2 = \bar{K}'_2 C_3 C_4. \tag{5.86b}$$

Thus

$$e_1 = S_1 \text{ and } e_2 = S_2 \tag{5.87}$$

provided

$$1 - \bar{K}_1 \bar{K}_2 \neq 0. \tag{5.88}$$

It may be noted from Figure 5-15 that the error-port signal of the coupler at the Antenna 2 line is used to control the Adaptive Canceller 1, the primary function of this canceller being to eliminate S_1 from the Antenna 2 line. From Eq. (5.85), however, it is seen that the only way S_1 is eliminated from the Antenna 2 line is to satisfy Eq. (5.86a), which is one of the equilibrium conditions. In other words, as the Adaptive Canceller 1 operates, the Antenna 2 line becomes progressively free of S_1 until such time as there is no S_1 left at the error port of the coupler in Antenna line 2. At this point, the signal control parameter $\bar{K}'_1$ becomes fixed, and Eq. (5.86a) is satisfied. Similarly, the error-port signal of the coupler at the Antenna 1 line controls the signal controller parameter $\bar{K}'_2$ of the Adaptive Canceller 2, and this canceller operation reaches an equilibrium when there is no S_2 signal left at the Antenna 1 line. From Eq. (5.85), it is seen that only when the Antenna 1 line signal, that is, e_1, becomes free of S_2 is when Eqs. (5.86a) and (5.86b) are satisfied. Also, as both Adaptive Cancellers operate, the reference-port signal of each canceller is "cleaned" by the other canceller, thereby making the convergence a faster process.

The arrangement shown in Figure 5-15 then separates two arbitrary signals S_1 and S_2 which are at the same frequency bands and which are completely arbitrary in levels. They may even be at the same level. An indirect benefit of the cross-connected adaptive cancellers, as shown in Figure 5-15, is that an interference cancellation is effected adaptively for both signals, leaving S_1 or S_2 free from contamination by the other. Thus, the adaptive canceller combination of Figure 5-15 becomes a powerful tool for interference cancellation which is not dependent on the high level of the interference.

The concept of signal separation can be extended where more than two signals are involved and a discrimination based on their magnitudes is not necessary for interference cancellation. We will consider such extensions in the following chapter while dealing with multiple interferences.

Although the cross-connected adaptive canceller, as shown in Figure 5-15, is a powerful tool to rid a class of difficult interferences, its implementation needs some careful engineering considerations. More specifically, we have noted in Eq. (5.88) that

$$1 - \bar{K}_1\bar{K}_2 = 1 - \bar{K}'_1\bar{K}'_2C_1C_2C_3C_4 \neq 0.$$

Equation (5.88), in fact, is the characteristic equation of the two loops in cascade which sets the condition for avoiding self-oscillation. To avoid such an oscillation, one may thus set the requirement as

$$|\bar{K}_1\bar{K}_2| < 1. \tag{5.89}$$

Equation (5.89) shows that there exist some constraints for the operation of the cross-connected adaptive cancellers. For example, if Antennas 1 and 2 are both omnidirectional vertical dipoles or monopoles over a plane ground base, it may not always be possible to maintain the relationship shown in Eq. (5.89). However, this constraint can be eliminated if, instead of two vertical dipoles, one uses a pair of cross-loop antennas, orthogonal to each other, to receive signals of the same polarization as the vertical dipole. For such a case, the factor $|\bar{K}_1\bar{K}_2|$ will be less than one unless S_1 and S_2 are absolutely codirectional. In other words, for such an antenna arrangement, at least $\bar{K}_1$ or $\bar{K}_2$, and usually both, will be less than one, and Eq. (5.89) will be satisfied.

Thus, an adaptive canceller or a combination of cancellers can be used to suppress or eliminate an arbitrary interference of remote origin under a variety of circumstances. In most common applications, the interference level will be overwhelmingly large in comparison with the desired signal at the receive-antenna line, and a reference antenna will be used to provide a sample of the interference. Also, for a high-gain receive antenna, the interference is most likely to be at the sidelobe of the antenna. However, as we have seen in some examples, the principal antenna main beam interference can also be suppressed by using the difference in polarization as a means for discrimination. In addition, although the level of interference is used for discrimination in most cases, one can suppress an arbitrary interference by employing a pair of cross-connected adaptive cancellers regardless of the interference level.

5.6 INTERFERENCE CANCELLATION AT RF AND IF STAGES

An adaptive canceller, shown schematically in Figure 4-1, can be designed for a wide operating frequency band, sometimes extending more than one octave in frequency range. This facilitates the use of the canceller since it need not be tailored for a particular receiver frequency or for a particular receiver. The capability of accommodating a wide operating frequency band, however, is not always an advantage. Let us consider, for example, the use of an adaptive canceller to suppress an interference of remote origin where a reference antenna is used to obtain a sample of the interference. Figure 5-16(a) shows the usual configuration of the canceller and the receiving and reference antennas. Let the receiver be intended for communications having only relatively small reception or information bandwidths. Let the frequency spectrum of the signals at the receiving antenna be what is shown in Figure 5-16(b). In this case, the receiver frequency is f_r and the reception bandwidth of interest is $f_r - \Delta$ to $f_r + \Delta$. As shown in the figure, there is a large interference at a frequency much removed from f_r. There is also an

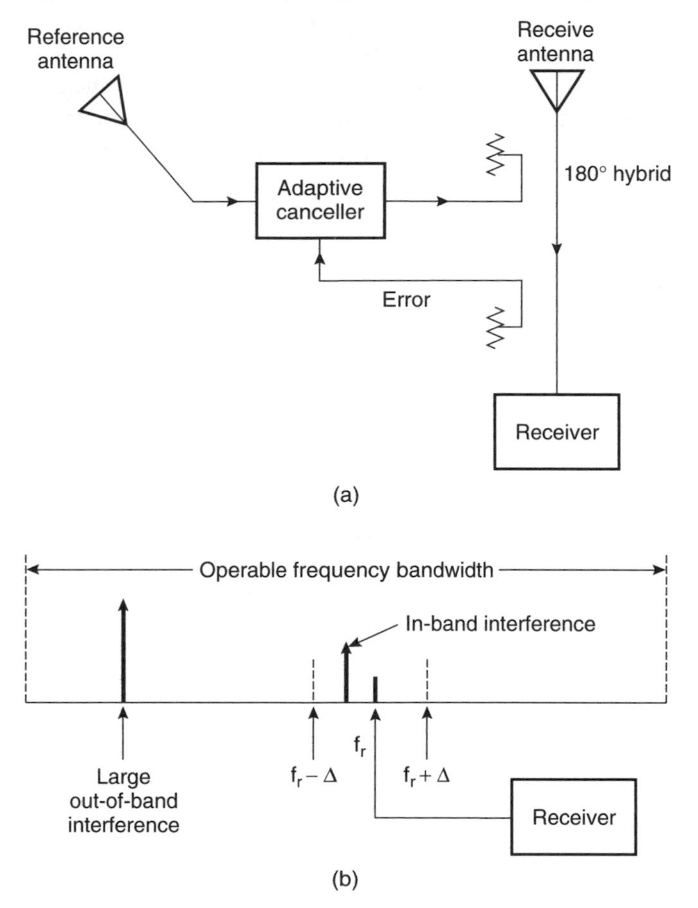

Figure 5-16 (a) Schematic illustration of the use of adaptive canceller. (b) Spectrum of received signal with a large out-of-band and a lesser in-band interference.

in-band interference, considerably lower in level than the out-of-band interference, but nonetheless objectionable. Since the adaptive canceller in Figure 5-16(a) has no *a priori* knowledge about either interference, it will try to eliminate both interferences, but succeed only in reducing the out-of-band interference because of its magnitude. If, however, no adaptive canceller is used for the situation shown in Figure 5-16(a) and (b), the out-of-band interference will most likely be rejected by the receiver selectivity. Thus, the use of an adaptive canceller is not helpful, particularly when the objectionable in-band interference is not addressed, whether or not the adaptive canceller is used.

This problem is remedied by providing the cancellation at an intermediate frequency, instead of at the RF, as shown in Figure 5-17. In this case, the signals at the receive antenna and reference antenna lines are heterodyned with a signal from a local oscillator, thereby down-converting the signals in both lines to a common intermediate frequency. If, now, the cancellation is effected by the adaptive canceller, addressing only the in-band interference by the canceller, there will be an effective use of the canceller automatically. All out-of-band interferences will be eliminated from the receiver due to the inherent receiver selectivity.

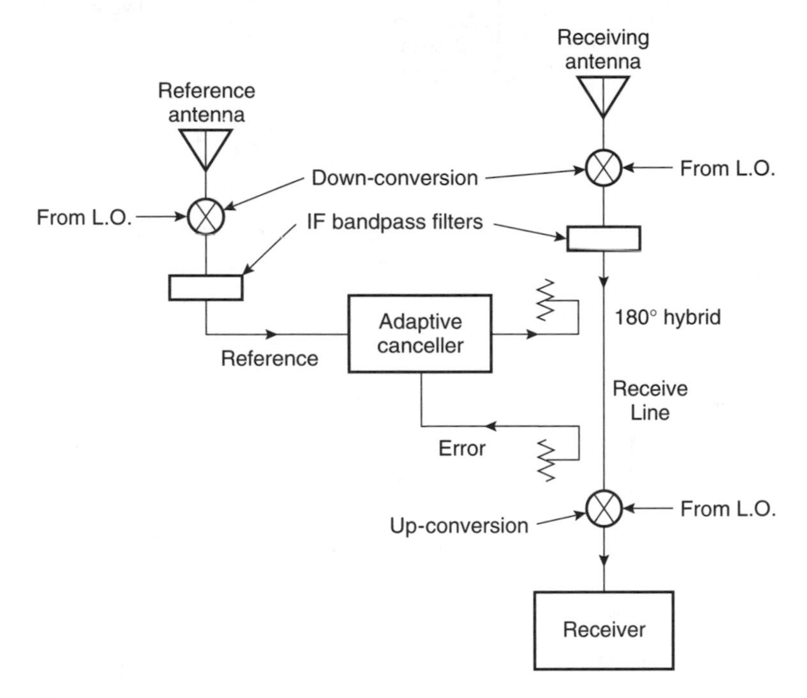

Figure 5-17 Adaptive cancellation at IF making use of receiver selectivity to reject out-of-band interferences.

Although the cancellation arrangement shown in Figure 5-17 addresses the correct interference of concern, its use in the form shown, however, is not desirable because of several reasons. First, the mixers at the reference and receiver antenna lines can generate intermodulation products which appear as additional interferences at the receive line that would not exist in the absence of mixers shown in Figure 5-17. This problem may be intolerable if the level of interference to be cancelled is high at the reference or receive antenna line or both. Second, the two bandpass filters at the reference and receive antenna lines may not track in amplitude and phase of the entire frequency band of the interference of concern. Consequently, the cancellation over the frequency band of the interference will not be uniform, regardless of how well the adaptive canceller performs. Third, following the cancellation, the desired signal at the receive antenna line will be at the down-converted IF. The receiver intended for the original use (in the absence of adaptive cancellation at IF) cannot be used unless the signals are up-converted again at the original frequencies of the received signals.

All these problems, however, can be eliminated if the signal-controller operation is maintained at RF, but the synchronous detectors for the canceller are operated at the IF, as shown in Figure 5-18. Thus, for example, there will be no intermodulation products at the receiver line since there is no mixer either at the signal controller or at the receive line. Also, nontracking bandpass filters at the reference and error ports of the synchronous detectors will not be harmful because of the closed-loop operation of the canceller that drives the error signal to zero. There will also be no need to up-convert the receive-line signal in frequency since

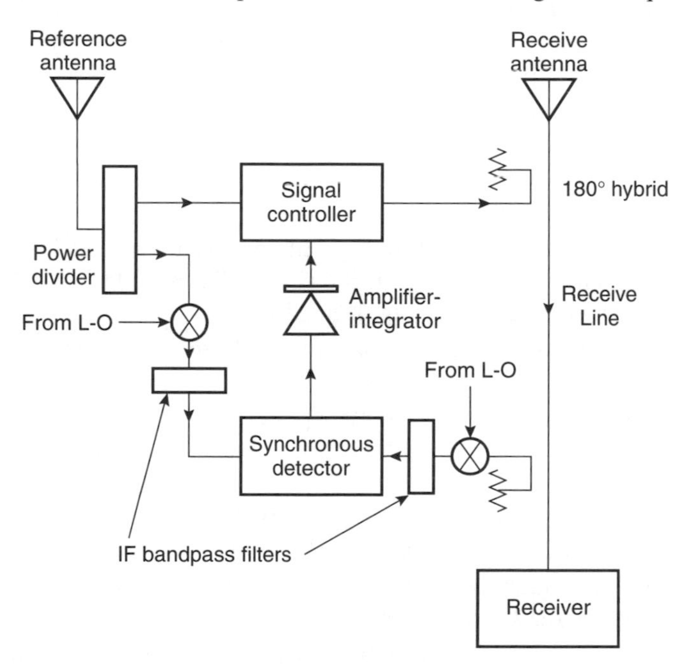

Figure 5-18 Adaptive cancellation with no frequency conversion at the receive line.

there is no down-conversion of the signal at the receive line requiring such a need. The frequency of the local oscillator (LO), shown in Figure 5-18, should be as close to the receiver frequency of interest as possible to permit interference cancellation most effectively for the receiver information bandwidth of concern.

It turns out that the ability to operate the synchronous detector of the adaptive cancellor at the narrow receiver frequency band of interest while maintaining linear paths for the receive and signal-controller lines, as shown in Figure 5-18, becomes a powerful tool for reducing the number of needed cancellers, particularly when one or a few receivers have to be protected from many interferences, and when the effect of propagation multipaths for the interference cannot be ignored. These problems are further discussed in the following chapters.

REFERENCES

[1] R. Ghose, "Theory and mathematical model of adaptive cancellation system," Tech Note 160004, Technology Research International, Inc., Calabasas, CA, June 1986.

[2] P. Howells, "Intermediate frequency side-lobe canceller," U.S. Patent 3,202,990, Aug. 24, 1965.

[3] R. Riegler and R. Compton, Jr., "An adaptive array for interference rejection," *Proc. IEEE*, vol. 61, pp. 748–758, June 1973.

[4] S. Applebaum, "Adaptive arrays," *IEEE Trans. Antennas Propagat.*, vol. AP-24, pp. 585–598, Sept. 1976.

[5] Special Issue on Adaptive Arrays, *IEEE Trans. Antennas Propagat.*, vol. AP-24, Sept. 1976.

[6] B. Widrow and S. Stearns, *Adaptive Signal Processing*. Englewood Cliffs, NJ: Prentice-Hall, 1985.

[7] N. Jablon, "Adaptive beamforming with the generalized sidelobe canceller in the presence of array imperfections," *IEEE Trans. Antennas Propagat.*, vol. AP-34, pp. 996–1012, Aug. 1986.

[8] R. Monzingo and T. Miller, *Introduction to Adaptive Array*. New York: Wiley, 1980.

[9] D. Morgan and A. Aridgides, "Adaptive sidelobe cancellation of wide-band multipath interference," *IEEE Trans. Antennas Propagat.*, vol. AP-33, Aug. 1985.

[10] B. Widrow, P. Mantey, L. Griffiths, and B. Goode, "Adaptive antenna systems," *Proc. IEEE*, vol. 55, p. 2143, Dec. 1967.

[11] L. Griffiths, "A simple adaptive algorithm for real-time processing in antenna arrays," *Proc. IEEE*, vol. 57, pp. 1696–1704, Oct. 1969.

[12] O. Frost, III, "An algorithm for linearly constrained adaptive array processing," *Proc. IEEE*, vol. 60, pp. 926–935, Aug. 1972.

[13] R. Ghose *et al.*, "Automatic separation system," U.S. Patent 4,466,131, Aug. 14, 1984.

[14] S. Haykin, Ed., "Array processing applications to radar," in *Benchmark Papers in Electrical Engineering and Computer Sciences*, vol. 22. Dowden, Hutchinson and Ross, 1980.

[15] Special Issue on Adaptive Processing Antenna System, *IEEE Trans. Antennas Propagat.*, vol. AP-34, no. 3, 1986.

6

SIMULTANEOUS CANCELLATION OF MULTIPLE INTERFERENCES

In many practical situations, there is often more than one interference or undesired signal that needs to be eliminated from the receive line or receiver simultaneously. Cancellations of multiple interferences may be effected by extending the concept of a single interference canceller, with multiple reference antennas providing the needed samples of the interferences so that corresponding "counterinterferences" can be synthesized for cancellation. Another approach for the cancellation of multiple interferences is to use a set of antenna elements in the form of an adaptive array that enhances the desired signal-to-noise ratio notwithstanding arbitrary multiple interferences.

Still another approach for effecting cancellations of multiple interferences is to separate the desired and undesired signals by a signal separation system that employs a number of adaptive cancellers. In most cases, multiple antennas are needed whenever multiple interferences need to be cancelled from the receive line or receiver.

6.1 ADAPTIVE CANCELLATION OF MULTIPLE INTERFERENCES

When a number of linearly independent reference inputs are available, one can extend the concept of the single interference cancellation to rid multiple interferences. Figure 6-1 shows, for example, an illustration of such an approach. Here, each adaptive canceller addresses one interference by arranging to sample that interference exclusively or predominantly by a reference sensor such as reference sensor 1, 2, or i, as shown in the figure. The monitoring of the error signal is effected following all the summing couplers or 180° hybrids used for various interferences at the receive line.

When the interference sources are collocated with the receiver such that they are accessible to obtain direct samples of the interferences as references, one

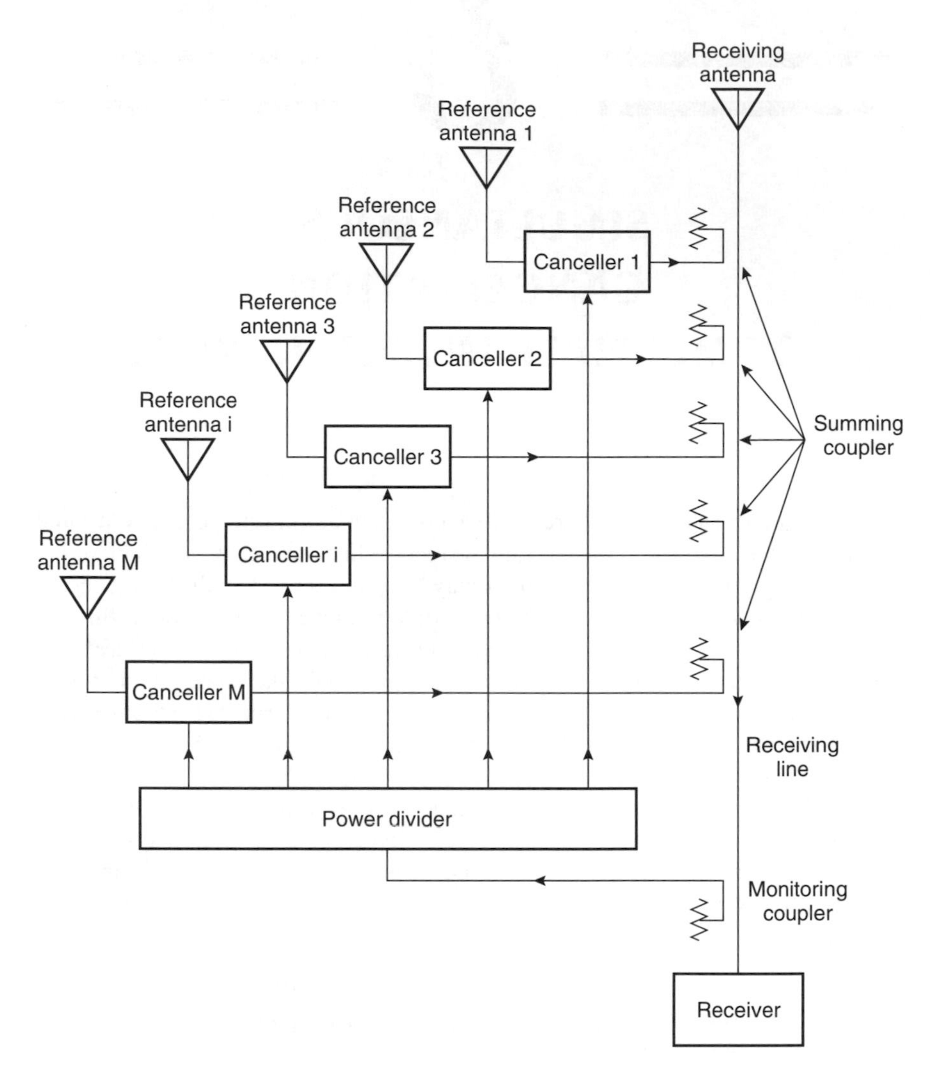

Figure 6-1 Illustration of simultaneous multiple interference cancellation with multiple references.

easily satisfies the requirement that each reference line contains one interference exclusively or as its dominant signal. However, if this is not the case, the samples of interferences have to be obtained by reference antennas placed close to the receiving antenna, and for such a case, provisions need to be made so that each reference antenna receives only one interference as its dominant received signal. The monitoring coupler, placed at the end of the summing couplers, at the receive line in Figure 6-1, permits the cancellation of a particular interference even when some components of the interference are introduced into the receive line through one or more other reference antennas. The degree of cancellation obtainable when the reference antennas receive more than one interference will depend on the degree of purity of the reference sensor output for a particular interference being addressed. The performance of any individual canceller and the factors affecting such a performance will be the same as those discussed earlier in connection with a single interference.

One can also extend the concept of the Wiener transfer function for a single canceller as discussed in Chapter 3 for the multiple reference canceller where one determines the weighting functions for various interferences to be eliminated. Figure 6-2 shows such a set of signal controllers providing such weighting functions and the associated reference sensors.

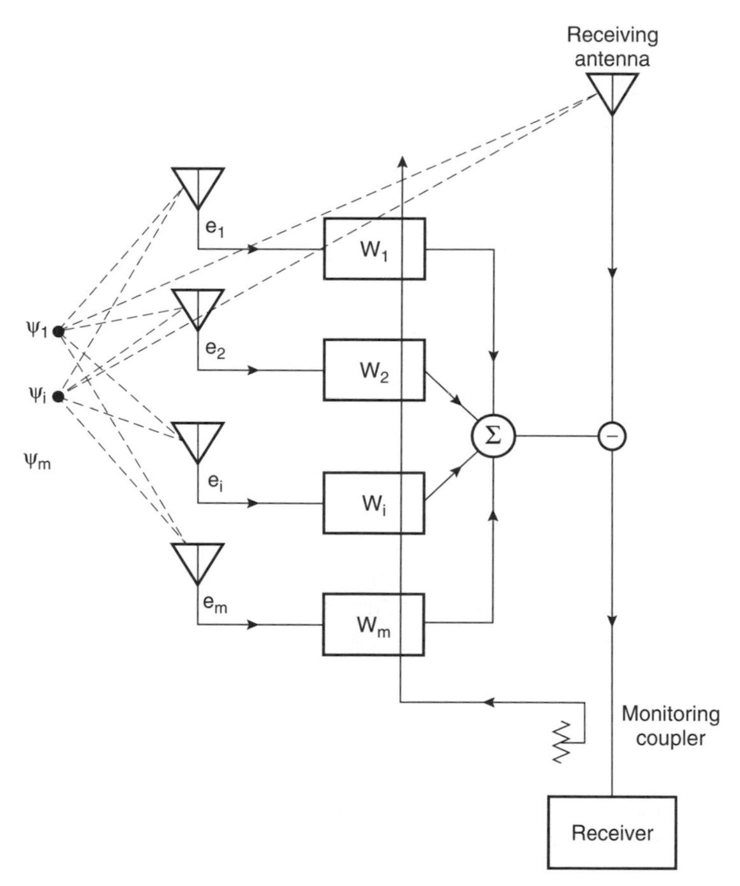

Figure 6-2 Illustration of multiple reference canceller. Here, each reference sensor receives all the interferences of concern.

Let $e_1, e_2, \cdots, e_i, \cdots, e_m$ denote the number of interferences received by the reference antennas 1, 2, $i, \cdots, m$, etc., as shown in the figure. Since e_1, e_2, etc., contain interferences from all sources ψ_1, ψ_2, ψ_3, etc., one may write

$$e_1 = \psi_1 F_{11} + \psi_2 F_{12} \cdots \psi_m F_{1m} \tag{6.1}$$

where

F_{11}, F_{12}, etc. = factors introduced by the propagation path between the reference antenna 1 and the respective sources.

The matrix F may be regarded as the transfer function matrix between the reference antennas and the interference sources.

Similarly

$$e_i = \psi_1 F_{i1} + \psi_2 F_{i2} \cdots \psi F_{im} \tag{6.2}$$

and in matrix notation

$$[e] = [F][\psi] \tag{6.3}$$

where

$$F = \begin{bmatrix} F_{11} & F_{12} \ldots F_{im} \\ \vdots & \\ F_{m1} & F_{m2} \ldots F_{mm} \end{bmatrix} \tag{6.4}$$

and the source matrix $[\psi]$ = a column matrix given by

$$\psi = \begin{bmatrix} \psi_1 \\ \psi_2 \\ \vdots \\ \psi_m \end{bmatrix}. \tag{6.5}$$

The power-density spectrum of the input interferences can be written as

$$[\phi_a] = [F^*][\psi^* \tilde{\psi}][\tilde{F}] \tag{6.6}$$

where

$\psi^* \tilde{\psi}$ = the source spectral matrix given by

$$[\phi_\psi] = \begin{bmatrix} \phi_{\psi_1 \psi_1} & 0 & 0 & 0 \\ \cdot & \phi_{\psi_2 \psi_2} & & \\ \cdot & & & \\ \cdot & \cdot & \cdot & \phi_{\psi_m \psi_m} \end{bmatrix}. \tag{6.7}$$

The cross-spectral vector between the reference inputs and the primary input or the receiving antenna input is given by

$$[\phi_c] = [F^*][\psi^* \tilde{\psi}][G] \tag{6.8}$$

where

G = the transfer function between the interference input sources and the primary input or the receiving antenna.

As discussed in Chapter 3, the Wiener filter function for the optimal weight vector, W^* (not conjugate of W) is then obtained as

$$[W^*] = \{[F^*][\phi_\psi][\tilde{F}]\}^{-1} \cdot [[F^*][\phi_\psi][G]]. \tag{6.9}$$

When $[F]$ is a square matrix as assumed in Eq. (6.4)

$$[W^*] = [\tilde{F}]^{-1} \cdot [G], \tag{6.10}$$

as may be easily verified by writing $G = \tilde{F} \cdot \tilde{F}^{-1} G$.

One may recall from Eq. (3.89) that for a single interference, the weighting function W^* is given by

$$W^* = Ke^{-j\omega T} = W$$

where the primary or receive-line input was assumed to be $KI(\omega)e^{-j\omega T}$ and the reference input was $I(\omega)$. From the definition of the reference input in Eq. (6.2)

$$e = F\psi = I(\omega) \tag{6.11}$$

where the suffix is omitted as there is only one interference.

Similarly, from the definition of the transfer function G

$$e_{\text{primary}} = G\psi = KI(\omega)e^{-j\omega T} \tag{6.12}$$

and

$$G/F = Ke^{-j\omega T}. \tag{6.13}$$

Thus, Eq. (6.10) is the matrix equivalent of the same result as given in Eq. (3.84).

To further illustrate the significance of Eq. (6.10) involving the Wiener filter function for a nontrivial case, one may consider two interferences ψ_1 and ψ_2 and two reference antennas 1 and 2, as shown in Figure 6-3. Let the interferences received by antennas 1 and 2 be e_1 and e_2, respectively. One may express

$$\begin{aligned} e_1 &= \psi_1 g_{11} \frac{e^{-j\beta r_{11}}}{r_{11}} + \psi_2 g_{12} \frac{e^{-j\beta r_{12}}}{r_{12}} \\ e_2 &= \psi_1 g_{21} \frac{e^{-j\beta r_{21}}}{r_{21}} + \psi_2 g_{22} \frac{e^{-j\beta r_{22}}}{r_{22}} \end{aligned} \tag{6.14}$$

where

g_{11}, g_{12}, etc. = factors due to the combined directivity gains of the source and reference antennas

$(e^{-j\beta r}/r)$ = the variation of the signal from the interference source with the distance r from the source. Here, β is assumed to be the common propagation constant.

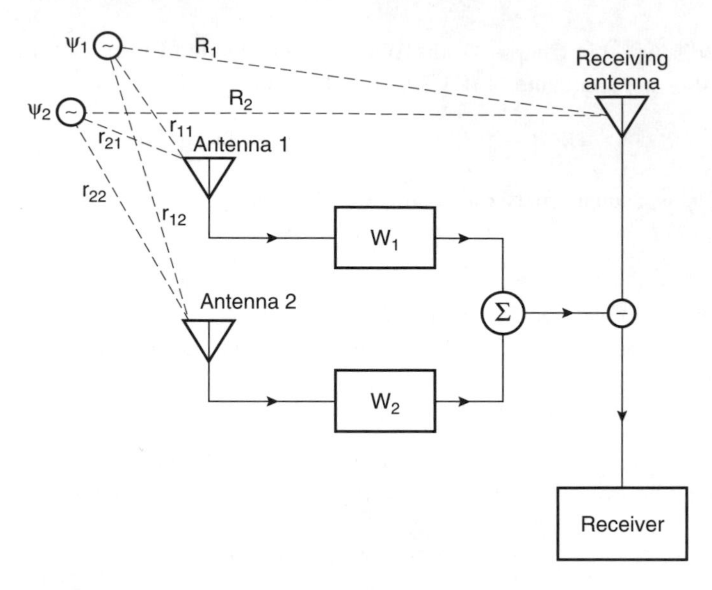

Figure 6-3 Cancellation of independent interferences Ψ_1 and Ψ_2 with Wiener filters.

The corresponding interferences as received by the receiving antenna may be expressed as

$$E_R = \psi_1 G_{11} \frac{e^{-j\beta R_1}}{R_1} + \psi_2 G_{12} \frac{e^{-j\beta R_2}}{R_2}. \tag{6.15}$$

In matrix notation

$$[e] = \begin{bmatrix} a & b \\ c & d \end{bmatrix} \begin{bmatrix} \psi_1 \\ \psi_2 \end{bmatrix} \tag{6.16}$$

where

$$a = g_{11} \frac{e^{-j\beta r_{11}}}{r_{11}}$$

$$b = g_{12} \frac{e^{-j\beta r_{12}}}{r_{12}}$$

$$c = g_{21} \frac{e^{-j\beta r_{21}}}{r_{21}}$$

$$d = g_{22} \frac{e^{-j\beta r_{22}}}{r_{22}}.$$

Similarly

$$E_R = \tilde{G}[\psi] \tag{6.17}$$

where

$$G = \begin{bmatrix} G_1 \dfrac{e^{-j\beta r_1}}{r_1} \\ G_2 \dfrac{e^{-j\beta r_2}}{r_2} \end{bmatrix} = \begin{bmatrix} e \\ f \end{bmatrix}.$$

From Eqs. (6.10), (6.16), and (6.17), the Wiener filter function matrix can be written as

$$[W] = \begin{bmatrix} \dfrac{de - cf}{ad - bc} \\ \dfrac{-be + af}{ad - bc} \end{bmatrix}. \tag{6.18}$$

It is easy to verify the appropriateness of Eq. (6.18) since the output of the filter element W_1 in Figure 6-3 is

$$E_1 = (a\psi_1 + b\psi_2)\frac{de - cf}{ad - bc}. \tag{6.19}$$

Similarly, the output of the filter element W_2 is

$$(c\psi_1 + d\psi_2)\frac{(-be + af)}{ad - bc} \tag{6.20}$$

and

$$E_1 + E_2 = e\psi_1 + f\psi_2 = E_R$$

from Eq. (6.17).

Thus, a subtraction of $(E_1 + E_2)$ from the receive-line signals will eliminate the interferences ψ_1 and ψ_2.

Although a solution exists for W_1 and W_2, their physical implementations, particularly in an adaptive manner, may not always be easy. The problem of implementation of the filter elements W_1 and W_2, by adaptive closed-loop controls, however, becomes considerably simplified if the reference antennas 1 and 2 could be so designed that each receives only one interference. Thus, if for example, the antennas are such that g_{12} and g_{21} are zero, b, and c in Eq. (6.18) will be zero and

$$W_1 = e/a \text{ and } W_2 = f/d. \tag{6.21}$$

Consequently, the outputs of the filter elements W_1 and W_2 will be $\psi_1 e$ and $\psi_2 f$, respectively, and their sum will be equal to E_R, as given in Eq. (6.20).

The ratio e/a is the ratio of the transfer function between the source antenna for ψ_1 and the receiving antenna to the transfer function between the same source antenna to the reference antenna 1. From Eqs. (6.16) and (6.17)

$$e/a = \left(\frac{G_1}{R_1}\frac{r_{11}}{g_{11}}\right)e^{-j\beta(R_1 - r_{11})} = K_1 e^{-j\omega T_1} \tag{6.22}$$

where
K_1 = an amplitude factor
T_1 = a time delay.

Similarly, the ratio f/d is the ratio of the transfer function between the source antenna for ψ_2 and the receiving antenna to the transfer function between the ψ_2 source antenna and the reference antenna 2, as shown in Figure 6-3. Also, from Eqs. (6.16) and (6.17)

$$f/d = \left(\frac{G_2}{R_2}\frac{r_{22}}{g_{22}}\right) e^{-j\beta(R_2 - r_{22})} = K_2 e^{-j\omega T_2} \tag{6.23}$$

where
K_2 = some other amplitude factor
T_2 = time delay.

A comparison between Eqs. (6.13) and (6.22) or between Eqs. (6.13) and (6.23) suggests that when the reference antennas can be designed such that each reference antenna receives one undesired signal either exclusively or predominantly, the cancellation of multiple undesired signals becomes the same as cancelling one undesired signal independent of any other canceller. Only the receiving antenna remains as the common element for the multiple undesired signal cancelling process.

6.2 MULTIPLE INTERFERENCE CANCELLATION BY AN ADAPTIVE ARRAY

The general considerations for interference cancellation by an array, where each array element is connected to a weighting network, W_1, W_2, etc., as shown in Figure 5-2 and the array output comprises a sum of the weighted signals, are outlined in Section 5.2. Referring to Figure 5-2 and Eqs. (5.12) and (5.13), we have seen that the array output for the desired signal and the noise that includes interferences can be written as

$$V_S = \alpha \sum_{K=1}^{N} S_K W_k \exp\left(-j\theta_{SK} - j\phi_{WK}\right)$$

$$V_n = \sum_{K=1}^{N} n_K W_K \exp\left(-j\phi_{WK}\right) \tag{6.24}$$

where, for the desired signal,

S_K = the amplitude
θ_{SK} = the phase angle
α = the level and time variation of the desired signal
n_K = the complex magnitude for the noise
W_K and ϕ_{WK} = the amplitude gain and phase introduced by the Kth weighting network, respectively.

In matrix notation, one may write

$$V_S = \alpha \tilde{W} S = \alpha \tilde{S} W$$
$$V_n = \tilde{W} n = \tilde{n} W \tag{6.25}$$

where

W, S, and n = column vectors representing the weighting network, the desired signal, and the noise at the array output, respectively.

The noise power P_n is obtained from the expected value of $|V_n|^2$, that is,

$$P_n = E\{|V_n|^2\}$$
$$= E\left\{\sum_{j=1}^{N}\sum_{k=1}^{N} n_j n_K^* W_j W_K(-j\phi_{Wj} + j\phi_{Wk})\right\} \tag{6.26}$$

where

E = the expectation operator

$*$ denotes the complex conjugate operation.

Since the expectation operator E affects only the noise term, one may write, from Eq. (6.26), in matrix form

$$P_n = \tilde{W}^* E\{N^* \tilde{N}\} W$$
$$= \tilde{W}^* M W \tag{6.27}$$

where

$M = E\{N^* \tilde{N}\} = [\mu_{jk}]$,

M being the covariance matrix of the noise components. If the noise components are uncorrelated, as will be the case, for example, in the absence of any particular interference, M will be a diagonal matrix, similar to that assumed in Eq. (5.20). When multiple interferences are present, there will, however, be nonzero entries in any position of the covariance matrix M, although M will still remain a Hermitian matrix since $\mu_{jK} = \mu_{Kj}^*$ and $\tilde{M} = M^*$.

Let us assume a transformation matrix A that transforms the signal and noise components from the array elements to S' and N' where

$$S' = AS$$
$$N' = AN. \tag{6.28}$$

This transformation corresponds to S and N from the array elements as inputs of a matrix transformation block such that the array output V_s and V_n can be written as

$$V_s = \alpha \tilde{W}' S' = \alpha \tilde{W}' A S$$
$$V_n = \tilde{W}' N' = \tilde{W}' A N \tag{6.29}$$

where
W' = the new weight vector.

Upon comparison of Eqs. (6.25) and (6.29), it is seen that combining the channels after the transformation matrix A with the weight vector W' is equivalent to the weight vector $\tilde{A}W'$ without the transformation matrix. Thus, for the equivalent outputs, we have

$$W = \tilde{A}W' = \tilde{W}'A. \tag{6.30}$$

From Eqs. (6.28) and (6.29), one may write

$$\begin{aligned} P_n &= E\{|\tilde{W}'N'|^2\} \\ &= E\{\tilde{W}'^* N'^* \tilde{N}' W'\} \\ &= \tilde{W}'^* E\{N'^* \tilde{N}'\} W'. \end{aligned} \tag{6.31}$$

Since the transformation matrix A decorrelates the noise components and equalizes their powers, the covariance matrix of the noise components after A transformation is simply the identity matrix of order N, N being the number of array elements. Thus,

$$E\{N'^* N'\} = p_o I_N \tag{6.32}$$

where
p_o = a constant

$$I_N = \begin{bmatrix} 1 & 0 & 0 & 0 & 0 & \\ . & 1 & . & . & . & \\ . & . & 1 & . & . & \\ 0 & 0 & 0 & . & . & 1 \end{bmatrix}. \tag{6.33}$$

From Eqs. (6.31) and (6.32)

$$P_n = p_o \tilde{W}'^* W' = p_o \|W'\|. \tag{6.34}$$

A comparison of Eqs. (5.21) and (6.34) shows that the output noise power in both cases, corresponding to the configuration shown in Figures 5-2 and 6-4, are identical in form, with $\|W'\|$ in Eq. (6.34) replacing $\|W'\|$ even when the off-diagonal elements of the covariance matrix M are not zero.

Now, from Eqs. (6.29), (6.30), (6.31), and (6.34)

$$\begin{aligned} P_n &= \tilde{W}^* M W \\ &= \tilde{W}'^* A^* M \tilde{A} W' \\ &= p_o \|W'\|. \end{aligned} \tag{6.35}$$

Thus

$$A^* M \tilde{A} = I_N \tag{6.36}$$

and

$$M = (\tilde{A} A^*)^{-1}.$$

We have seen in Section 5.2 of Chapter 5 that the optimum weighting vector, W_{opt}, where the noise components n_k have equal power and are uncorrelated, is given by

$$W_{\text{opt}} = \mu S^* \tag{6.37}$$

where,
$\mu =$ a constant, as given in Eq. (5.29).

From Eqs. (6.28), (6.30), and (6.37)

$$\begin{aligned} W_{\text{opt}} &= \tilde{A} W'_{\text{opt}} \\ &= \tilde{A} \mu S' \\ &= \mu \tilde{A} A^* S^*. \end{aligned} \tag{6.38}$$

And, from Eqs. (6.37) and (6.38)

$$W_{\text{opt}} = \mu M^{-1} S^* \tag{6.39}$$

or

$$M W_{\text{opt}} = \mu S^*. \tag{6.40}$$

The covariance matrix M in Eq. (6.40) is the sum of $M_0 + M_J$ where M_0 is the covariance matrix of the noise in the absence of the jammers. Thus, Eq. (6.39) shows the required weighting vectors, corresponding to Figure 5-2, that will maximize the SNR regardless of the number of jammers or interferences as long as it is less than N.

Since the initial formulations of Howells–Applebaum approaches on adaptive beamforming by an antenna array to reject multiple interferences, while maintaining the array directivity in the desired look direction, in the 1970s, much has been explored on the implementation of such arrays, and on overcoming various drawbacks of such arrays such as slow convergence, possibility of the desired signal null under certain circumstances, etc. In general, a sidelobe canceller addressing cancellation of multiple interferences could be made effective when the interference levels are relatively much higher than the desired signal level. In the absence of special strategies, an adaptive beamforming could be hypersensitive to array imperfection [3] when the desired signal-to-receiver-noise ratio exceeds a certain threshold. Various algorithms, including the constrained least mean square [4], have been developed to constrain adaptive beamforming such that the array output for the desired look direction remains unity while interferences from other directions are nulled. Publications relating to many such activities are noted in the References section at the end of this chapter. They are not discussed here further.

Before we conclude our discussions on adaptive beamforming antenna arrays for multiple interference rejections, we should consider some approaches, instead of algorithms, that ensure the array directivity in the desired look direction while nulling the array output for interferences arriving at the array from other directions, as outlined in the following section.

6.3 SPATIAL INTERFERENCE NULLING WITH A PILOT SIGNAL

In the simplest arrangement for multiple interference cancellation, shown in Figure 6-1, it is assumed that an adaptive canceller is dedicated to each interference, where the reference antenna or sensor for the canceller receives the particular interference to be cancelled as its dominant signal. In some cases, however, the particular interference being the dominant signal for the reference antenna of the dedicated canceller is not enough, even when the canceller eliminates the intended interference at its output almost completely. This is because the interference of the same specie as that cancelled completely by a dedicated adaptive canceller may be reintroduced[1] into the receive line by other cancellers as long as their reference antennas contain the interference of the specie under consideration. An arrangement, similar to that shown in Figure 6-1, except where each adaptive canceller addresses the desired signal instead of the different interferences, can sometimes be more effective, particularly when the interferences and their directions of arrival at the antenna array are too numerous to make separate provision for their cancellations. To illustrate this concept, let us suppose that a pilot signal, representing the look direction of the desired signal and its frequency spectrum, is available to the antenna array where each array element except one (Ant. 1) is connected to an adaptive canceller, as shown in Figure 6-4. In this arrangement, the receiving antenna in Figure 6-1 is replaced by one of the array elements (Ant. 1) that can receive both the desired and undesired signals. Let us suppose that the pilot signal is the dominant signal for each array element, connected to an adaptive canceller such that each adaptive canceller eliminates the desired signal from the array element 1, as shown in Figure 6-4, in an adaptive manner. The countersignals created at the outputs of the signal controllers can then be summed as shown, the summed signal being the array output. When the reference signal for each adaptive canceller in Figure 6-4 contains the pilot signal as the dominant signal, the countersignal created at the output of each canceller will contain no or negligible interferences, regardless of their directions of arrival at the array. Consequently, the array output also will contain no or negligible interferences.

It should be noted that, although the pilot signal at the reference port of each adaptive canceller will have a different phase, depending on the location of the

[1] A common monitoring coupler following the subtraction of all cancelling signals at the receive line lessens the effect of the reintroduction of any interference because of the closed-loop controls.

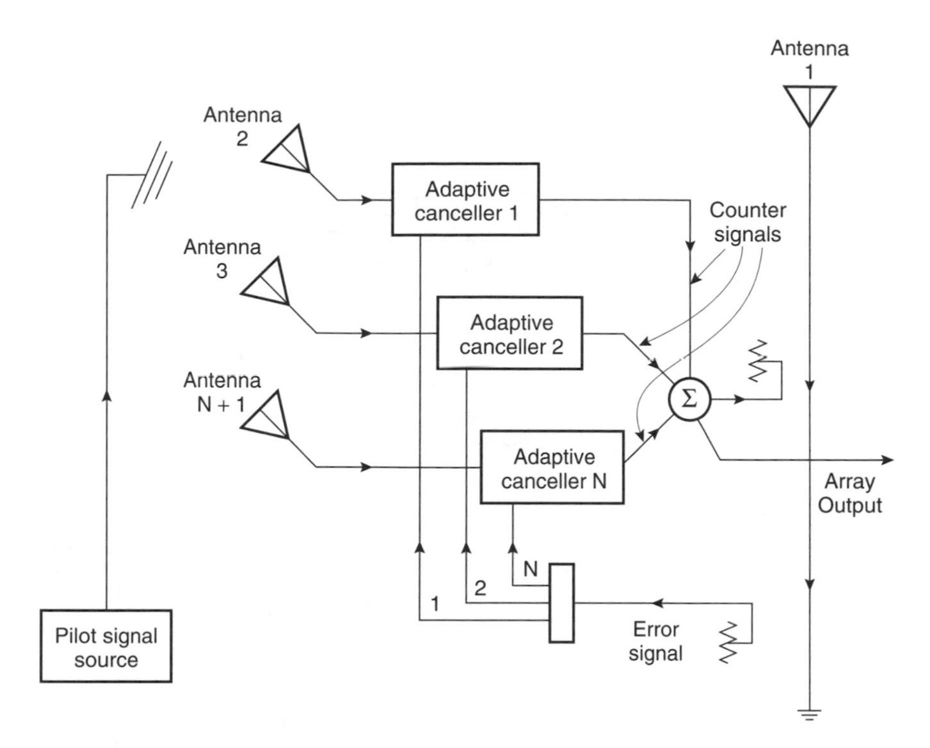

Figure 6-4 Illustration of multiple interference cancellations by addressing desired signal only.

array element with respect to the pilot signal source, the countersignals at the output of each signal controller will have the same phase as that of the pilot signal as each countersignal cancels the same pilot signal at the array element 1. The summer shown in the figure, then, sums all cophased pilot signals which become the array output, the amplitude of the summed signal being N times the signal input at the array element 1.

When the signal controller of each adaptive canceller, shown in Figure 6-4, consists of two electronic attenuators, phase separated by 90° as discussed in Chapter 3, the adaptive cancellers can effectively accommodate a relatively narrowband pilot signal. When the desired and hence pilot signal is a wideband signal, each adaptive canceller in Figure 6-4 can be replaced by an adaptive transversal filter as discussed in Section 2.3 of Chapter 2. Such an approach is illustrated in Figure 6-5.

Ideally, for the arrangements shown in Figures 6-4 and 6-5, the pilot signal should be the desired signal having the look direction and spectral characteristics of those of the desired signal. In reality, the desired signal is seldom known that well, as otherwise there would be no need for a receiver or a receiving antenna array. However, adaptive cancellers are effective for a band of frequencies, and

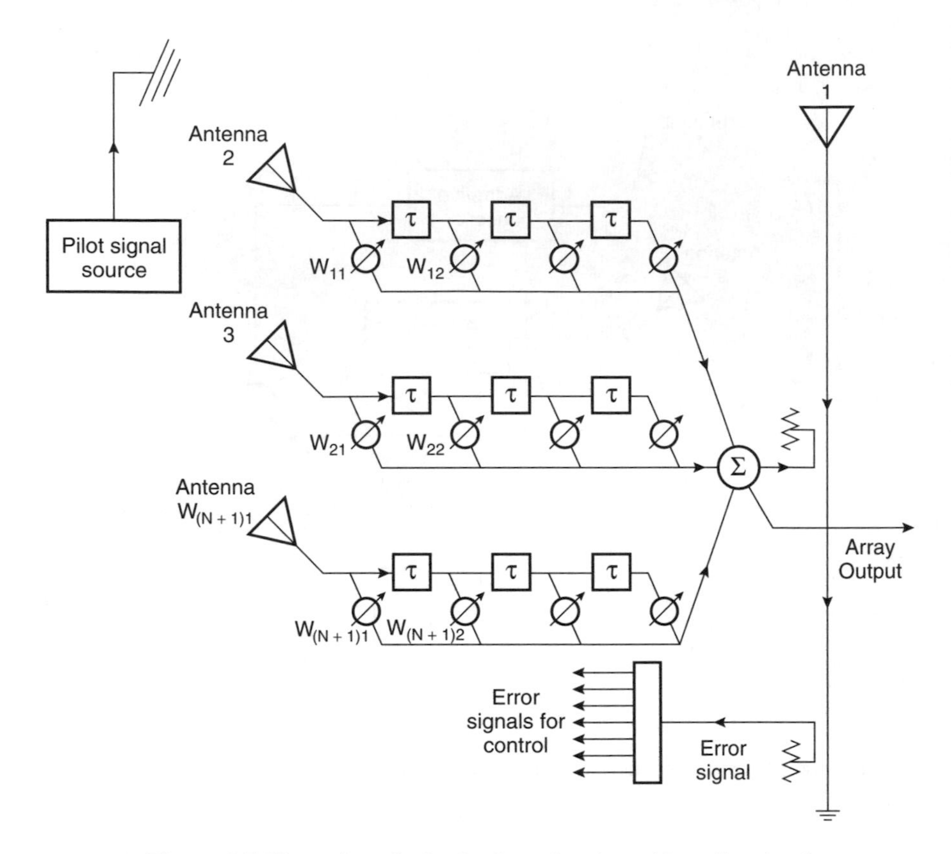

Figure 6-5 Illustration of adaptive beamforming with a pilot signal.

if the pilot signal is chosen from that frequency band, such that the pilot signal frequency and the look angle approximate those of the desired signal, then a control system that operates on the pilot signal will, in general, set up signal-controller parameters that are almost the same as those needed for the desired signal. For such cases, then, the controls of the adaptive cancellers effected by the pilot signals will be very nearly the same as needed to receive the desired signal optimally. When the implementation of such a pilot signal is feasible, one can also make the pilot signal the dominant signal for the reference antenna for each adaptive canceller shown in Figure 6-4 or 6-5. Finally, the receiver selectivity to discriminate against the pilot signal can be utilized to receive the desired signal only from the array output while discarding the pilot signal.

Widrow *et al.* [6] suggested a different configuration for utilizing the pilot signal for the constrained linear adaptation as referred to earlier. Figure 6-6 illustrates such an approach. There are two adaptation modes for this arrangement. During the "*A*-mode" operation, the inputs to the processor are connected to actual antenna element outputs, as shown in the figure, and the adaption with the pilot signal is absent. During the "*P*-mode" operation, on the other hand, a set of delayed signals derived from the pilot signal becomes the inputs to the processor. The time delays $\delta_1 \cdots \delta_k$ are chosen so as to obtain a set of input signals identical to those that would appear if the array were actually receiving a distant radiated plane-wave pilot signal from the desired look direction, that is, the direction intended for the main lobe of the receiving directivity pattern.

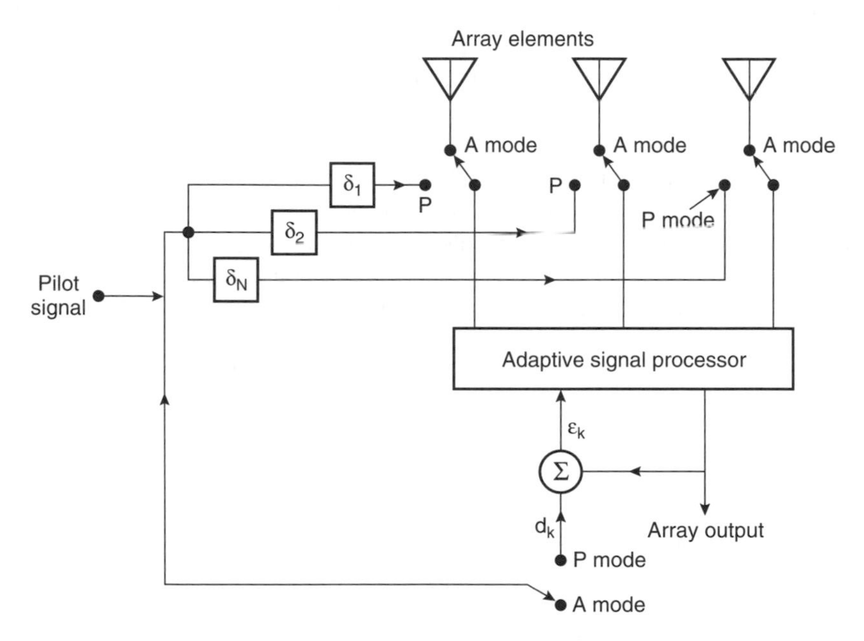

Figure 6-6 Two-mode operation of an array and signal processor. This arrangement maintains a directivity at the desired look direction, while cancelling one or more interferences.

During the "*P*-mode" operation, the desired response is set to zero, as shown in the figure, and hence the adaptation process proceeds to eliminate all received signals. Obviously, a continuous operation in "*A*-mode" will cause all weights to be zero, causing the system to shut itself off. However, by alternating frequently between "*P*-mode" and "*A*-mode," and causing only small changes in weights during each mode of adaptation, one may maintain a unit-gain beam approximately in the desired look direction, and minimize, approximately at the same time, the reception of incident noise or interference power. Also, the pilot signal can be chosen as the sum of several signals at different frequencies, so that the adaptation in "*P*-mode" will constrain the antenna gain and phase in the look direction to have specific values at each of these frequencies. Additionally, if several pilot signals, corresponding to different directions of arrival, are added together, it is possible to constrain the array gain simultaneously for various look angles and frequencies when adapting in "*P*-mode." Thus, one may obtain some control of the bandwidth and beamwidth in various look directions.

In the two-mode adaptation operation as illustrated in Figure 6-6, the beam is formed and supported during the "*P*-mode," while the interferences or noise are rejected in the least-square sense during the "*A*-mode" operation, subject to the constraints created during the "*P*-mode" operation. Thus, a signal reception is impossible during the "*P*-mode" operation as the processor is not connected to the array elements. Reception can therefore take place only during the "*A*-mode" operation. This problem can be eliminated by employing two signal processors, one slaved by the other, as shown in Figure 6-7. As shown in the figure, the pilot signals and the received signals through the array elements enter into the adaptive signal processor. For this processor, the desired response is the pilot signal. A second, slaved signal processor generates the actual array output signals, but it performs no adaptation. Its input signals do not contain the pilot signal. It is slaved to the first adaptive signal processor in such a way that its weights are exact copies of the adapting system so that it does not need to receive the pilot signal.

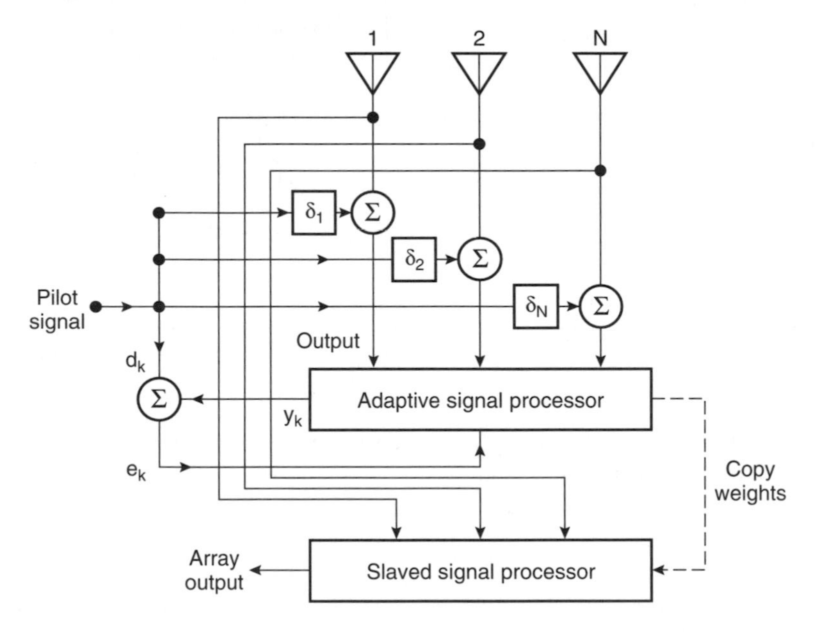

Figure 6-7 Illustration of a single-mode operation with a pilot signal. (Adapted from B. Widrow *et al.*, "Adaptive antenna systems," *Proc. IEEE*, Dec. 1967.)

In the arrangement shown in Figure 6-7, the pilot signal is on all the time. An adaptation to minimize the mean-square error forces the adaptive processor to reproduce the pilot signal as closely as possible, and reject at the same time all signals received by the array elements which are not correlated with the pilot signal, at least in the mean-square sense. Thus, the adaptive process creates a directivity-pattern main lobe toward the desired look direction and the desired signal passband, as caused by the pilot signal, and creates nulls in the direction of interferences and noise and their frequency bands. Such a constrained beamforming to reject multiple interferences while maintaining directivity toward the desired signal source, however, is feasible only when the approximate direction of the desired signal and its approximate frequency passband are known so that the pilot signal can be generated accordingly. For a spatial scanning receiving antenna, various desired signal look directions can be assumed and the corresponding pilot signal can be simulated. The presence of the desired signal may then be sought along the pilot signal look direction while rejecting multiple interferences from other directions during the scanning operation.

The pilot-signal-driven adaptive beamformer as discussed above places a "soft constraint" for signal reception in the pilot signal look direction. Thus, a weak desired signal arriving at the pilot signal look direction would have no effect on the sensitivity of the adaptive beamformer to this signal. However, a strong desired signal would tend to be partially rejected, even if it were to arrive exactly in the pilot-signal look direction. This difficulty is overcome by the adaptive beamformer devised by Frost [4], which places a "hard constraint" in the look direction. With this hard constraint, the sensitivity of Frost's adaptive beamformer is fixed on the look direction, regardless of the strength of the desired signal from this direction.

The degree of cancellation usually feasible with a constrained beamforming antenna array that attempts to reject multiple interferences simultaneously is more limited than it would be if dedicated adaptive cancellers are used to reject particular interferences. Some such arrangements are discussed later in this chapter.

6.4 ARRAY PATTERN SYNTHESIS FOR NULLING MULTIPLE INTERFERENCES

Most adaptive beamforming antenna arrays considered earlier attempt to maximize the desired signal-to-noise ratio, as in the case of the Applebaum [2] array, for example. In the case where the jammers are the dominant noise sources, this process automatically places pattern nulls in the direction of jammers. In the constrained beamforming array also, such as suggested by Frost [4], a directivity gain is maintained along the desired signal look direction while minimizing the noise or interference power received by the array from other directions. In either case, the role of the antenna pattern is not clear. An alternative approach for multiple interference suppression by an antenna array could be a direct pattern synthesis which includes specific pattern nulls along the specific look directions of the interferences or jammers. For example, a spatial narrowband interference can be suppressed by imposing a signal null in the antenna pattern at the proper angle. A spatial wideband jammer that covers an angular sector may require an antenna pattern null over the entire sector. This can be achieved either by imposing signal pattern nulls on a set of closely spaced directions over the sector of concern or by imposing nulls in the pattern and its derivative at the middirection of the sector. Such a pattern synthesis is analogous to Chebyshev and Butterworth type alternatives in filter design. Steyskal [13] has suggested an array pattern synthesis based on least-mean-square approximation.

To illustrate the approach, let us consider a situation where an antenna array is being illuminated by desired signals and by highly dominant interference signals from certain discrete directions. We assume that the optimum array pattern in this case is the desired pattern in the absence of the interferences, which we will refer to as the quiescent pattern, but modified so as to form pattern nulls in the look directions of the interferences of concern. When the interference rejection is the primary objective, the degrees of freedom or design flexibility available during the array pattern synthesis will be used to form the required pattern nulls first, the remaining degrees of freedom being used for approximating the quiescent pattern.

The antenna pattern synthesis suggested by Steyskal consists of determining the closest approximation of the array pattern, P_a, to the given quiescent pattern, p_o, subject to a set of null constraints due to a set of discrete interferences. For simplicity, we consider a linear array of N isotropic antenna elements with uniform half-wavelength spacing, as shown in Figure 6-8. The far-field antenna pattern can be written as

$$p(u) = \sum_{1}^{N} x_n e^{-jn\pi u} \tag{6.41}$$

where
$u = \sin\theta$, θ being defined in the figure
$x_n =$ the complex excitation of the nth array element.

The synthesis process can now be stated as follows: Minimize $\epsilon(p_a)$ in the least-mean-square sense where

$$\epsilon(p_a) = 1/2 \int_{-1}^{1} |p_o(u) - p_a(u)|^2 \, du \tag{6.42}$$

subject to the constraint

$$\frac{d^n}{du^n} p_a \left[(u_m)_1^{M_1} \right] = 0 \qquad m = 1, 2, \cdots, M_1 \qquad n = 0, 1, 2, \cdots, M_2 \tag{6.43}$$

where
$\{u_m\}_1^{M_1} =$ the angular locations of M_1 interference sources.

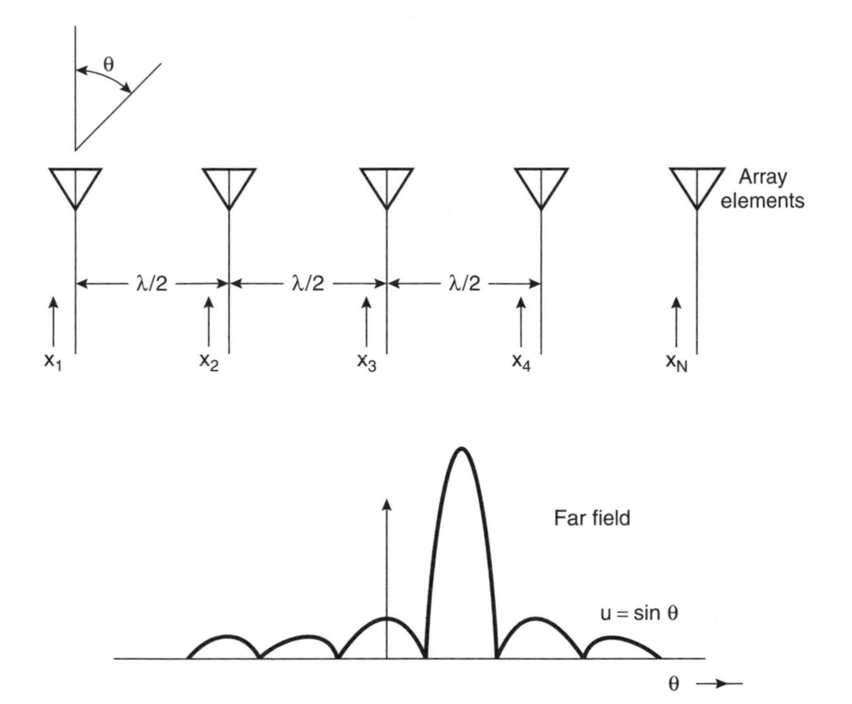

Figure 6-8 Antenna array and far field.

Since the total number of constraints must not exceed the total number of free variables, it is required that

$$M_1(M_2 + 1) \leq N. \tag{6.44}$$

Let it be assumed that the desired quiescent pattern is given as a sum of N weighted harmonics as given in Eq. (6.41). For the general case where $p_o(u)$ has any known functional form, p_o may be approximated by the first N terms of its Fourier-series expansion.

Substituting Eq. (6.41) into Eq. (6.42), one may write

$$\epsilon(P_a) = 1/2 \int_{-1}^{1} \left| \sum_{n=1}^{N} x_{on} e^{-jn\pi u} - \sum_{n=1}^{N} x_{an} e^{-jn\pi u} \right|^2 du \tag{6.45}$$

where

x_{on} and x_{an} = the excitations corresponding to p_o and p_a, respectively.

It should be noted that only the excitation functions could be different for any n, $\exp(-jn\pi u)$ being the same for the same array element spacing.

Combining the two series in Eq. (6.45), one obtains

$$\begin{aligned}\epsilon(P_a) &= 1/2 \int_{-1}^{1} \left| \sum_{n=1}^{N} (x_{on} - x_{an}) e^{-jn\pi u} \right|^2 du \\ &= 1/2 \sum_{n=1}^{N} \sum_{m=1}^{N} (x_{on} - x_{an})(x_{om}^* - x_{am}^*) \int_{-1}^{1} e^{-jn\pi u} e^{jm\pi u}. \end{aligned} \tag{6.46}$$

Because of the orthogonal relation of the two terms in the integrand of Eq. (6.46), when m, n are integers

$$\begin{aligned} 1/2 \int_{-1}^{1} e^{-j(n-m)\pi u}\, du &= 1 \qquad \text{for } n = m \\ &= 0 \qquad \text{for } n \neq m. \end{aligned} \tag{6.47}$$

Thus, from Eqs. (6.46) and (6.47)

$$\epsilon(p_a) = \sum_{n=1}^{N} |x_{on} - x_{an}|^2. \tag{6.48}$$

The array excitations $\{x_{on}\}_1^N$ and $\{x_{an}\}_1^N$ can be represented as vectors

$$\begin{aligned} \bar{x}_o &= (x_{o1}, x_{o2}, \cdots, x_{oN}) \\ \bar{x}_a &= (x_{a1}, x_{ao2}, \cdots, x_{aN}). \end{aligned} \tag{6.49}$$

Then, from Eqs. (6.48) and (6.49)

$$\begin{aligned} \epsilon(p_a) &= \tilde{x}^* x \\ &= \|x\| \end{aligned} \tag{6.50}$$

where
$x = \bar{x}_0 - \bar{x}_a$.

As noted earlier, the pattern synthesis involves minimizing $\|x\|$ subject to the constraints imposed by the nulls. Equation (6.43) relates to the needed constraints. For the single or zero-order null, we may write from Eq. (6.43)

$$\sum_{1}^{N} x_n e^{j(\Psi - n\pi)u_m} = 0 \tag{6.51}$$

in which a phase factor is introduced in Eq. (6.41) to shift the phase center of the pattern to the array center, Ψ being given as

$$\Psi = \frac{N+1}{2}. \tag{6.52}$$

Similarly, from Eq. (6.43), one obtains for the mth-order null

$$\sum_{1}^{N} x_n (j[\Psi - n\pi])^m e^{j(\Psi - n\pi)u_m} = 0. \tag{6.53}$$

Defining the constraint vector

$$\begin{aligned}\bar{y}(m) = ([j(\Psi - \pi)]^m e^{j(\Psi-\pi)u_m}, \\ [j(\Psi - 2\pi)]^m e^{j(\Psi-2\pi)u_m} \\ \cdots [j(\Psi - N\pi)^m] e^{j(\Psi-N\pi)u_m})\end{aligned} \tag{6.54}$$

one obtains the orthogonality condition for the array excitation parameters as given by

$$(\bar{x}, \bar{y}_m^{(0)}) = (\bar{x}, \bar{y}_m^{(1)}) = \cdots (\bar{x}, \bar{y}_m^{(M)}) = 0 \tag{6.55}$$

where

$(\bar{x}, \bar{y}_m^{(0)})$, $(\bar{x}, \bar{y}_m^{(1)})$, etc. = the inner product such that

$$(\bar{x}, \bar{y}) = \sum x_n y_n^*,$$

y_n^* = the complex conjugate of y_n.

From Eqs. (6.48) and (6.55), the array synthesis problem can be expressed as

$$\epsilon = \|\bar{x}_0 - \bar{x}_a\|^2 = \text{ minimum} \tag{6.56}$$

and

$$(\bar{x}_a, \bar{y}_m) = 0 \qquad m = 1, 2, 3, \cdots, M \tag{6.57}$$

where

$\bar{y}_m = \bar{y}_m^{(0)}, \bar{y}_m^{(1)}$, etc.

$\bar{x}_0$ and $\bar{x}_a$ = the unconstrained and constrained array excitation, respectively.

Equation (6.57) shows that the desired array excitation $\bar{x}_a$ is orthogonal to the constraint vector $\{\bar{y}_m\}_1^M$.

Now, since any vector $\bar{x}$ can be decomposed into a pair of mutually orthogonal components, that is,

$$\bar{x} = \bar{y} + \bar{z} \tag{6.58}$$

in which $\bar{y}$ is contained in subspace Y and z is contained in subspace Z and $\bar{z}$ is normal to Y, one may then decompose $\bar{x}_0$ and $\bar{x}_a$ and obtain from Eqs. (6.56) and (6.57)

$$\epsilon = \|\bar{x}_0 - \bar{x}_a\|^2 = \|\bar{y}_0 - \bar{y}_a\|^2 + \|\bar{z}_0 - \bar{z}_a\|^2 = \text{ minimum}. \tag{6.59}$$

When $\bar{x}_a$ is resolved into two mutually orthogonal components, $\bar{y}_a$ and $\bar{z}_a$, with $\bar{y}_a$ contained in subspace Y as noted above, one may write

$$\bar{x}_a = \bar{y}_a + \bar{z}_a$$

and the inner product

$$(\bar{x}_a, \bar{y}_m) = (\bar{y}_a, \bar{y}_m) + (\bar{z}_a, \bar{y}_m).$$

But $\bar{z}_a$ being orthogonal to y_a and y_a and y_m being in the subspace Y

$$(\bar{z}_a, \bar{y}_m) = 0.$$

Thus

$$(\bar{x}_a, \bar{y}_m) = 0 = (\bar{y}_a, \bar{y}_m). \tag{6.60}$$

Again, the only way $(\bar{y}_a, \bar{y}_m)$ could be zero in this case, both being in the same subspace, is

$$\bar{y}_a = 0. \tag{6.61}$$

This makes $\bar{x}_a$ confined to the subspace Z, as illustrated in Figure 6-9.

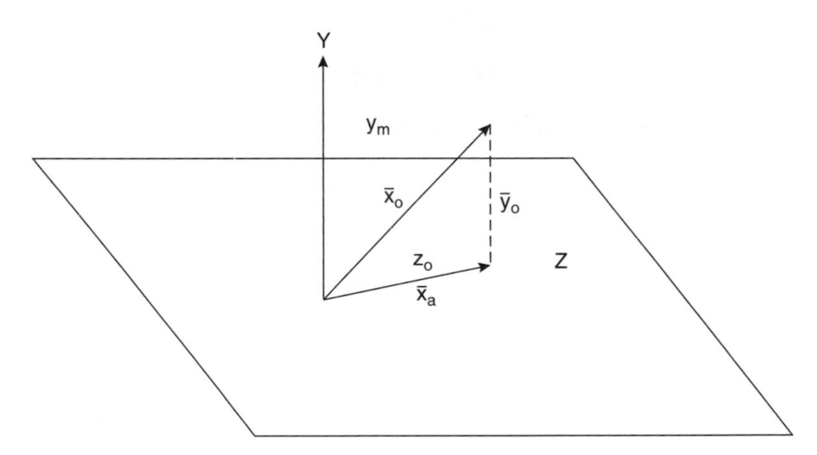

Figure 6-9 Illustration of the approximation of $\bar{x}_a$.

From Eq. (6.59), it is seen that ϵ is minimized when one sets $\bar{z}_0 = \bar{z}_a$, that is,

$$\epsilon_{\min} = \|\bar{x}_0 - \bar{x}_a\|^2 = \|\bar{y}_0\|^2 \tag{6.62}$$

and

$$\bar{x}_a = \bar{y}_a + \bar{z}_a = \bar{z}_0 = \bar{x}_0 - \bar{y}_0. \tag{6.63}$$

It is seen from Figure 6-9 that when $\bar{x}_a$ is confined to the subspace Z, the best approximation to $\bar{x}_0$ is obtained when $\bar{x}_a = \bar{z}_0$ since of all elements $\bar{z}$ contained in the subspace Z; this point is the closest to $\bar{x}_0$.

Equations (6.62) and (6.63) constitute the solution of the array excitation problem. Expressing, now, $\bar{y}_0$ as a linear combination of $\bar{y}_m$ such that

$$\bar{y}_0 = \sum_{1}^{N} \alpha_m \bar{y}_m \tag{6.64}$$

where $\alpha_m s$ = coefficients to be determined.

One may write

$$\bar{x}_a = \bar{x}_0 - \sum_{1}^{M} \alpha_m \bar{y}_m. \tag{6.65}$$

Equation (6.65) suggests that the sought excitation $\bar{x}_a$ consists of the quiescent excitation $\bar{x}_0$ and a weighted sum of the vector $\bar{y}_m$. It also follows from Eq. (6.65) that the constrained pattern $p_a(u)$ is the quiescent pattern $p_o(u)$ with M number of beams superimposed on the quiescent pattern. In other words

$$p_a(u) = p_0(u) - \sum_{1}^{M} \alpha_m F_m(u) \tag{6.66}$$

where F_m is some function of u.

For the case of M single nulls corresponding to $m = 0$ in Eq. (6.43), Steyskal [13] has shown that

$$p_a(u) = p_o(u) - \sum_{1}^{M} \alpha_m \frac{\sin N\pi(u - u_m)/2}{\sin \pi(u - u_m)/2}. \tag{6.67}$$

When the number of array elements N is large, Eq. (6.67) can be approximated as

$$p_a(u) = p_o(u) - N \sum_{1}^{M} \alpha_m \operatorname{sinc}\left[N\pi(u - u_m)\right].$$

Thus, the pattern p_a is simply the quiescent pattern p_0 and M number of sinc beams. It may be noted that a similar result is obtained from Applebaum analysis such as given in Eq. (5.60) where $G_0(\theta)$ corresponds to $p_0(u)$ and $\bar{s}(x)$ in Eq. (5.59) corresponds to $\sin(Nv/2)/\sin(v/2)$, v being equal to $\pi(u - u_m)$.

To complete the solution for $p_0(u)$, one still needs to determine the coefficients α_m, for $m = 1, 2$, etc. From Eqs. (6.65) and (6.57), one may write

$$\begin{bmatrix} (\bar{y}_1, \bar{x}_0) \\ (\bar{y}_2, \bar{x}_0) \\ \vdots \\ (\bar{y}_m, \bar{x}_0) \end{bmatrix} = \begin{bmatrix} \bar{y}_1, \bar{y}_1 & \bar{y}_1, \bar{y}_2 & \cdots & \bar{y}_1, \bar{y}_m \\ \bar{y}_2, \bar{y}_1 & \bar{y}_2, \bar{y}_2 & \cdots & \bar{y}_2, \bar{y}_m \\ \vdots & & & \\ \bar{y}_m, \bar{y}_1 & \bar{y}_m, \bar{y}_2 & \cdots & \bar{y}_m, \bar{y}_m \end{bmatrix} \begin{bmatrix} \alpha_1 \\ \alpha_2 \\ \vdots \\ \alpha_m \end{bmatrix}. \tag{6.68}$$

The coefficients α_1, α_2, etc., are obtained from the column matrix involving α which, in turn, is obtained from the matrix equation given in Eq. (6.68).

To illustrate the effectiveness of this alternative method for optimum pattern synthesis by modifying the quiescent pattern, one may consider, for example, the computer simulated patterns for 21 equally spaced antenna elements in an array, as shown in Figure 6-10. Here, Figure 6-10(a) shows the quiescent pattern, with the main beam corresponding to $\sin\theta = 0$ denoting the direction of the desired signal. Figure 6-10(b) shows the modified pattern with a zero order null at $\sin\theta = 0.22$ for an assumed interference direction. Similarly, Figure 6-10(c) and (d) shows, respectively, the pattern modification for a first-order and a second-order null. As noted earlier, a higher than zero-order null is desired when a pattern notch over a sector is to be synthesized, as in the case shown in Figure 6-10(c) and (d). Also, in all cases of quiescent pattern modification, there is a slight reduction of maximum directivity. In Figure 6-10(b), (c), and (d), this reduction is 0.06, 0.11, and 0.59 dB, respectively.

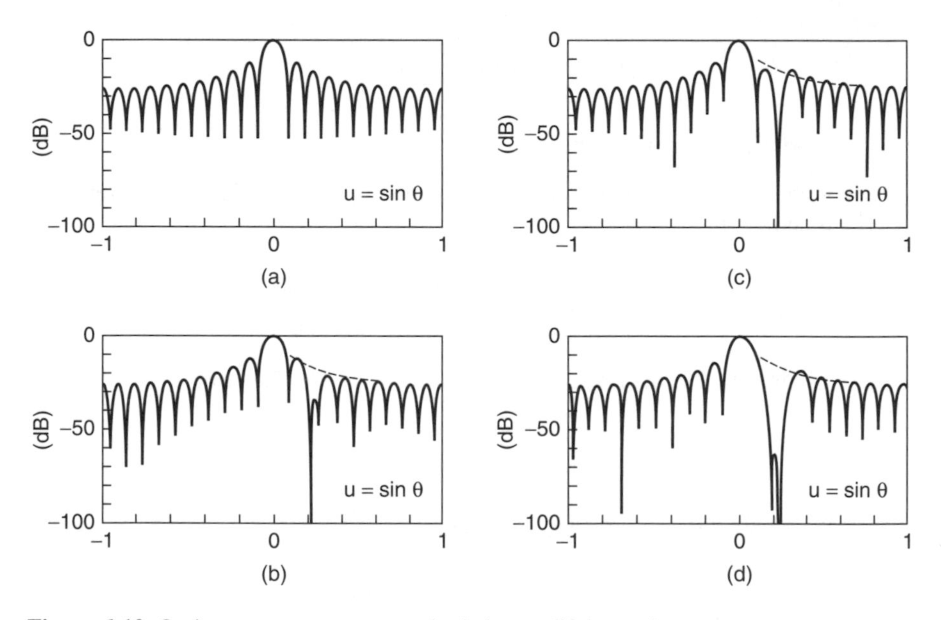

Figure 6-10 Optimum array pattern synthesis by modifying quiescent pattern. (a) Quiescent pattern for a 21-element array. (b) Modified pattern with zero-order null at $\sin\theta = 0.22$. (c) Modified pattern with first-order null at $\sin\theta = 0.22$. (d) Modified pattern with second-order null at $\sin\theta = 0.22$.

In reviewing the array beamforming approaches to rid interferences as advanced by Applebaum and the alternative approach introduced by Steyskal, it is seen that they both attempt to modify the quiescent pattern p_0 that would be desired in the absence of any interference. Both attempt to superimpose the interference nulling pattern on the quiescent pattern, although the steps involved in each case are different. Also, unlike an adaptive cancellation, both approaches have the disadvantage of adversely affecting the signal of interest when the interference direction is too close to the desired signal direction. In many cases, the concept of multiple-order null for blocking out reception from an angular sector in the pattern, as is available in the Steyskal approach, is a distinct advantage. In contrast, an adaptive canceller that does not rely on spatial discrimination means for desired and undesired signal directions is often preferred, particularly when the desired and undesired signals' directions are too close to one another.

Byrnes and Newman [14] investigated the multiple null steering problem by an antenna array using "phase" only adjustments for the array-element circuits. The synthesis of such an array is based on the one-to-one correspondence between polynomials and linear arrays, with commensurable separations between elements. It is seen that this type of array can also be synthesized to create multiple nulls along the direction of multiple interferences or jammers.

Although there have been many investigations to solve the multiple interference problems for a receiving antenna by an antenna array, the degree of success, as measured by the magnitude of cancellation of completely arbitrary interferences, has been limited. Additionally, an array requiring a large number of elements is also not feasible for many practical applications, when the available space may be limited. Other means are therefore needed to solve the multiple interference problem, particularly when the interferences are completely arbitrary.

6.5 MULTIPLE INTERFERENCE CANCELLATIONS WITHOUT THEIR REINTRODUCTION

We have noted earlier that a multiple interference cancelling system with one adaptive canceller dedicated to one interference only can sometimes be an unsatisfactory remedy, even when each adaptive canceller provides a cancellation on the order of 60 dB or more for the particular interference it addresses. This is because the interference which is cancelled by one canceller can be reintroduced into the receive line through other adaptive cancellers. When a very high degree of cancellation is required, the constrained beamforming array, attempting to cancel multiple interferences from various different directions, also is often not adequate, particularly when the desired signal is also strong. It is possible, however, to obtain a high degree of cancellation of multiple interferences without the fear of reintroduction of any interference into the receive line at the expense of additional auxiliary antenna elements and additional cancellers. Such an arrangement is shown schematically in Figure 6-11. Here, each particular interference is cancelled by a row of antenna elements and corresponding cancellers so that at the second row of cancellers, there is no input interference of the same specie as cancelled in the first row. Similarly, for the third row of cancellers, there are no input interferences of the species

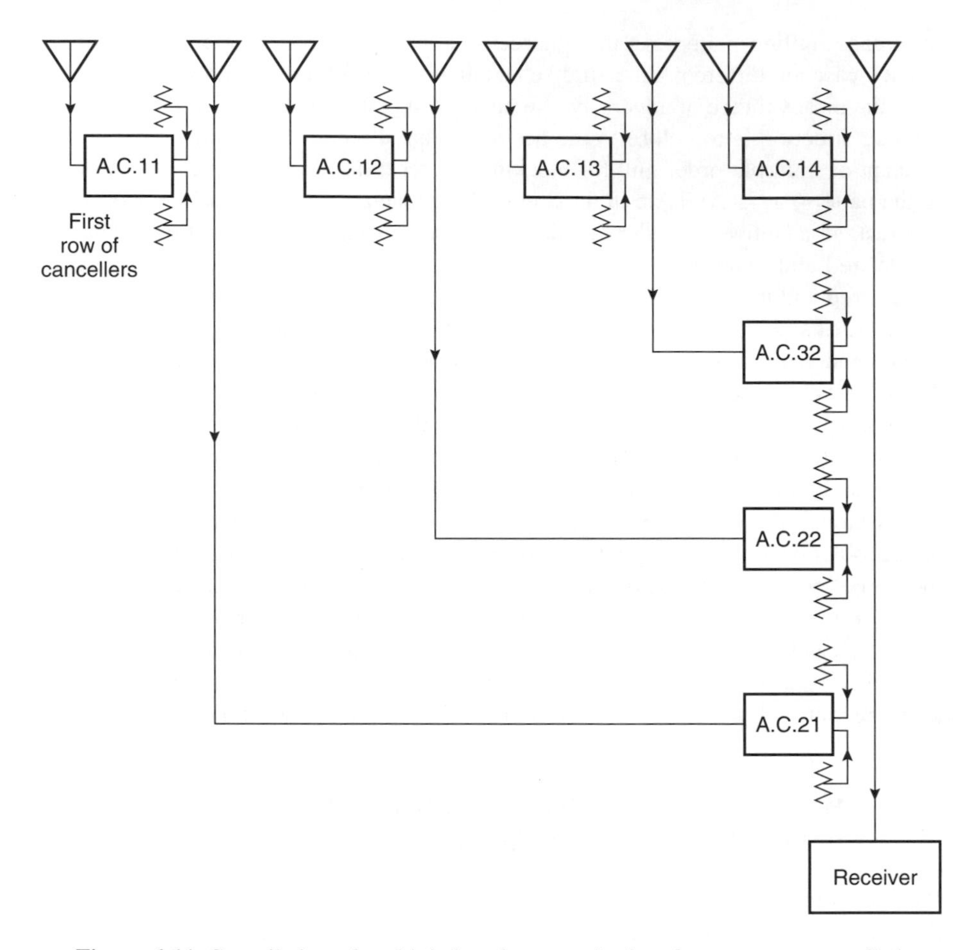

Figure 6-11 Cancellation of multiple interferences. An interference, once cancelled, cannot be reintroduced.

cancelled at the first and second rows of cancellers. More specifically, at the first row, each pair of array elements and associated cancellers address the same interference such that each of the four cancellers, shown in Figure 6-11, cancels the same interference, say interference No. 1, assuming it to be the strongest. For the second row of cancellers, then, there is no interference No. 1 present at any of the canceller inputs or reference ports. The second row of adaptive cancellers then cancels the second or next strongest interference after the No. 1 interference. The process is continued until all interferences of concern are eliminated. As in the case of all adaptive cancellers considered in Chapters 3 and 4, the cancellers address the interferences only, the desired signal flow through the receive line suffering, in most cases, only an insertion loss. The biggest drawback of the arrangement shown in Figure 6-11 is that a large number of array elements and cancellers are needed, although the degree of cancellation of each interference can be very high. In the arrangement shown in Figure 6-11, three independent interferences and a number of desired signals are assumed, the number of required auxiliary antenna elements being 2^3 or 8 and the number of cancellers being $2^3 - 1$ or 7. For N number of interferences, then, one needs 2^N antenna elements and $2^N - 1$ number of adaptive cancellers. Obviously, as N becomes large, such an arrangement may become very involved and perhaps prohibitively expensive. Also, as N increases, the through-line insertion loss due to the cancelling couplers at the receive line also may be high. Notwithstanding these problems, the arrangement shown in Figure 6-11 is very useful for the cancellation of completely arbitrary undesired signals and interferences when the number of independent interferences, N, is 3 or 4.

6.6 MULTIPLE INTERFERENCE CANCELLATION BY SIGNAL SEPARATION

We have seen in Chapter 5 that one way to rid an interference from a receive line containing a desired signal along with the interference is to separate the desired and undesired signals and use the desired signal only following the signal separation. In Figure 5-15, for example, we have seen how two arbitrary signals, in the same frequency band and with unknown and time-varying signal levels, can be separated from each other by two cross-connected adaptive cancellers. For such a case, two antennas are used, each receiving the same two arbitrary signals, although in different proportions, where one of which may be labeled as the desired signal, while the other is an interference. We have seen that the two cancellers operate in such a way that following cancellation, one and only one signal appears at the output of one antenna line, while the output of the other antenna contains the second signal only. This concept of adaptive signal separation to achieve the same objective as ridding an arbitrary interference can be extended to separate multiple signals, and hence to rid multiple interferences.

Figure 6-12 shows schematically the extension [12] of the concept of signal separation to three signals, instead of two signals, as shown in Figure 5-15. Each of the antennas shown in Figure 6-12 receives all three arbitrary signals, S_1, S_2, and S_3. For each antenna line, two adaptive cancellers are used to cancel two signals from the other two antenna lines such that, following cancellation, each antenna line contains only one signal at its output. Thus, following the cancellation of two signals in each antenna line, one obtains a separation of the signals, S_1, S_2, and S_3, only one signal appearing at the output of each line. An essential criterion for the signal separation concept is that for each adaptive canceller associated with an antenna line, Antenna Line 1 for example, the reference port is coupled to the other line, that is, Antenna Line 2 or Antenna Line 3, whereas the error and output ports of the canceller are coupled to the same line, that is, Antenna Line 1. Once the signal separation is effected, one can choose only the desired signal for the receiver, creating the same effect as the cancellations of the other two undesired signals.

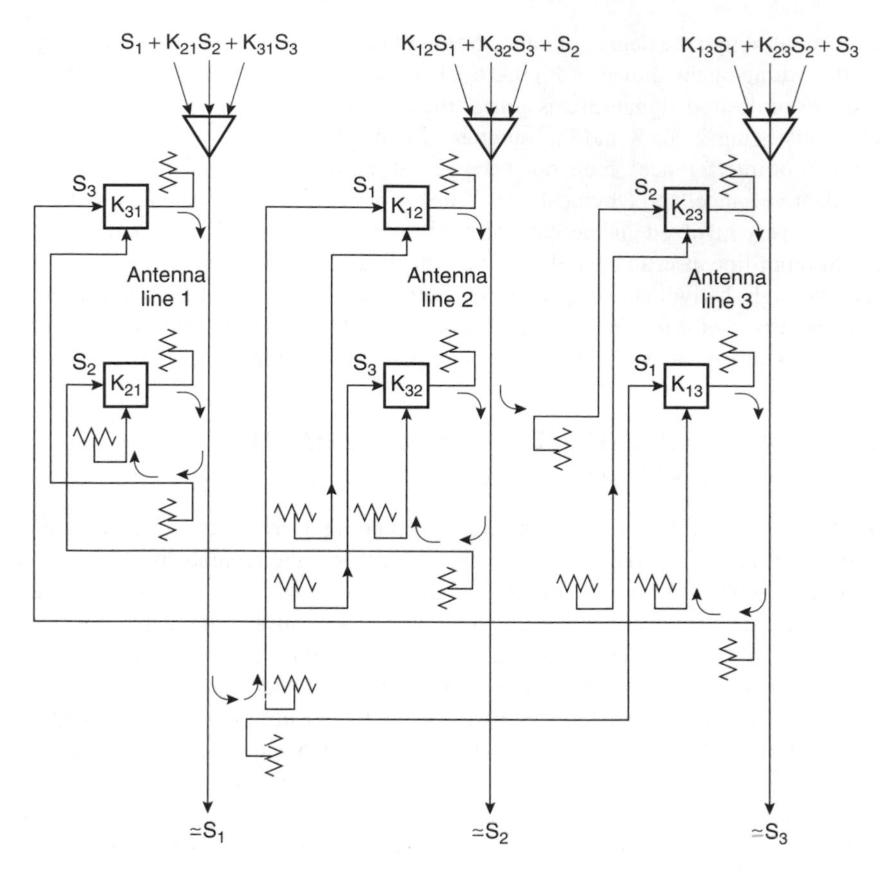

Figure 6-12 Illustration of signal separations of three signals. Square boxes denote adaptive cancellers.

Although an analysis of the dynamics of the simultaneous operations of six cancellers shown in Figure 6-12 is involved, one may easily verify the steady-state requirement and feasibility of the signal separation arrangement. Thus, for example, let the signal inputs to the Antenna lines be as follows:

$$
\begin{aligned}
\text{Antenna line 1} &= S_1 + K_{21}S_2 + K_{31}S_3 \\
\text{Antenna line 2} &= K_{12}S_1 + K_{32}S_3 + S_2 \\
\text{Antenna line 3} &= K_{13}S_1 + K_{23}S_2 + S_3
\end{aligned}
\tag{6.69}
$$

where

$K_{21}, K_{12}, K_{31}, K_{13}, K_{32}, K_{23}$ = complex quantities, each involving an amplitude and a phase,

S_1, S_2, S_3 = transforms of three arbitrary signals to be separated.

The transfer functions of the signal controllers of various cancellers are shown in the figure. Since such transfer functions are obviously, physically realizable, one obtains, in the steady state, the signal S_1 in Antenna Line 1, S_2 in Antenna Line 2, and S_3 in Antenna Line 3.

As noted by Ghose *et al.*, the signal separation approach, as described above, is not limited to the three signals considered in Figure 6-12. In fact, the concept can be extended to an arbitrary number of signals, as long as there are as many antennas available as there are signals to be separated. The principal disadvantage of the multiple interference rejection through a signal separation is that a large number of cancellers is needed for such a case. As may be inferred from Figure 6-12, one needs $N(N-1)$ number of adaptive cancellers and N number of independent antennas when the total number of desired signal and interferences to be separated is N. Again, notwithstanding this disadvantage, effective cancellations of multiple interferences by the signal separation means is a powerful tool for signal processing in a multiple interference environment when no meaningful constraints can be imposed on the types and characteristics of signals involved, and when the number of available antennas is limited.

REFERENCES

[1] B. Widrow, J. Glover, Jr., J. McCool, J. Kaunitz, C.Williams, R. Hearn, J. Zeidler, E. Dong, Jr., and R. Goodlin, "Adaptive noise cancelling principles and applications," *Proc. IEEE*, vol. 63, pp. 1692–1716, Dec. 1975.

[2] S. Applebaum, "Adaptive arrays," *IEEE Trans. Antennas Propagat.*, vol. AP-24, pp. 585–598, Sept. 1976.

[3] N. Jablon, "Adaptive beamforming with the generalized sidelobe canceller in the presence of array imperfections," *IEEE Trans. Antennas Propagat.*, vol. AP-34, pp. 996–1012, Aug. 1986.

[4] O. Frost, III, "An algorithm for linearly constrained adaptive array processing," *Proc. IEEE*, vol. 60, pp. 926–935, Aug. 1972.

[5] B. Widrow and S. Steans, *Adaptive Signal Processing*. Englewood Cliffs, NJ: Prentice-Hall, 1985.

[6] B. Widrow, P. Mantey, L. Griffiths, and B. Goode, "Adaptive antenna systems," *Proc. IEEE*, vol. 55, Dec. 1967.

[7] J. Mayhan, "Nulling limitations for a multiple-beam antenna," *IEEE Trans. Antennas Propagat.*, vol. AP-24, pp. 769–779, Nov. 1976.

[8] J. Hudson, *Adaptive Array Principles*. London: Peter Peregrinus Ltd., 1981.

[9] R. Monzinyo and T. Miller, *Adaptive Beamforming*. New York: Wiley, 1980.

[10] L. Griffiths, "A simple adaptive algorithm for real-time processing in antenna arrays," *Proc. IEEE*, vol. 57, pp. 1696–1704, Oct. 1969.

[11] R. Haupt, "Adaptive nulling in monopulse antennas," *IEEE Trans. Antennas Propagat.*, vol. AP-36, pp. 202–208, Feb. 1986.

[12] R. Ghose and W. Sauter, "Automatic separation system," U.S. Patent 4,466,131, Aug. 14, 1984.

[13] H. Steyskal, "Synthesis of antenna patterns with prescribed nulls," *IEEE Trans. Antennas Propagat.*, vol. AP-30, pp. 273–279, Mar. 1982.

[14] J. Byrnes and D. Newman, "Null steering employing polynomials with restricted coefficients," *IEEE Trans. Antennas Propagat.*, vol. 36, Feb. 1988.

[15] W. Gabriel, "Adaptive arrays—An introduction," *Proc. IEEE*, vol. 64, pp. 239–272, 1976.

[16] S. Applebaum, "Adaptive arrays with main beam constraints," *IEEE Trans. Antennas Propagat.*, vol. AP-24, pp. 650–662, 1972.

[17] Special Issue on Adaptive Procesing Antenna System, *IEEE Trans. Antennas Propagat.*, vol. AP-34, no. 3, 1986.

MULTIPATH INTERFERENCE CANCELLATION

Often, an interference, arriving from its source to a receiving antenna through a number of propagation multipaths, is no less of a problem than multiple interferences originating from different sources as considered in the previous chapter. From the viewpoint of providing a remedy, one should note that the interferences are uncorrelated with one another in the case of multiple interferences, while the interferences from a common source but arriving at the receiving antenna through different propagation paths could be highly correlated with one another, as they are, after all, the same interference, only traveling through different propagation paths. Another important difference from the remedial viewpoint is that an interference propagating through a number of multipaths may experience a fading where the magnitude of the interference may vary widely with time. A fading, however, cannot occur with multiple interferences which are uncorrelated with one another. Still another distinction between the problems of multiple interferences and multipath interferences is that each interference in the case of multiple interferences is uncorrelated with the desired signal, whereas a component of the desired signal itself, propagating through multipaths, can be an interference, as it is in the case of the television "ghost" signal or in a direction-finding system where the desired signal travels through different multipaths having different arrival directions. Adaptive cancellers can be used to remedy the multipath interference problem also, as will be seen in this chapter.

7.1 CANCELLATION OF NARROWBAND MULTIPATH INTERFERENCES

To examine whether or how an adaptive canceller can be useful in a multipath environment, one may first notice the characteristic difference between the cancellation requirements for a single-path and multipath interference, as illustrated in Figure 7-1(a) and (b). Figure 7-1(a) shows, for example, a single propagation path

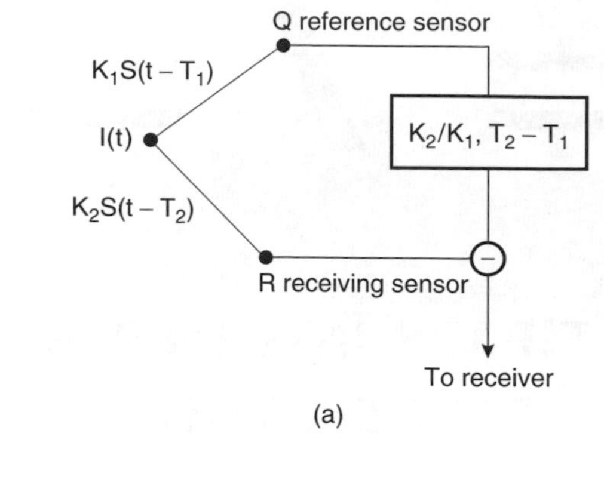

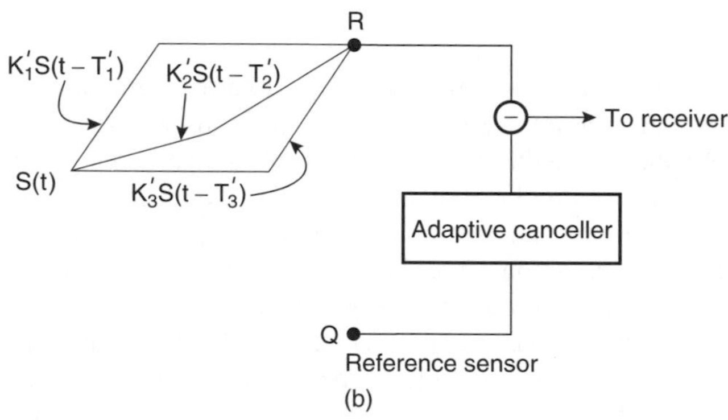

Figure 7-1 Illustration of the difference between a single-path and multipath interference. (a) A single propagation path characteristic and the signal-controller requirement. (b) Multiple propagation path characteristic and cancellation requirement.

between the receiving antenna and the source of interference $S(t)$. If the interference at the reference and receiving sensors is denoted, respectively, as $K_1 S(t - T_1)$ and $K_2 S(t - T_2)$, where K_1 and K_2 are amplitude factors and T_1 and T_2 are time delays introduced by the propagation paths as shown in the figure, then the requirement for the signal controller to cancel the interference will be to modify the magnitude of the signal-controller input by a factor (K_2/K_1) and to delay the input by $(T_2 - T_1)$.

Figure 7-1(b) shows that the interference $I(t)$ arrives at the receiving antenna by more than one propagation path, each path introducing an amplitude reduction factor K_1', K_2', K_3', etc., and a time delay T_1', T_2', T_3', etc. It appears that more than one signal controller may now be needed, one corresponding to each path to effect cancellations of all multipath interferences. Even when multiple signal controllers are available, it is not obvious how the adaptive cancellations will be effected as the interferences in signal controllers and cancellation loops will be highly correlated as it is the same interference, and hence interactions among loops cannot be easily avoided. In general, one may obtain a reasonable degree of interference cancellation in most cases involving a multipath environment by employing cancellation loops much fewer in number than the number of multipaths. Clearly, for a CW interference, one cancellation loop will be adequate since the combined effect of all multipaths will be equivalent to one propagation path, and the situation will not be very much different for near CW or narrowband interferences.

The feasibility of cancellation of a narrowband multipath interference can, perhaps, be further explained by an example. Let the interference of concern be an amplitude-modulated signal of the form

$$I(t) = A(t)\sin(\omega t - \phi) \tag{7.1}$$

where

$A(t)$ = a time-varying amplitude, typical of a pulse modulated radar

ω = angular frequency

t = time

ϕ = some arbitrary phase angle.

If the interference arrives at the receiving antenna through a number of multipaths as shown in Figure 7-2, the interference received at the receiving antenna can be written as

$$I_R(t) = \sum_{i=1}^{N} \Gamma_i I(t - T_i) \tag{7.2}$$

where

Γ_i = the amplitude reduction factor

T_i = the time delay introduced at the ith propagation path, there being N number of paths.

The reference signal obtained for the adaptive canceller can be written as

$$I_{\text{Ref}}(t) = KI(t) \tag{7.3}$$

since it is obtained by coupling directly from the interference transmitter line, K being a coupling constant.

From Eqs. (7.1) and (7.2), one may write

$$I_R = \sum_{i=1}^{N} \Gamma_i A(t - T_i)\sin(\omega[t - T_i] - \phi). \tag{7.4}$$

If the amplitude variation with time t is such that the amplitude does not change from one path to another during the transit time at the propagation paths, that is, if

$$A(t - T_1) \simeq A(t - T_2) \simeq A(t - T_N) \simeq A(t - T), \tag{7.5}$$

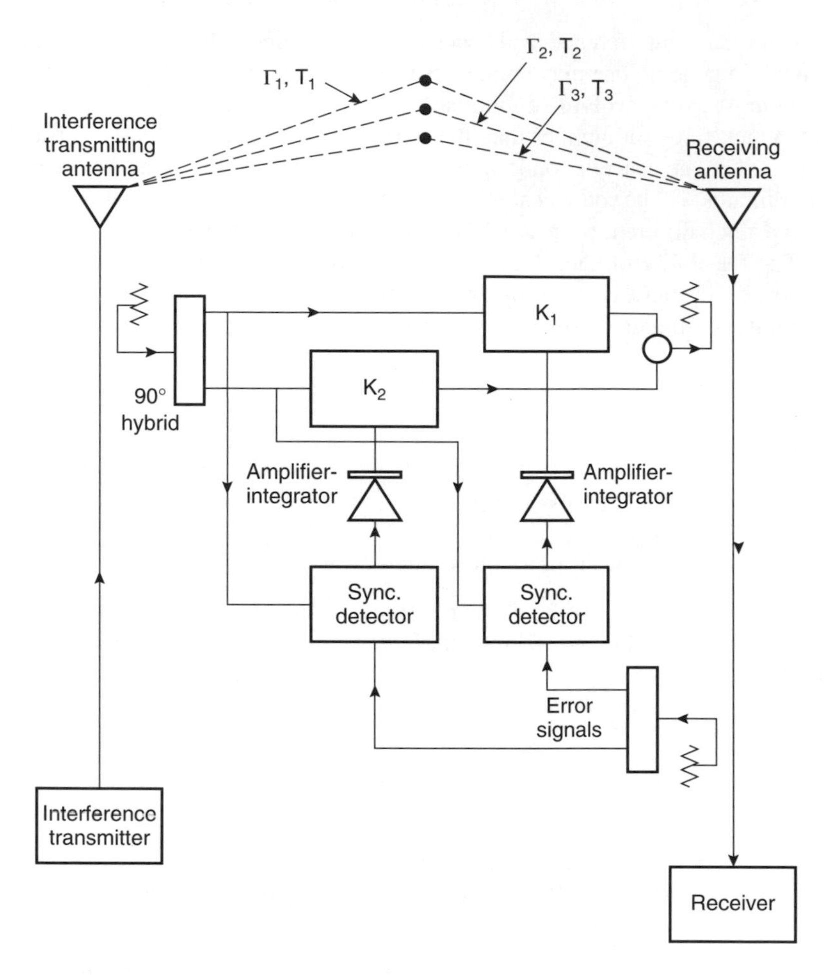

Figure 7-2 Illustration of narrowband multipath interference cancellation by a single canceller.

then

$$
\begin{aligned}
I_R(t) &= A(t-T)\sin(\omega t-\phi)\sum_{i}^{N}\Gamma_i\cos\omega T_i \\
&\quad - A(t-T)\cos(\omega t-\phi)\sum_{i}^{N}\Gamma_i\sin\omega T_i \\
&= KA(t-T)\sin(\omega t-\theta-\phi) \qquad (7.6)
\end{aligned}
$$

where

$K^2 = K_1^2 + K_2^2$

$$K_1 = \sum_i^N \Gamma_i \cos \omega T_i$$

$$K_2 = \sum_i^N \Gamma_i \sin \omega T_i$$

$\tan \theta = K_2/K_1,$

K_1 and K_2 not being functions of time.

Since, at any instant of time, there is only one set of values of K_1 and K_2, which according to Eq. (7.6) are functions of the angular frequency ω, it is seen that for the cancellation to be effective, the required values of K_1 and K_2 must be the same or very nearly the same for all frequency components of the interference. This usually implies that the interference cancellation approach shown in Figure 7-2 is confined to only narrowband interferences.

Even when the interference is not narrowband, the desired signal for the receiver could, sometimes, be a very narrowband signal, and since the receiver needs protection against the interference only for the desired-signal bandwidth of the receiver, one may examine the characteristics of broadband multipath interference when observed at the narrow bandwidth corresponding to that of the desired signal. The arrangement shown in Figure 5-18, where the synchronous detector path is down-converted and filtered with the filter bandwidth corresponding to the desired-signal bandwidth, can then be used to eliminate the multipath interference where it needs to be eliminated, that is, at the receiver bandwidth of interest for the desired signal.

For this purpose, let the interference of concern at its source be denoted as

$$I(t) = A(t) \sin[\omega t + \phi(t)] \tag{7.7}$$

where

$A(t)$ and $\phi(t)$ = a time-varying amplitude and phase, respectively.

The sum of the interference arriving through N number of multipaths can be written as

$$E_R(t) = \sum_{K=1}^{N} \Gamma_K A(t - T_K) \sin(\omega_0[t - T_K] + \phi[t - t_K]) \tag{7.8}$$

where

Γ_K = the equivalent amplitude reduction factor

T_K = a time delay introduced by the Kth propagation path.

If the multipath interference $E_R(t)$ flows through a narrowband filter centered at the angular frequency ω_0, one may write the output of the filter as the transform of the product of the filter-transfer function $F(\omega)$ and the transform of $E_R(t)$. Thus, the output $E_0(\omega)$ can be written as

$$E_0(t) = \frac{1}{2\pi}\int_{-\infty}^{\infty} F(\omega)E_R(\omega)e^{j\omega t}\,d\omega \tag{7.9}$$

where
$F(\omega)$ = the filter-transfer function
and

$$E_R(\omega) = \int_{-\infty}^{\infty} E_R(t)e^{-j\omega t}\,dt. \tag{7.10}$$

Let us approximate the transfer function of the narrowband filter as

$$\begin{aligned} F(\omega) &= M \qquad \text{for } \omega - \Delta \le \omega \le \omega_0 + \Delta \\ &= 0 \qquad \text{elsewhere.} \end{aligned} \tag{7.11}$$

Thus, from Eqs. (7.9) and (7.11)

$$\begin{aligned} E_0(t) &= \frac{M}{2\pi}\int_{\omega_0-\Delta}^{\omega_0+\Delta} e^{j\omega t}\,d\omega \int_{-\infty}^{\infty} E_R(t)e^{-j\omega t}\,dt \\ &= \frac{M}{2\pi}\sum_{K=1}^{N}\int_{\omega_0-\Delta}^{\omega_0+\Delta} \Gamma_K \\ &\quad \cdot e^{j\omega[t-T_K]}\,d\omega \int_{-\infty}^{\infty} A(t-T_K)\sin(\omega_0[t-T_K] \\ &\quad + \phi[t-T_K]e^{-j\omega[t-T_K]}\,d[t-T_K]) \\ &= \frac{M}{2\pi}\sum_{K=1}^{N}\int_{\omega_0-\Delta}^{\omega_0+\Delta} \Gamma_K e^{j\omega[t-T_K]}E_R(\omega)\,d\omega. \end{aligned} \tag{7.12}$$

If, now, the half-bandwidth Δ of the filter is so small that the integrand may be approximated by its value at $\omega = \omega_0$, then

$$\begin{aligned} E_0(t) &\simeq \frac{M}{2\pi}\sum_{K=1}^{N}\Gamma_K(\omega_0)E_R(\omega_0)e^{j\omega_0[t-T_K]}\int_{\omega_0-\Delta}^{\omega_0+\Delta} d\omega \\ &= \frac{M}{\pi}E_R(\omega_0)e^{j\omega_0 t}P\,\Delta \end{aligned} \tag{7.13}$$

and

$$P(\omega_0) = \sum_{K=1}^{N} \Gamma_K(\omega_0) e^{-j\omega_0 T_K}. \tag{7.14}$$

Since $P(\omega_0)$ is independent of the running index K corresponding to the multipaths, $E_0(t)$ will also be independent of K. Equation (7.13) suggests that the multipath signal observed through a very narrowband filter will be like a CW or near CW signal. A cancellation arrangement shown in Figure 7-2 will, then, be adequate to cancel the interference at the receiver, even when it arrives at the receiver through an arbitrary number of multipaths.

When the multipath characteristics are such that the approximation shown in Eq. (7.13) is not permissible, one may still examine the adequacy or inadequacy of a single adaptive canceller, even though such multipaths may suggest the need for multiple cancellers.

The frequency-domain expression for the received multipath interference corresponding to $E_0(t)$ can be written as

$$E_0(\omega) = E_R(\omega) \sum_{K=1}^{N} \Gamma_K e^{j\omega T_K}. \tag{7.15}$$

Assuming a cancellation approach as shown in Figure 7-2, we may write the reference interference from which the countermultipath interference has to be synthesized as

$$E_{\text{Ref}}(\omega) = K e^{j\theta} E_R(\omega). \tag{7.16}$$

Let us suppose that an adaptive canceller can be built for a perfect cancellation of the interference at $\omega = \omega_0$. For the cancellation of $E_0(\omega)$, then, one needs

$$K e^{j\theta} E_R(\omega_0) T_{SC} = E_R(\omega_0) \sum_{K=1}^{N} \Gamma_K e^{-j\omega_0 T_K} \tag{7.17}$$

where

T_{SC} = is the transfer function of the signal controller of the canceller.

In Eq. (7.17), Γ_K is assumed to be independent of ω, at least for the limited range of ω of concern. Thus

$$T_{SC} = \frac{1}{K e^{j\theta}} \cdot \sum_{K=1}^{N} \Gamma_K e^{-j\omega_0 T_K}. \tag{7.18}$$

To find the degradation of cancellation at any other angular frequency ω, we may write the residual interference following the cancellation as

$$
\begin{aligned}
E_{\text{Res}}(\omega) &= E_R(\omega)\left\{\sum_{K=1}^{N}\Gamma_K e^{-j\omega T_K} - K e^{j\theta}\cdot T_S\right\} \\
&= E_R(\omega)\left\{\sum_{K=1}^{N}\Gamma_K e^{-j\omega T_K} - \sum_{K=1}^{N}\Gamma_K e^{-j\omega_0 T_K}\right\}.
\end{aligned} \tag{7.19}
$$

Assuming a small deviation $\delta\omega$ from ω_0, one may write $\omega = \omega_0 + \delta\omega$, and

$$
\sum_{K=1}^{N}\Gamma_K e^{-j\omega T_K} = \sum_{K=1}^{N}\Gamma_K e^{-j\omega_0 T_K} - i\,\delta\omega\sum_{K=1}^{N}\Gamma_K T_K e^{-j\omega_0 T_K} \tag{7.20}
$$

and

$$
\left|\frac{E_{\text{Res}}}{E_R}\right| \simeq \left|\frac{\sum_{K=1}^{N}\Gamma_K(\delta\omega T_K)e^{-j\omega_0 T_K}}{\sum_{K=1}^{N}\Gamma_K e^{-j\omega_0 T_K}}\right|. \tag{7.21}
$$

If the magnitudes of $\delta\omega$ and T_K, for any K, are such that

$$
(\delta\omega T_K) \le P \tag{7.22}
$$

P being real, then

$$
\left|\sum_{K=1}^{N}\Gamma_K(\delta\omega T_K)e^{-j\omega_0 T_K}\right| \le P\left|\sum_{K=1}^{N}\Gamma_K e^{-j\omega_0 T_K}\right| \tag{7.23}
$$

and

$$
\left|\frac{E_{\text{Res}}}{E_R}\right| \le P. \tag{7.24}
$$

Thus, for example, if $\delta\omega$ is 10^6 and the time delay in the largest propagation path is 100 n seconds, then P will be (1/10), and from Eq. (7.24), $|E_{\text{Res}}/E_R|$ will be at least -20 dB regardless of the number of multipaths. On the other hand, there may be very little cancellation when $\delta\omega$ is $2\pi \cdot 1.5 \times 10^6$ and T is 100 ns, T being the delay at the largest multipath. Figures 7-3 and 7-4 show, for example, the expected variation of the degree of cancellation with one adaptive canceller for two different multipath environments, where the cancellation is effected at about 30 MHz intermediate frequency (IF) band, the carrier frequency of the interference being at approximately 10 GHz. The cancellation arrangement and the multipath environment are similar to those shown in Figure 7-2, except that the reference and receive-line interferences are down-converted to the 30 MHz IF band as noted above.

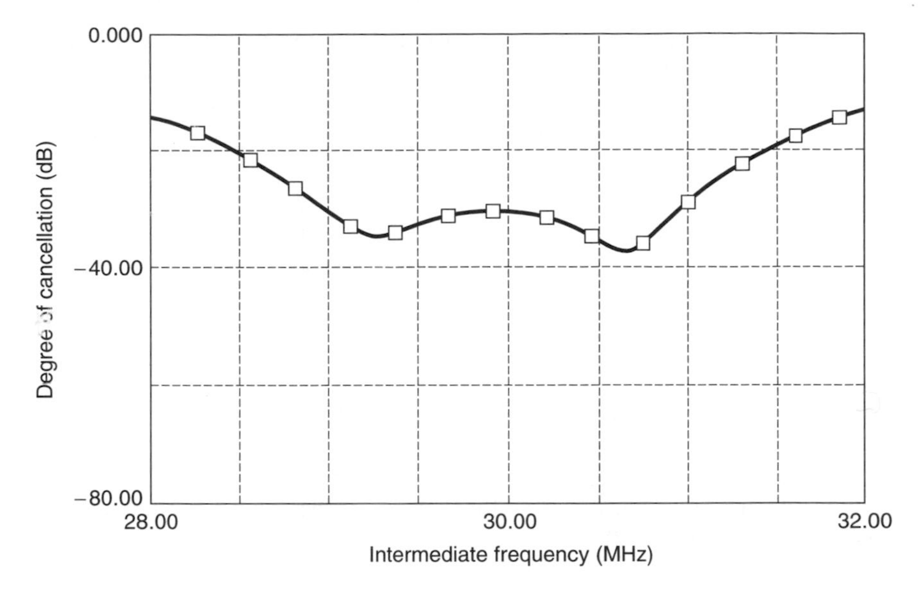

Figure 7-3 Degree of cancellation for a multipath interference. Relative amplitudes of four propagation paths are 1, 0.7, 0.9, and 0.5. Corresponding time delays in paths are 100, 230, 130, and 110 ns. Reference line delay is 129 ns and is optimized for above multipath parameters.

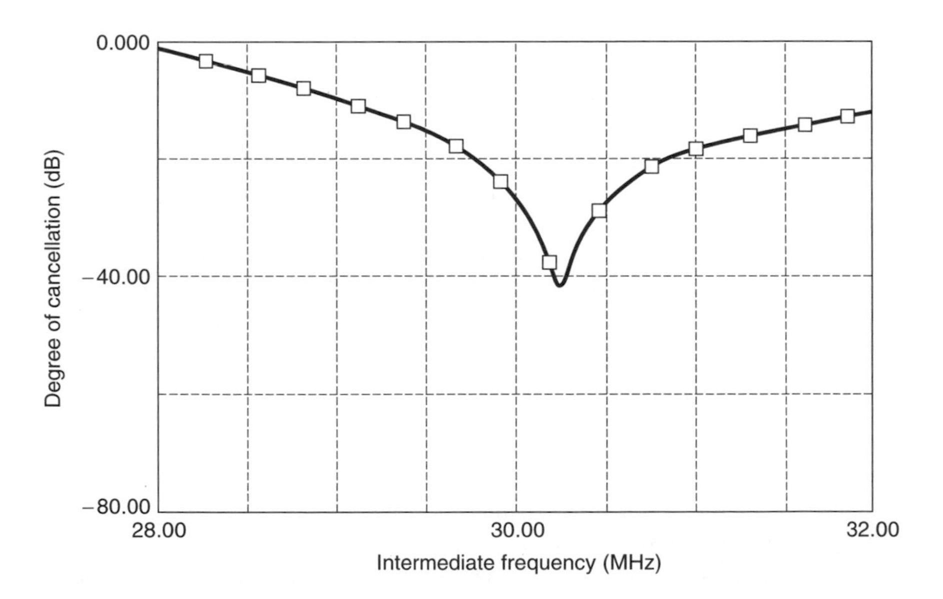

Figure 7-4 Degree of cancellation for multipath interference. Relative amplitudes and time delays are the same as in Figure 7-3, but the reference line delay is kept fixed and not optimized.

The amplitude reduction factors and associated time delays, corresponding to Figure 7-3, are as follows:

Multipath *K*	**Γ**	***T*(ns)**
1	1	100
2	0.7	125
3	0.5	150
4	0.6	200

A total time delay of 129 ns, optimized for best cancellation for the given multipath characteristics, is assumed to have been introduced at the reference line. As may be seen from Figure 7-3, a cancellation on the order of −40 dB for a bandwidth of 1.75 MHz on either side of the center frequency of 30 MHz is realizable with a single canceller.

Figure 7-4 shows the expected degree of cancellation with the same canceller and the reference-line delay as used for Figure 7-3, the corresponding multipath characteristics being as follows:

Multipath *K*	**Γ**	***T*(ns)**
1	1	100
2	0.7	230
3	0.9	130
4	0.5	110

As seen from Figure 7-4, the expected degree of cancellation is much worse for the same bandwidth of the interference. The degradation of interference cancellation is primarily due to the fact that the time delay introduced at the reference line has been optimized for the multipath environment considered in Figure 7-3, while there is no time-delay optimization for the environment associated with Figure 7-4.

The rationale for expecting any reasonable cancellation of an arbitrary multipath interference by a single adaptive canceller is that with the optimized time delay at the reference line of the canceller, one attempts to approximate an equivalent single propagation path for the interference that will have, in effect, the same propagation path characteristics as that of the combined multipaths. This is necessary since a single canceller can effectively accommodate a single propagation path for the interference. Usually, such an approximation is possible for a given set of fixed multipaths, and a limited bandwidth, and the degree of cancellation of the interference for such a case depends on how closely it is possible to simulate the equivalent single propagation path over the frequency band of concern.

To estimate the required optimized time delay, let ω_0 be the angular frequency at which the maximum cancellation is desired. From Eqs. (7.15) and (7.16)

$$Ke^{j\theta} = \sum_{K=1}^{N} \Gamma_K e^{-j\omega_0 T_K} \tag{7.25}$$

and

$$\theta = \tan^{-1}\left\{\frac{\sum_{K=1}^{N} \Gamma_K \sin \omega_0 T_K}{\sum_{K=1}^{N} \Gamma_K \cos \omega_0 T_K}\right\}. \tag{7.26}$$

But $\theta = \omega_0 T_{\text{op}}$, where T_{op} is the optimized time delay that approximates the equivalent single-path time delay to provide ideal cancellation at ω_0. Thus, one may write

$$T_{\text{op}} = \frac{1}{\omega_0} \tan^{-1}\left\{\frac{\sum_{K=1}^{N} \Gamma_K \sin \omega_0 T_K}{\sum_{K=1}^{N} \Gamma_K \cos \omega_0 T_K}\right\}. \tag{7.27}$$

It is evident from Eq. (7.27) that T_{op} will change as the multipath parameters Γ_K and T_K for any K change. Consequently, if the reference-line delay is kept fixed at T_{op} as obtained from Eq. (7.27) for one set of multipath environments, it will not, in general, be the time delay required for another set of multipaths, and the cancellation may not be effective, as seen in Figure 7-4.

7.2 MULTIPLE CANCELLERS FOR MULTIPATH INTERFERENCE

We have seen in the previous section that a single adaptive canceller can be effective if the multipath propagation characteristics for the interference can be approximated by a single equivalent propagation path, at least for a very narrow frequency band. For a wider band cancellation requirement, one is tempted to use more than one canceller. Also, by making the frequency for which the cancellation is maximum different for different cancellers, one may avoid the interactions not atypical among cancellers, while correcting a common error signal. Again, the rationale for using multiple cancellers is to approximate two or more propagation paths which will be equivalent to the multipath environment of concern for a somewhat wider frequency band than what was considered earlier.

Figure 7-5 shows an arrangement employing two adaptive cancellers with two different time delays at the reference ports of the cancellers. Here, a common

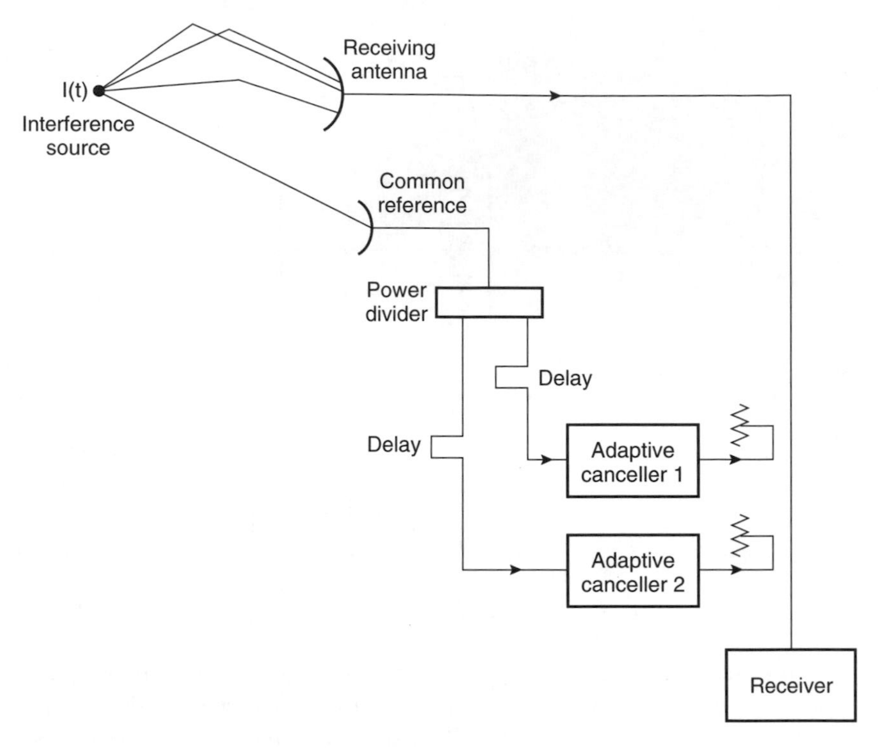

Figure 7-5 Illustration of approximating multipath characteristics by two equivalent paths and two cancellers.

reference sensor is used to provide the inputs to two cancellers. The optimized delay lines associated with the two cancellers are determined from Eqs. (7.25) and (7.27).

$$T_{\mathrm{OP1}} = \frac{1}{\omega_{01}} \tan^{-1} \left\{ \frac{\sum_{K=1}^{N} \Gamma_K \sin \omega_{01} T_K}{\sum_{K=1}^{N} \Gamma_K \cos \omega_{01} T_K} \right\} \tag{7.28}$$

and

$$T_{\mathrm{OP2}} = \frac{1}{\omega_{02}} \tan^{-1} \left\{ \frac{\sum_{K=1}^{N} \Gamma_K \sin \omega_{02} T_K}{\sum_{K=1}^{N} \Gamma_K \cos \omega_{02} T_K} \right\} \tag{7.29}$$

where

ω_{01} and ω_{02} = the two angular frequencies for which the two adaptive cancellers will provide maximum cancellations regardless of the multipath environment.

Usually, ω_{01} and ω_{02} are chosen such that

$$\begin{aligned} \omega_{01} &= \omega_0 - \delta \\ \omega_{02} &= \omega_0 + \delta \\ \delta &= 2\pi \Delta f \end{aligned} \tag{7.30}$$

where

ω_0 and Δf = the angular frequency corresponding to the receiver center frequency and half-bandwidth, respectively, over which the multipath interference cancellation is desired.

Figure 7-6 shows, for example, the expected cancellation with two adaptive cancellers when Δf is 1 MHz and the multipath environment is what is shown in the figure. As may be seen from Figure 7-6, a cancellation of at least 20 dB is realizable for a bandwidth of ±2 MHz for an IF bandwidth of 30 MHz where the cancellation is effected. The center angular frequency ω_0 in Figure 7-6 is $2\pi \times 30 \times 10^6$ rad/s.

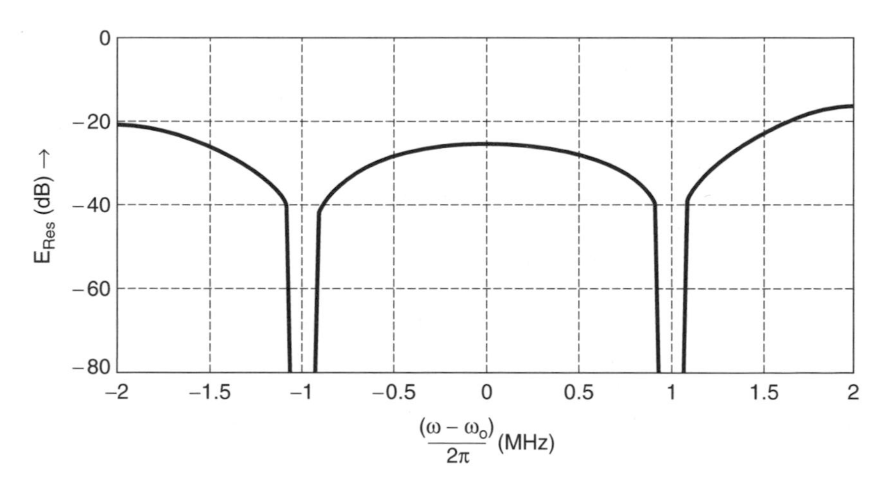

Figure 7-6 Degree of cancellation of a multipath interference by two cancellation loops. By design, $E_{\text{Res}} = 0$ at $\omega = \omega_0 \pm \delta$, $\delta = 2\pi(1 \times 10^6)$ rad/s. Multipaths: amplitude factors Γ_1, Γ_2, etc., are 0.9, 0.7, and 0.5. Time delays are 100, 130, and 150 ns, respectively.

Similarly, Figure 7-7 shows the degree of multipath interference cancellation possible with two cancellers arranged as shown in Figure 7-5. Here, $\omega_0 = 2\pi \times$

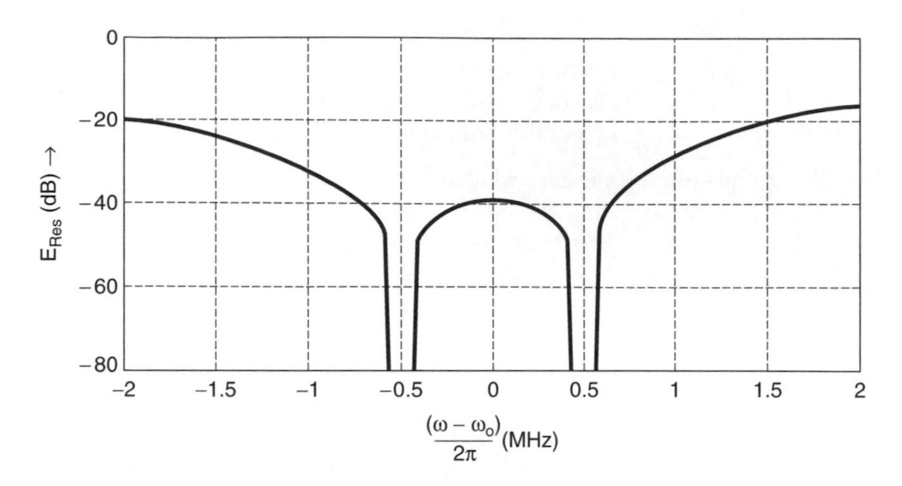

Figure 7-7 Cancellation of multipath interference by two cancellers. By design, $E_{\text{Res}} = 0$ at $\omega = \omega_0 \pm \delta$, $\delta = 2\pi(0.5 \times 10^6)$ rad/s. Multipaths: amplitude factors Γ_1, Γ_2, etc., are 0.9, 0.7, and 0.5. Corresponding time delays are 100, 130, and 150 ns.

30×10^6 rad/s and Δf is 0.5 MHz. Although the degree of cancellation is more at or near the center frequency, the realizable cancellation bandwidth is less than that when Δf is 1 MHz, as shown in Figure 7-6. In both cases shown in Figures 7-6 and 7-7, the number of adaptive cancellers is less than that of the multipaths. Also, since in both cases the two frequencies of maximum cancellation are different, there is no troublesome interaction between two cancelling loops to affect the overall cancellation of the multipath interference.

The concept of introducing an optimized time delay at the reference port of each canceller assumes that the multipath environment is not time varying, that is, Γ_K and T_K, for the propagation path K, do not change significantly with time. This assumption, however, may not always be valid. For example, when a radar or an electronic countermeasure (ecm) equipment is the source of interference, and this source is collocated with a communication-receiving antenna on a ship, there could be a time-varying multipath environment as the radar or ecm transmitting antenna scans over a range of angular directions. In this case, scatterings from shipboard objects and ocean surface that create the multipaths will change with time. One approach for adaptive cancellation of such a multipath interference is to employ adaptive cancellers where the signal controllers provide a variable attenuation and a time delay, so that the effective time delay at the reference port can be changed in accordance with the multipath environment. We have already discussed such variable time-delay signal controllers in Chapter 3.

Figure 7-8 shows schematically an arrangement that can accommodate time-varying multipaths as required for a multipath interference cancellation. As in

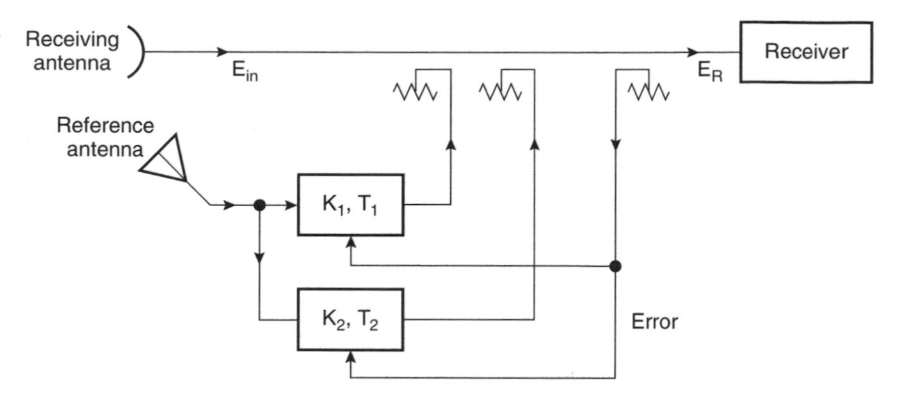

Figure 7-8 Use of two time-varying signal controllers for adaptive multipath interference cancellation.

Eqs. (7.15) and (7.16), we assume that the transform of the multipath interference, as received by the receiving and reference antennas shown in Figure 7-8, can be written as

$$E_0(\omega)E_R(\omega)\sum_{K=1}^{N}\Gamma_K e^{-j\omega T_K}$$

and

$$E_{\text{Ref}}(\omega) = Ke^{j\theta}E_R(\omega)$$

where
$E_R(\omega) =$ the transform of the interference at its source.

As the signal controllers in Figure 7-8 provide attenuation K_1 and K_2, respectively, and time delays T_1 and T_2, respectively, for the cancellers 1 and 2, one requires for the cancellation of the multipath interference

$$\sum_{K=1}^{N}\Gamma_K e^{-j\omega T_K} = K_1' e^{-j\omega T_1} + K_2' e^{-j\omega T_2} \tag{7.31}$$

where
$K_1' = K_1 K e^{j\theta}$
and $K_2' = K_2 K e^{j\theta}$.

Since it may not be possible for Eq. (7.31) to be valid for all ω or frequencies of interest within the interference band, let it be assumed that one obtains a perfect cancellation at the angular frequency ω_0; let it further be assumed that the derivative of the residual interference at ω_0 is also zero. One may thus write

$$\sum_{K=1}^{N} \Gamma_K e^{-j\omega_0 T_K} = K_1' e^{-j\omega_0 T_1} + K_2' e^{-j\omega_0 T_2}$$

$$\sum_{K=1}^{N} T_K \Gamma_K e^{-j\omega_0 T_K} = T_1 K_1' e^{-j\omega_0 T_1} + T_2 K_2' e^{-j\omega_0 T_2}. \quad (7.32)$$

That is,

$$K_1' e^{-j\omega_0 T_1} = \frac{P_1 T_2 - P_2}{T_2 - T_1}$$

$$K_2' e^{-j\omega_0 T_2} = \frac{P_1 T_1 - P_2}{T_1 - T_2} \quad (7.33)$$

where

$$P_1 = \sum_{K=1}^{N} \Gamma_K e^{-j\omega_0 T_K}$$

$$P_2 = \sum_{K=1}^{N} T_K \Gamma_K e^{-j\omega_0 T_K}. \quad (7.34)$$

The degree of cancellation D, defined as the ratio of the residual multipath interference following cancellation by the arrangement shown in Figure 7-8 to the same interference before cancellation, can now be expressed as

$$D = 1 - \frac{(P_1 T_2 - P_2)e^{-j\Delta T_1} - (P_1 T_1 - P_2)e^{-j\Delta T_2}}{(T_2 - T_1)\sum_{K=1}^{N} \Gamma_K e^{-j\omega T_K}} \quad (7.35)$$

where
Δ = the deviation angular frequency from ω_0, that is, $\Delta = (\omega - \omega_0)$.

It is seen from Eq. (7.35) that as $\Delta \to 0$, $D \to 0$, regardless of the values of T_1, T_2, P_1, and P_2. At any other frequency, D can be determined when the values of P_1, P_2, T_1, and T_2 are obtained from the multipath-characteristic parameters such as Γ_K and T_K for the propagation path K and the center angular frequency ω_0. They are constants for a given ω_0 and the given propagation environment. To determine the parameters T_1 and T_2 for a given set of P_1 and P_2, one may impose an additional criterion that $D = 0$ at $\Delta = \pm\delta$, where δ is chosen depending on the cancellation requirement over the interference bandwidth. This leads to the following equations:

$$(P_1 T_2 - P_2)e^{-j\delta T_1} - (P_1 T_1 - P_2)e^{-j\delta T_2} = (T_2 - T_1)\sum_{K=1}^{N} \propto_K e^{-j\delta T_K}$$

$$(P_1 T_2 - P_2)e^{j\delta T_1} - (P_1 T_1 - P_2)e^{j\delta T_2} = (T_2 - T_1)\sum_{K=1}^{N} \propto_K e^{j\delta T_K} \quad (7.36)$$

where

$\propto_K = \Gamma_K e^{-j\omega_0 T_K}$.

When T_1 and T_2 are thus determined for a given set of values of P_1, P_2, and ω_0, and the results are substituted in Eq. (7.35), one obtains the degree of cancellation of any value of ω. Figures 7-9 and 7-10 show the expected values of $|D|$ for various multipath characteristics as noted in the figures and for various values of Δ.

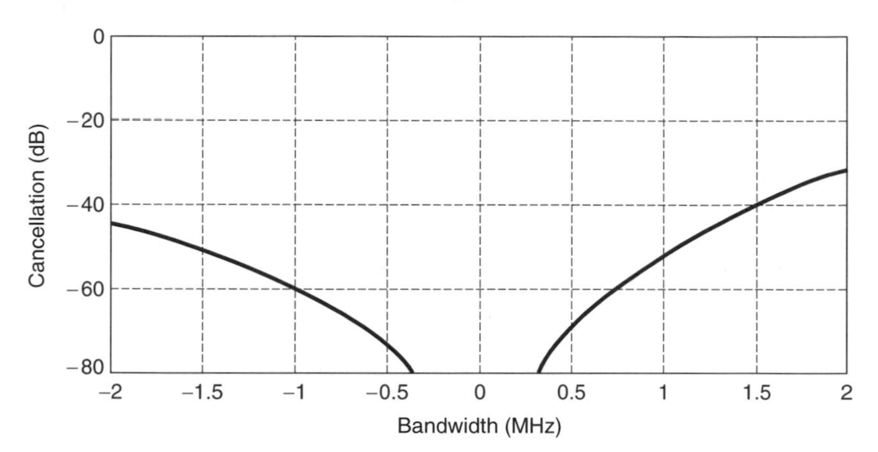

Figure 7-9 Cancellation of three-multipath interference by two time-varying signal controllers. Multipaths: amplitude factors Γ_1, Γ_2, etc., are 0.9, 0.7, and 0.5, with corresponding time delays of 50, 75, and 100 ns.

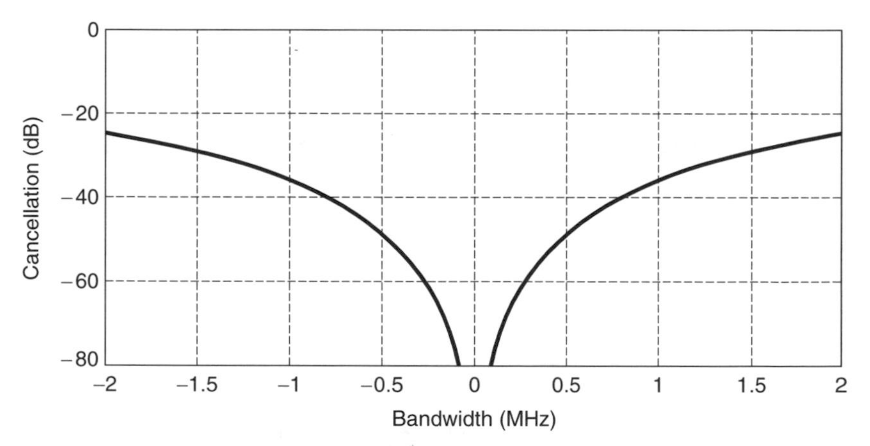

Figure 7-10 Cancellation of four-multipath interference by two time-varying signal controllers. Multipaths: amplitude factors, Γ_1, Γ_2, etc., are 0.9, 0.7, 0.5, and 0.6, with corresponding time delays of 50, 75, 100, and 150 ns.

7.3 MULTIPATH FADING AND ITS AVOIDANCE

So far in this chapter, we have considered the cancellation of an unwanted interference arriving at the receiving antenna through a number of propagation paths to avoid a drastic reduction of the signal-to-noise ratio for the desired signal. Sometimes, however, propagation multipaths can cause a drastic reduction of the signal-to-noise ratio even when there is no specific high-level interference. One such case arises when propagation multipaths cause a fading of the desired signal. It will be desirable to explore whether the adaptive cancellation concept discussed earlier could somehow be used to avoid a serious desired signal fading.

Usually, a fading occurs when the sum of the signals arriving through multipaths from the same source is reduced in amplitude because of destructive phase relations. The simplest example of fading is experienced when a relatively narrowband signal arrives at the receiving antenna through a direct propagation path and a ground-reflected path, as shown in Figure 7-11, the phases of the signals in the two paths being approximately 180° from each other. In this case, the signal from either path may be adequate, while the combined signals may not be.

Another common occurrence of fading is observed for communication and broadcast signals involved in the ionospheric propagation when the so-called "sky waves" and "ground waves" interfere with each other to create a nearly self-cancelling desired signal. The fading is also not uncommon for the line-of-sight propagation [1] at ultra-high and microwave frequencies as in a microwave relay line, for example, where the variation of received signal strength is due to multipath transmission. The fading for such cases may be divided into two main types. One type of fading is relatively rapid, and is caused by interference between two or more "rays" arriving at the receiving antenna by slightly different propagation paths. This type is referred to as the atmospheric multipath fading. The second type of fading is less rapid, and is due to interference between the direct and reflected rays similar to that shown in Figure 7-11. The second type is known as reflection-multipath fading. In general, the number of fades per unit time due to atmospheric multipath increases with path length between the transmitting and receiving antennas. The duration of a fade of given depth, however, tends to decrease with increasing path length.

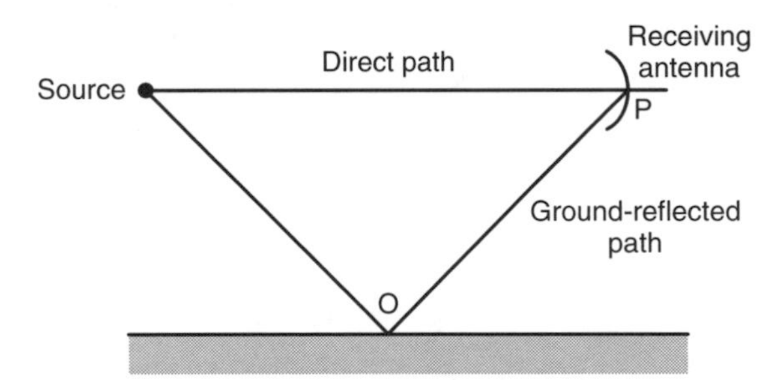

Figure 7-11 Illustration of fading of a narrowband signal. Here, the signal in any one path reduces that due to the other.

For the line-of-sight propagation, where there is one reflected ray combining with the direct ray at the receiving point P, as shown in Figure 7-11, the resulting field strength, neglecting the difference in angles of arrival and assuming a perfect reflection at 0, is related to the direct-ray field intensity by

$$E_R = 2E_d \sin\left(\frac{\pi\delta}{\lambda}\right) \tag{7.37}$$

where
E_R = resulting field strength
E_d = direct-ray field strength
δ = geometrical length difference between direct and reflected paths
λ = wavelength of the field.

The geometrical length can be approximated as

$$\delta \simeq 2H_t H_r / d \tag{7.38}$$

where
H_t and H_r = the transmitting and receiving antenna heights, respectively, above a reflecting plane tangent to the effective earth surface
d = the direct path length.

To illustrate the consequences of fading, one may note from Eqs. (7.37) and (7.38) that

$$\begin{aligned} E_R &= 0 && \text{when} \quad H_t H_r = d\lambda/2 \\ E_R &= 2E_d && \text{when} \quad H_t H_r = d\lambda/4 \\ E_R &= E_d && \text{when} \quad H_t H_r = d\lambda/12. \end{aligned} \tag{7.39}$$

Also, when $H_t = H_r = H$

$$\begin{aligned} E_R &= 0 && \text{when} \quad H = (d\lambda/2)^{1/2} \\ E_R &= 2E_d && \text{when} \quad H = (d\lambda/4)^{1/2} \\ E_R &= E_d && \text{when} \quad H = (d\lambda/12)^{1/2}. \end{aligned}$$

It is seen from Eq. (7.37) that when H_r is varied, the field strength varies accordingly. Thus, the use of two receiving antennas at different heights provides a means of compensating, to a certain extent, for changes in electrical path differences between the direct and reflected rays. Such a means, known as space diversity reception, is often used to avoid drastic fading and a consequential reduction of the desired signal-to-noise ratio.

To avoid fading for a line-of-sight propagation path, the spacing between the two receiving antennas for diversity reception should be approximately such as to give a $\lambda/2$ variation between geometric-path differences in the two cases. An approximate value[1] of the spacing is given by $(\Delta d) = \lambda d/4H_t$, all quantities being in the same units. Thus, when λ is 4 cm, d is 20 mi, and H_t is 50 ft, the spacing (Δd) is 69.33 ft.

For the reflection-multipath fading, the variation in strength of the desired signal is, in general, frequency selective, that is, signals on different frequencies, even within the same frequency band, may exhibit at any instant radically different behavior.[2,3] This aspect of multipath propagation characteristics may also be used to mitigate transmission impairments due to fading. For the purpose of this frequency diversity, the carrier is shifted to an alternative and less vulnerable frequency.

The purpose of both space and frequency diversity is to avoid a drastic reduction of the desired signal-to-noise ratio in a fading environment, even when there is no specific interference which is commonly responsible for the signal-to-noise ratio reduction. Once the alternative satisfactory means for receiving the desired signal is obtained, one may switch the appropriate antenna for receiver manually or automatically. An adaptive signal combining approach [6] that can automatically sum the magnitude of the signal received by each antenna that exceeds a preset threshold level may also be used to avoid a fading when two or more diversity antennas are employed. The provision of the preset threshold level excludes those diversity antennas which contribute more to the noise than the desired signal from the usable "receive" signals. The adaptive signal combining approach can also be used in the frequency diversity system when different carrier frequencies with the same modulation are employed to avoid a fading.

There is still another set of circumstances where the multipath propagation constitutes a problem, even when there is no specific interference and there may not even be any fading of the desired signal. In a direction-finding system, for example, the directions of arrival of the signals for different multipaths are usually not the same, thus creating an ambiguity of the true direction of the source of signal as measured by the direction-finding system. Multipath propagations may also affect the radar resolution [3], and may cause undesirable phase errors in tracking systems [4].

As noted earlier, our motivation of addressing such problems here is to examine whether adaptive cancellers could be used directly or in combination with other approaches to provide a remedy for problems relating to propagation multipath. Clearly, when the signals in different propagation paths can be viewed as

[1] E. Jordan, Ed.-in-Chief, *Reference Data for Engineers.* IN: Howard W. Sours & Co., 1985.

[2] A. Crawford and W. Jakes, Jr., "Selective fading of microwaves," *Bell Syst. Tech. J.,* Jan. 1952.

[3] R. Kaylor, "A statistical study of selective fading of super-high frequency radio signals," *Bell Syst. Tech. J.,* Sept. 1953.

independent signals, each of which can be accommodated by a separate adaptive canceller, there will be a straightforward remedy, the number of required cancellers in such cases being the same as the number of objectionable multipaths. Since the propagation path characteristics almost invariably change with time, any canceller, to be effective, has to be adaptive.

When a reference antenna can be devised such that it receives the signal from one and only one propagation path or receives one propagation path signal predominantly, and the signals in different propagation multipaths are not highly correlated, one can effectively use an adaptive canceller to rid the particular signal received from this path, thus preventing its contribution toward fading. Thus, for example, in the case of ionospheric propagation fading, if one can arrange a reference antenna that receives the "sky wave" signal arriving at the receiving antenna at a higher elevation angle than the "ground wave" signal exclusively or predominantly, the signal thus received by the reference antenna can be used in an adaptive canceller to eliminate the "sky wave" contribution of the received signal and thereby eliminate the consequential probability of fading. In the absence of fading, the "ground wave" signal is often found to be adequate. Also, the needed decorrelation of the "sky wave" and "ground wave" signals is often found, particularly for frequency-modulated and wideband signals. Besides the adaptive canceller, the signal separation system, discussed in Section 6.6 of Chapter 6, can also be used to separate the ground and sky wave signals, and the stronger of the two signals may be used to ensure a signal reception without a reduction of the fading-related signal-to-noise ratio. It should be noted here that the use of an adaptive canceller or signal separation system is possible only when separate reference signals are obtainable for one or more propagation paths and a sufficient decorrelation among multipaths signals exist. Under such circumstances, propagation multipath signals become equivalent to independent multiple interferences which can be accommodated by various means, as discussed in Chapter 6. Similar approaches are also available to resolve ambiguities for the true direction of a signal as determined by a direction-finding system. In particular, when multipath signals are sufficiently decorrelated because of the wide bandwidth of the signals, the signal separation approach as discussed in Section 6.6 of Chapter 6 can be used first to separate the multipath signals, and then to find the direction for one or more selected propagation paths.

Sussman [7] has discussed a detection system employing matched filters to mitigate the adverse effect of propagation multipaths for radio teletype communications transmitting "Mark" or "Space" symbols. Although there is no interference cancellation involved, the signal-to-noise ratio is improved at the receiver.

A similar use of matched filters to avoid the adverse effects of multipath propagation has been suggested by Turin [8], [9]. If, for example, a matched filter can be designed for each dominant propagation path based on the approach discussed in Section 2.2 of Chapter 2, then the contribution of the kth path signal at the output of the ith path matched filter will approach zero in the absence of noise at time $t = t_1$, when the output of the ith path matched filter is maximum, provided $i \neq k$. Such a situation provides a means for discriminating the signal contributions of various multipaths, and thus may provide a means to avoid their destructive interferences that cause fading.

Other types of linear filters to minimize the undesirable effects on the desired signal due to multipaths for special cases have also been investigated. Balakrishnan [10], for example, explored the feasibility of physically realizable corrective filters to minimize distortion of television signals due to multipaths. Here, the distortion is defined through the correlation coefficient ρ so that

$$\rho = \frac{E[(x(t) - m_1)(y(t) - m_2)]}{[E(x(t) - m_1)^2 E(y(t) - m_2)^2]^{1/2}} \tag{7.40}$$

where

$x(t)$ = signal received in the absence of multipaths
$y(t)$ = actual signal received
m_1, m_2 = mean values of $x(t)$ and $y(t)$, respectively
E = is the expectation operator.

Optimum linear filter solutions for amplitude- and frequency-modulated signals have been derived by Balakrishnan to minimize distortion.

We have seen in earlier chapters that, besides adaptive cancellers, an adaptive array or an adaptive filter can also be used to remedy the multiple interference problems. It turns out that such approaches could also be helpful in some situations to minimize adverse effects of multipath fading in a receiver. More specifically, in digital mobile communications, the use of high-speed data widens the required instantaneous signal frequency bandwidth, a 270 kb/s data rate already being used in Europe and a data rate in excess of 1 Mb/s being envisioned for the near future. This wideband signal requirement, in turn, results in much greater susceptibility of system performance due to multipath fading than what is experienced at a lower data speed. An adaptive equalizer [11], [12] is sometimes used for reducing the effect of multipath fading by means of compensating the channel frequency response. However, the range of time delay that can be compensated is limited by the total line lengths of the taps. In contrast, adaptive arrays have no such limit. Additionally, an adaptive array can cancel undesired signals by directing nulls in an array pattern toward them, and has the potential to cancel a cochannel interference.

The implementation of one such array particularly addressing the multipath fading problem for mobile communications in an urban area has been described by Ohgane *et al.* [13], [14]. This array comprises four elements, each connected to a weighting network that performs adaptation. The weighting network parameters are updated in accordance with a constant modulus algorithm[4] (CMA), which is different from the LMS algorithm usually employed with the adaptive array. No reference is needed for CMA, but signals in one or several multipaths are automatically selected as the desired signal. Also, CMA does not need to know the arrival timing of the incident rays when the array weights are updated. For the array implemented by Ohgane *et al.*, weighting network control functions are effected by digital signal processing. Figure 7-12 shows an example of performance characteristics of an array in terms of bit error rate. Here, the bit error rate as a function of E_b/N_o is compared when the array is used and not used, E_b and N_o being, respectively, the energy per bit and the noise density. The figure corresponds to slow Rayleigh fading, the Doppler shift being about 5 Hz. A remarkable improvement in bit-error-rate performance is thus achievable with the adaptive array in a reasonably expected fading environment.

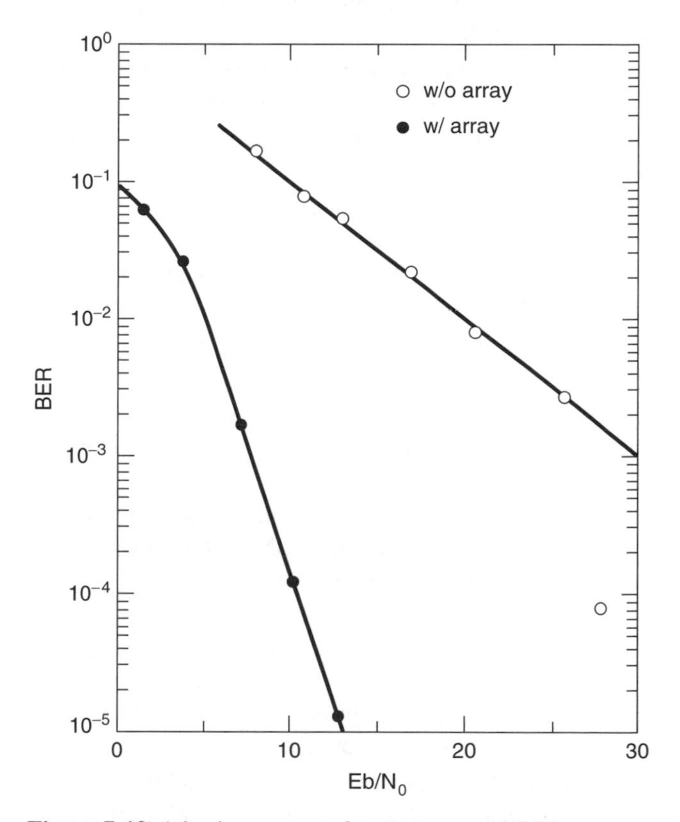

Figure 7-12 Adaptive array performance toward BER improvement in a Rayleigh fading environment. E_b = energy per bit. N_0 = Noise density.

[4]The CMA Algorithm was suggested by Treichler *et al.* [15]. Its advantages are further explained in [13].

References

[1] K. Bullington, "Radio propagation fundamentals," *Bell Syst. Tech. J.*, vol. 36, no. 3, pp. 593–626, 1957.

[2] A. Crawford and W. Jakes, Jr., "Selective fading of microwaves," *Bell Syst. Tech. J.*, Jan. 1952.

[3] D. Barton, *Radar Resolution and Multipath Effects*. Dedham, MA: Artech House, 1975.

[4] T. Sollenberger, "Multipath phase errors in CW-FM tracking systems," *IRE Trans. Antennas Propagat.*, vol. AP-3, pp. 185–192, Oct. 1955.

[5] R. Ghose *et al.*, "Automatic separation system," U.S. Patent 4,466,131, Aug. 14, 1984.

[6] R. Ghose, "Adaptive multifrequency signal combining system," U.S. Patent 5,014,061, May 7, 1991.

[7] S. Sussman, "A matched filter communication system for multipath channels," *IRE Trans. Inform. Theory*, p. 367, June 1960.

[8] G. Turin, "An introduction to matched filter," *IRE Trans. Inform. Theory*, pp. 311–329, June 1960.

[9] G. Turin, "Communication through noisy, random multipath channels," in *1956 IRE Conv. Rec.*, pt. 4, pp. 154–166.

[10] A. Balakrishnan, "Multipath distortion of TV signal and design of corrective filter," in *1956 IRE Conv. Rec.*, pt. 4, pp. 167–175.

[11] A. Lender, "Decision directed digital adaptive equalization techniques for high-speed data transmission," *IEEE Trans. Commun.*, vol. COM-18, pp. 625–632, Oct. 1970.

[12] G. Forney, Jr. "The Viterbi algorithm," *Proc. IEEE*, vol. 61, pp. 268–278, Mar. 1973.

[13] T. Ohgane, T. Shimura, N. Matsuzawa, and H. Sasaoka, "An implementation of CMA adaptive array for high-speed GMSK transmission in mobile communications," *IEEE Trans. Vehicular Technol.*, vol. 42, pp. 282–288, Nov. 1993.

[14] T. Ohgane, N. Matsuzawa, T. Shimura, and M. Mizuno, "BER performance of CMA adaptive array for high-speed GMSK mobile communication—A description of measurements in central Tokyo," *IEEE Trans. Vehicular Technol.*, vol. 42, pp. 484–490, Nov. 1993.

[15] J. Treichler and B. Agee, "A new approach to multipath correction of constant modulus signals," *IEEE Trans. Acous., Speech, Signal Processing*, vol. ASSP-31, pp. 459–472, Apr. 1983.

8

OTHER APPLICATIONS OF ADAPTIVE CANCELLERS

The underlying principle of adaptive cancellation is the adaption or learning where, for example, any error in the synthesis of the counterinterference is continuously driven toward zero. This implies a closed-loop control, the primary function of which is to minimize error. It is not surprising, therefore, that the basic mechanism of the adaptive canceller will find uses in a variety of applications where a closed-loop control is involved. We have already considered one such application in the case of a signal separation system where two or more cancellers are used to separate two or more uncorrelated signals received by the antennas, even when the signals are in the same frequency band and are comparable in levels. Other possible applications are various types of adaptive filters, adaptive direction finding, adaptive means for enhancing the radar cross section of an object, etc. The use of adaptive cancellers by themselves or in combination with other systems for a few such diverse applications is considered in this chapter.

8.1 ADAPTIVE LINEAR COMBINER AS A FILTER

An adaptive linear combiner is a tapped delay line with a set of bias weights that can be adaptively adjusted to provide the function of a filter. We have already considered the construction of such a linear combiner in Section 2.3 of Chapter 2 without the discussion of the adaptive algorithm. We will consider in this section the rationale for the adaptive bias weights adjustment process.

Figure 8-1 shows the essential elements of the linear combiner with bias weights and closed-loop controls for weight adjustments. Let x_K be the kth element of the time series representing the input signal such that $x_K = x(KT)$, T being the time step or interval between samples. The input signal vector may thus be defined as

$$x_K = \begin{Bmatrix} x_{0K} \\ x_{1K} \\ \vdots \\ x_{NK} \end{Bmatrix}. \tag{8.1}$$

As shown in Figure 8-1, the input signal components appear simultaneously on all input lines at discrete times indexed by the subscript K. The weighting coefficients or multiplying factors $W_0, W_1, W_2, \ldots, W_N$ are adjustable. One may write the weight vector as

$$W = \begin{bmatrix} W_0 \\ W_1 \\ W_2 \\ \vdots \\ W_N \end{bmatrix}. \tag{8.2}$$

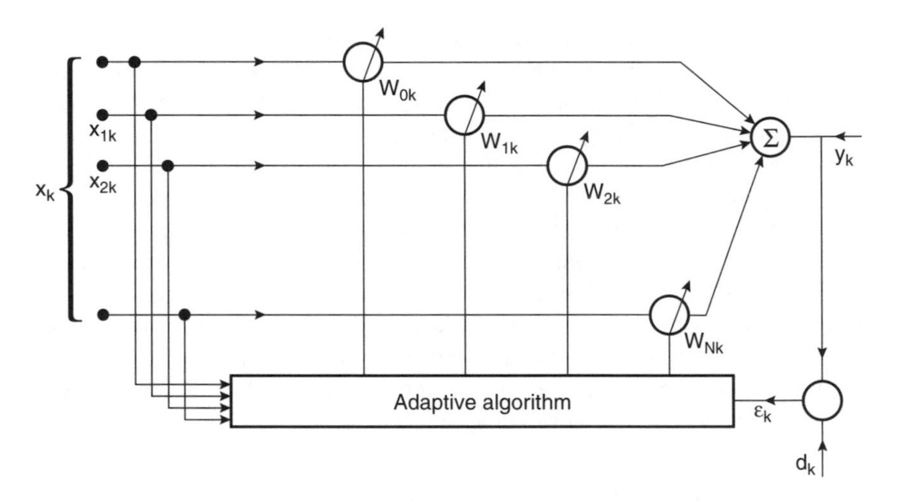

Figure 8-1 An adaptive combiner with a closed-loop control.

The output y_K, as shown in the figure, is the sum of the products of $x_{iK}W_i$ for each $i = 1, 2, 3$, etc. Thus

$$y_K = \sum_{i=1}^{N} x_{iK}W_i = \sum_{I=1}^{N} W_i x_{iK}. \tag{8.3}$$

In matrix form, y_K is the inner product of x_K and W, that is,

$$y_K = \tilde{x}_K W = \tilde{W} x_K \tag{8.4}$$

where

$\tilde{x}_K$ and $\tilde{W}$ = the transposed matrices of x_K and W, respectively.

Let ϵ_K be defined as the difference between the desired response and the actual response y_K as noted in the figure. If the linear combiner is to behave as a filter, d_K will represent the desired response of the filter. Again, in matrix notation

$$\epsilon_k = d_K - \tilde{x}_K W = d_K - \tilde{W} x_K$$
$$= d_K - \sum_{i=1}^{N} x_{iK} W_i. \tag{8.5}$$

The adaptive algorithm that adjusts the weights of the linear combiner to minimize the error ϵ_K in the least-mean-square (LMS) sense is referred to as the LMS algorithm. A general expression for the mean-square error, as a function of the weight values, assuming that the input signals and the desired response are statistically stationary and the weights are fixed, can be obtained from Eq. (8.5). Thus

$$\epsilon_K^2 = \left(d_K - \sum_{i=1}^{N} x_{iK} W_i\right)^2 = d_K^2 - 2d_K \sum_{i=1}^{N} x_{iK} W_i + \left(\sum_{i=1}^{N} x_{iK} W_i\right)^2. \tag{8.6}$$

But since

$$\left(\sum_{i=1}^{N} x_{iK} W_i\right)^2 = \sum_{i=1}^{N} \sum_{j=1}^{N} x_{iK} W_i x_{jK} W_j$$
$$= \sum_{i=1}^{N} \sum_{j=1}^{N} W_i (x_{iK} x_{jK}) W_j \tag{8.7}$$

$$\epsilon_K^2 = d_K^2 - 2d_K \sum_{i=1}^{N} x_{iK} W_i + \sum_{i=1}^{N} \sum_{j=1}^{N} W_i (x_{iK} x_{jK}) W_j. \tag{8.8}$$

Again, in matrix notation

$$\epsilon_K^2 = d_K^2 - 2d_K \tilde{x}_K W + \tilde{W} x_K \tilde{x}_K W. \tag{8.9}$$

Taking the expected values of both sides of Eq. (8.8) or (8.9) yields

$$E[\epsilon_K^2] = E[d_K^2] - 2E\left[d_K \sum_{i=0}^{N} x_{ik} W_i\right] + E\left[\sum_{i}^{N} \sum_{j}^{N} W_i (x_{iK} x_{jK}) W_j\right]$$
$$= E[d_K^2] - 2E[d_K \tilde{x}_K] W + \tilde{W} E[x_K \tilde{x}_K] W. \tag{8.10}$$

If, now, one defines P as the cross correlation between the desired response and the input signal such that

$$P \triangleq E[d_K x_K] \triangleq E\left[d_K \sum_{i=0}^{N} x_{iK}\right] \tag{8.11}$$

and R as the input correlation matrix such that

$$\begin{aligned} R &= E\begin{bmatrix} x_{0K}x_{0K} & x_{0K}x_{1K} & \cdot & \cdot \\ x_{1K}x_{0K} & x_{1K}x_{1K} & \cdot & \cdot \\ x_{2K}x_{0K} & x_{2K}x_{1K} & \cdot & \cdot \\ x_{nK}x_{0K} & \cdot & \cdot & x_{nK}x_{nK} \end{bmatrix} \\ &= E[x_K \tilde{x}_K], \end{aligned} \tag{8.12}$$

R being a symmetric matrix, one may write from Eqs. (8.9) through (8.12)

$$E[\epsilon_K^2] = E[d_K^2] - 2\tilde{P}W + \tilde{W}RW. \tag{8.13}$$

It is seen from Eqs. (8.10) and (8.13) that the expected value of the error is always positive and is a quadratic function of the weights. Adjusting the weights to minimize the error during adaptation involves descending along a concave hyperboloidal surface defined by Eqs. (8.10) or (8.13). As noted by Widrow *et al.* [1], gradient methods are commonly used for minimizing the error.

The gradient ∇ of the error function is obtained by differentiating Eqs. (8.10) and (8.13) with respect to W. Thus

$$\nabla = \begin{bmatrix} \dfrac{\partial E[\epsilon_k^2]}{\partial W_0} \\ \dfrac{\partial E[\epsilon_k^2]}{\partial W_1} \\ \vdots \\ \dfrac{\partial E[\epsilon_k^2]}{\partial W_N} \end{bmatrix}. \tag{8.14}$$

The optimal weights are obtained by setting the gradient of the mean-square-error function to zero.

Now, from Eq. (8.10)

$$\nabla E[d_K^2] = 0 \tag{8.15}$$

since d_K is not a function of the weights.

Similarly

$$\nabla E\left[d_K \sum_{i=0}^{N} x_{iK} W_i\right] = E\begin{bmatrix} d_K x_{0K} \\ d_K x_{1K} \\ \vdots \\ d_K x_{NK} \end{bmatrix} = P. \tag{8.16}$$

Also,

$$\nabla E\left[\sum_{i=1}^{N} x_{iK} W_i\right]^2 = E\left[\sum_{i=0}^{N} x_{iK} W_i x_j\right] + E\left[\sum_{j=0}^{N} x_{jK} W_j x_i\right]. \tag{8.17}$$

But

$$\sum_{i=0}^{N} x_{iK} W_i x_j = \sum_{j=0}^{N} x_{jK} W_j x_i.$$

Thus

$$\begin{aligned}\nabla E\left[\sum_{i=0}^{N} x_{iK} W_i\right]^2 &= 2\sum_{i=0}^{N} x_{iK} x_{jK} W_i \\ &= 2E\begin{bmatrix} x_{0K}x_{0K} & \cdot & \cdot & x_{0K}x_{NK} \\ x_{1K}x_{0K} & \cdot & \cdot & \cdot \\ \cdot & & & \\ x_{NK}x_{0K} & \cdot & \cdot & x_{NK}x_{NK}\end{bmatrix}\begin{bmatrix} W_0 \\ W_1 \\ \cdot \\ W_N\end{bmatrix} \\ &= 2RW. \end{aligned} \tag{8.18}$$

When

$$\begin{aligned} E[\epsilon_K^2] &= 0 \\ -2P + 2RW &= 0 \end{aligned} \tag{8.19}$$

and hence

$$\overline{W} = R^{-1} P \tag{8.20}$$

where
$\overline{W}$ = the optimal weight vector, usually referred to as the Wiener weight vector.

The LMS adaptive algorithm is a practical method for finding a close approximate solution of Eq. (8.20). The algorithm does not require explicit measurements of correlation functions, and no matrix inversion as implied in Eq. (8.20) is involved. The method used to implement the algorithm comprises iteration of the weight vector W where each iteration occupies a unit time period. If W_K denotes the weight vector of the Kth iteration, then the next weight vector W_{K+1} is obtained from the relation

$$W_{K+1} = W_K - \mu \nabla_K \tag{8.21}$$

where
μ = a factor that controls the stability and rate of convergence
∇_K = the gradient as defined in Eq. (8.14) for the Kth iteration.

The LMS algorithm estimates an instantaneous gradient in an approximate manner by assuming that the error vector ϵ_K^2, that is, the square of a single error sample, is an estimate of the mean-square error and by differentiating ϵ_K^2 with respect to W. Let the true and estimated gradients ∇_K and $\bar{\nabla}_K$, respectively, be defined as

$$\nabla_K \triangleq \begin{bmatrix} \dfrac{\partial}{\partial W_0} E[\epsilon_K^2] \\ \dfrac{\partial}{\partial W_1} E[\epsilon_K^2] \\ \vdots \\ \dfrac{\partial}{\partial W_N} E[\epsilon_K^2] \end{bmatrix}_{W=W_K} \tag{8.22}$$

and

$$\bar{\nabla}_K = \begin{bmatrix} \dfrac{\partial \epsilon_K^2}{\partial W_0} \\ \dfrac{\partial \epsilon_K^2}{\partial W_1} \\ \vdots \\ \dfrac{\partial \epsilon_K^2}{\partial W_N} \end{bmatrix}_{W=W_K} \tag{8.23}$$

One may also write

$$\bar{\nabla}_K = 2\epsilon_K \begin{bmatrix} \dfrac{\partial \epsilon_K}{\partial W_0} \\ \dfrac{\partial \epsilon_K}{\partial W_1} \\ \vdots \\ \dfrac{\partial \epsilon_K}{\partial W_N} \end{bmatrix}_{W=W_K} \tag{8.24}$$

and from Eq. (8.5), one obtains

$$\frac{\partial \epsilon_K}{\partial W_i} = -x_{iK}. \tag{8.25}$$

In matrix form

$$\begin{bmatrix} \frac{\partial \epsilon_K}{\partial W_0} \\ \frac{\partial \epsilon_K}{\partial W_1} \\ \vdots \\ \frac{\partial \epsilon_K}{\partial W_N} \end{bmatrix} = -x_K. \tag{8.26}$$

Thus, from Eqs. (8.24) and (8.26)

$$\bar{\nabla}_K = -2\epsilon_K x_K \tag{8.27}$$

and from Eqs. (8.21) and (8.27)

$$W_{K+1} = W_K + 2\mu\epsilon_K x_K. \tag{8.28}$$

From Eq. (8.28), the individual weight change can be written as

$$W_{i,K+1} = W_{i,K} + 2\mu\epsilon_K x_i. \tag{8.29}$$

The LMS algorithm is simple and easy to implement. Also, it does not require squaring, averaging, or differentiation as the name implies, although it makes use of the gradients of the mean-square error functions.

The gradient estimate used in the LMS algorithm yields weights, W_1, W_2, etc., that converge to the weights as required in Eq. (8.20), when the input signal vectors x_K are uncorrelated over time, that is, when

$$E[x_K x_{K+m}] = 0 \qquad m \neq 0. \tag{8.30}$$

Starting with an arbitrary initial weight vector W_i, the algorithm will converge in the mean, and will remain stable as long as the parameter μ is greater than 0 and is bounded by M which depends on the correlation matrix R as given in Eq. (8.12), such that

$$M > \mu > 0. \tag{8.31}$$

The time constant associated with the initially chosen weight vector W_i converging to that required in Eq. (8.20) is seen to be inversely proportional to the equivalent gain factor μ and the number of weights [1].

The adaptive linear combiner may be used with a tapped delay line to form the LMS adaptive filter. The schematic approach is shown in Figure 8-2. In this

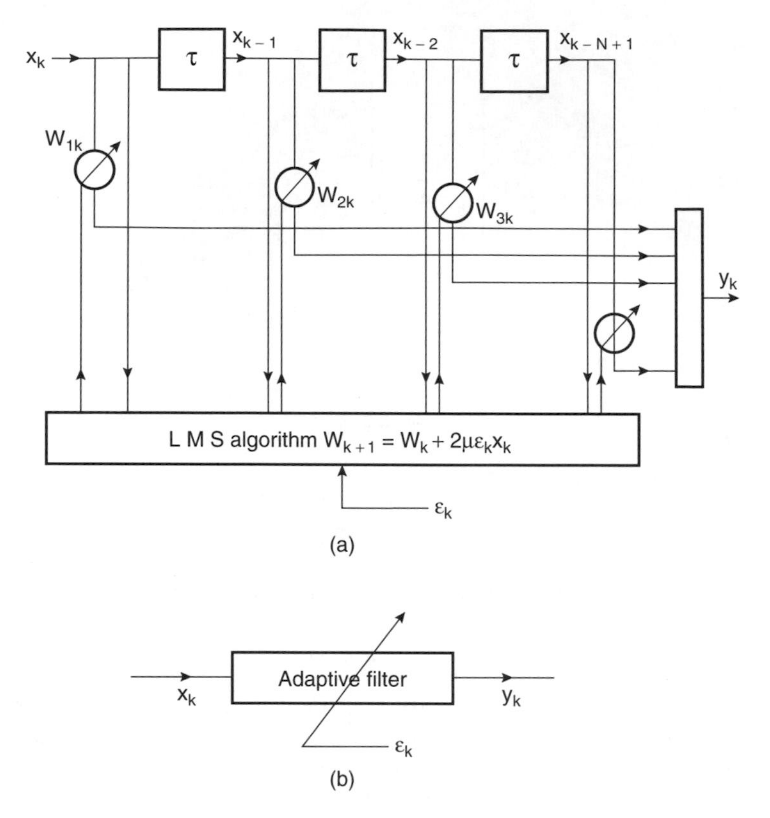

Figure 8-2 Adaptive filter employing LMS algorithm. (a) Block diagram. (b) Symbolic representation.

case, various input signal components are delayed versions of the input signal x_K. Thus, the input signal vector may be written as

$$x_K = \begin{bmatrix} x_K \\ x_{K-1} \\ \vdots \\ x_{K-N+1} \end{bmatrix}. \tag{8.32}$$

This kind of filter permits the adjustment of gain and phase at many frequencies simultaneously, and is useful in adaptive wideband signal processing. Simplified design rules are:

(a) Tap spacing at the delay line must be as short as the reciprocal of twice the signal bandwidth.

(b) Total real-time length of the delay line is determined by the reciprocal of the filter frequency resolution.

The number of weights required is generally equal to twice the ratio of the total signal bandwidth to the frequency resolution of the filter.

8.2 NOTCH FILTER

In certain situations, the undesired signal or interference is sinusoidal in nature, and the conventional method of eliminating such interference is through the use of a notch filter. An adaptive canceller as discussed in earlier chapters can be effectively used as a notch filter. To illustrate such feasibility, one may consider the situation shown in Figure 8-3, where the receiving antenna rceeives the desired signal $s(t)$, contaminated by an overwhelmingly large, sinusoidal, undesired signal, denoted as $K\cos(\omega_0 t - \phi)$, K and ϕ being an arbitrary amplitude and an arbitrary phase, respectively. The angular frequency ω_0 is $2\pi f_0$, where f_0 is the frequency of the undesired signal.

As shown in the figure, the reference signal for the adaptive canceller is $A\cos(\omega_0 t - \phi_R)$ where A and ϕ_R are some arbitrary amplitude and phase, respectively. When the frequency of the undesired signal f_0 is known with certainty,

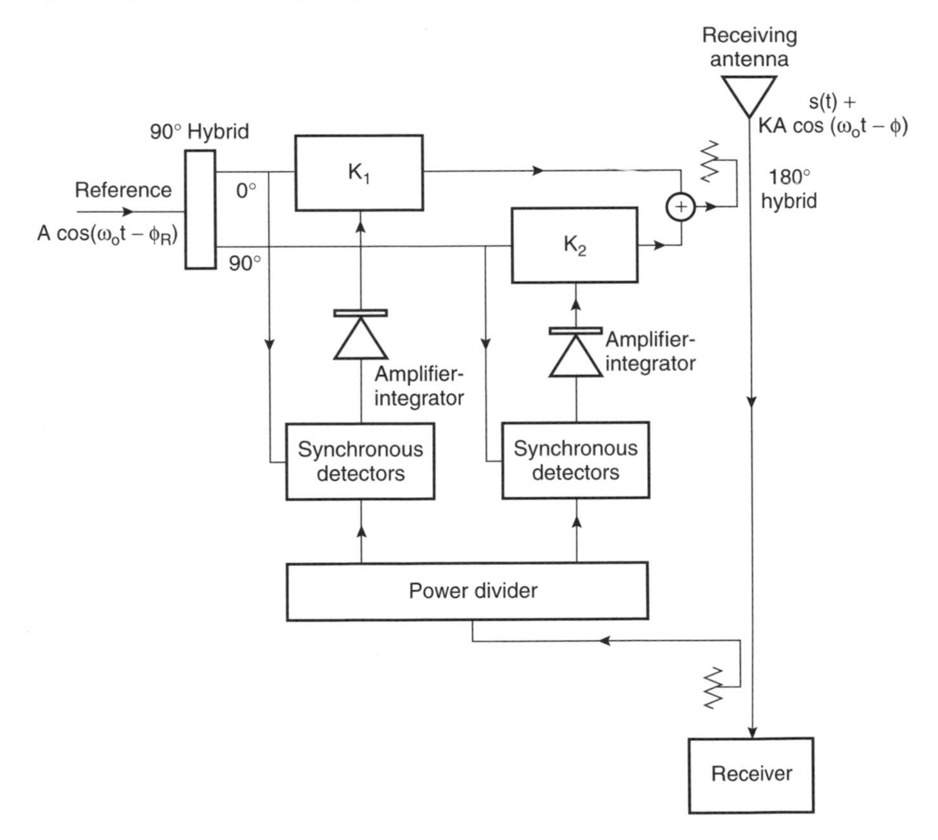

Figure 8-3 Illustration of the use of an adaptive canceller as a notch filter.

power line frequency, 60 Hz for example, the reference signal could be provided by any signal source, unrelated to the source of the undesired signal. Alternatively, the reference signal could be obtained from a voltage-controlled oscillator (VCO) of a phase-lock circuit that receives its input from a reference antenna tuned to f_0, as shown in Figure 8-4. The advantage of this alternative approach is to ensure a good reference for the adaptive canceller, when the undesired signal changes or drifts in frequency with time or when the exact frequency of the undesired signal is not known. Referring to Figure 8-3 it is readily seen that the 0° and 90° signal controllers adjust the values of K_1 and K_2 such that

$$K_1^2 + K_2^2 = K^2 \quad \text{and} \quad \tan^{-1} K_2/K_1 = \phi_R - \phi. \tag{8.33}$$

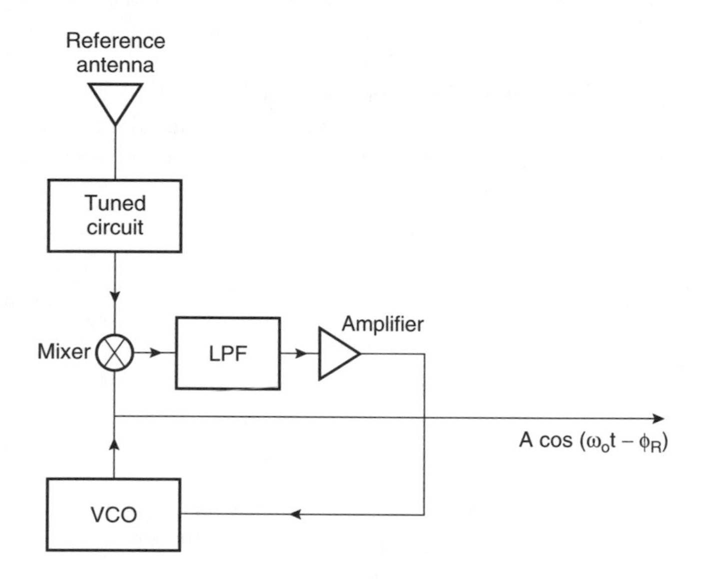

Figure 8-4 An alternative means for obtaining a reference signal for a notch filter.

The amplifier–integrator assembly shown in Figure 8-3 can be arranged to hold the interference-cancelling values for the signal controllers as given in Eq. (8.33) for a reasonable time so as to accommodate an intermittent, CW interference. It should also be noted that if there is more than one CW interference present at the receiving antenna, each one of them may be cancelled by providing the appropriate reference signals to the adaptive cancellers used to eliminate multiple CW interferences. Additionally, when the CW interferences are harmonically related, one may be able to cancel almost all harmonics with a single adaptive canceller, as we will see later in this chapter.

Besides interference cancellation, a notch-filter-like application may be encountered when one needs to measure very low-level spurious signals and noise output of a transmitter, particularly at the adjacent channels, where the presence of a high-level transmitter carrier limits such measurements. It is not atypical, for example, for the sideband power level of a transmitter to be 100 dB below the level of its carrier, while the dynamic range of a typical spectrum analyzer used for such measurements is limited to approximately 70 dB. Usually, the frequency separation between the spurious output and the main carrier is not adequate for conventional passive filters to be used during such measurements to sufficiently suppress the carrier. Thus, the problem of accurate measurements of spurious and other noise outputs becomes difficult when their power levels are more than 70 dB below the carrier level. An adaptive canceller [2] may be used to enhance the effective dynamic range to more than 120 dB where the transmitter carrier level is reduced by 55 dB.

Figure 8-5 shows a block diagram of the arrangement that permits an effective increase of the dynamic range of a spectrum analyzer while measuring spurious transmission or noise, particularly in adjacent channels, of a transmitter output by suppressing the transmitter carrier. As shown in the figure, the signal from the transmitter is divided into two branches, A and B, where the branch A feeds an adaptive canceller, while the branch B is capable of introducing a variable attenuation and a time delay. The adaptive canceller primarily cancels the carrier, but any time delay introduced in branch B "spoils" the cancellation at adjacent frequencies, the degradation of cancellation being a function of the time delay. To better understand the consequence of the delay in one branch and the adaptive canceller at the other, let us assume that the input signals from the transmitter to branches A and B are

$$\begin{aligned} S_A &= A \sin(\omega t + \phi_A) \\ S_B &= B \sin(\omega t + \phi_B). \end{aligned} \tag{8.34}$$

Let a time delay T_1 be introduced in branch B, and the adaptive canceller be used to cancel the signal from branch B at summer No. 2. The cancellation at the angular frequency ω_0 requires that

$$KA \sin(\omega_0 t + \phi_A + \theta) = B \sin(\omega_0 [t - T_1] + \phi_B) \tag{8.35}$$

where

K and θ = the amplitude factor and phase delay, respectively, caused by the signal controllers, as shown in Figure 8-5.

Here, the cancellation is assumed to be effected at the angular frequency $\omega_0 = 2\pi f_0$, f_0 being the carrier frequency, since it is the dominant signal that controls the adaptive canceller.

It is evident from Eq. (8.35) that

$$\phi_A - \phi_B + \theta = -\omega_0 T_1. \tag{8.36}$$

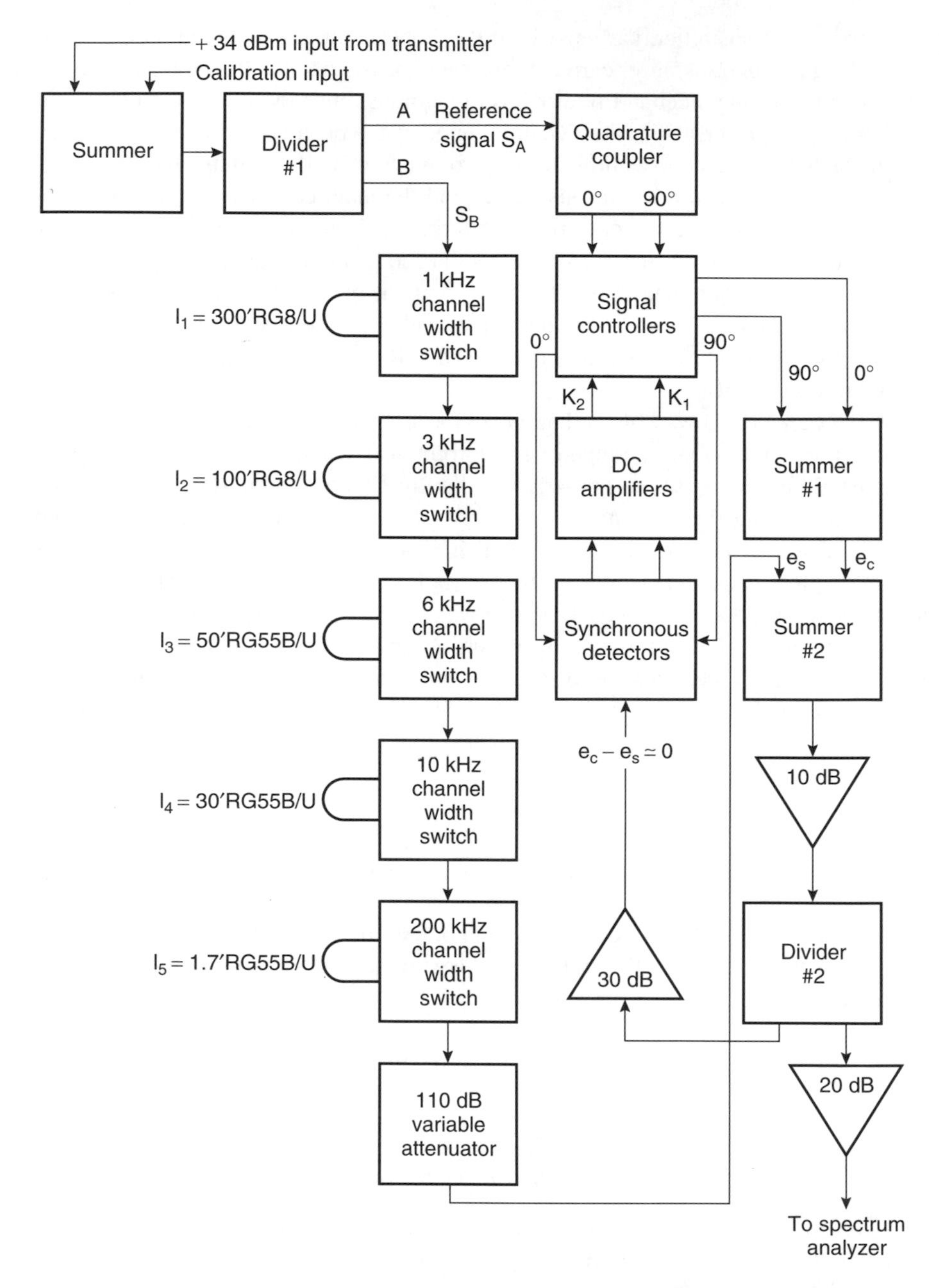

Figure 8-5 Illustration of the use of adaptive canceller to enlarge the dynamic range of spectrum analyzer while measuring adjacent channel noise.

At any other angular frequency ω_1, adjacent to ω_0, there will not, in general, be a perfect cancellation, and consequently, there will be a residual signal instead, given by

$$e_R = B\sin(\omega_1(t - T_1) + \phi_B) - KA\sin(\omega_1 t + \phi_A + \theta).$$

And, since $B = KA$

$$\begin{aligned} e_R &= 2B\sin\left[\frac{\theta + \phi_A - \phi_B + \omega_1 T_1}{2}\right]\sin(\omega_1 t + \psi) \\ &= 2B\sin([\omega_1 - \omega_0](T_1/2))\sin(\omega_1 t + \psi) \end{aligned} \quad (8.37)$$

where
ψ = a phase shift.

From Eqs. (8.36) and (8.37), the cancellation becomes

$$20\log_{10}\left|\frac{e_R}{S_B}\right| \simeq 20\log_{10}(\Delta\omega T_1)\ \text{dB} \quad (8.38)$$

where

$$\begin{aligned} \Delta\omega &= (\omega_1 - \omega_0) \\ \sin(\Delta\omega T_1/2) &\simeq \Delta\omega T_1/2. \end{aligned}$$

Thus, by introducing a time delay in branch B, as in Eq. (8.38), one can cause a degradation of cancellation at a frequency $\Delta\omega/2\pi$ away from the carrier while still maintaining a high degree of cancellation for the carrier. This ability to cancel the high-level carrier while not cancelling as much at the adjacent frequencies away from the carrier permits an examination of the relatively low-level spurious transmissions or noise at the adjacent frequencies, even when such a level is 100–120 dB below the level of the carrier.

As an example, suppose the cancellation provided by the adaptive canceller is 75 dB or more at the carrier frequency, and it is desired that the corresponding cancellation for the same signal controller settings does not exceed 52 dB at a frequency 10 kHz away from the carrier. From Eq. (8.38), T in this case should be approximately 40 ns. If a section of cable is used to create this time delay, and the relative velocity of electromagnetic waves in the cable is 0.7 with respect to that in free space, the length of the cable will be approximately 28 ft. Figure 8-6 shows the effect of cancellation of the carrier while not cancelling the sidebands as much. As may be seen from this figure, low-level spurious outputs of the source, which could not have been measurable otherwise, are now visible on the spectrum analyzer screen.

It may be noted that the same concept as inserting a delay line in branch B of Figure 8-5 can be used to create a notch-filter-like characteristic, particularly

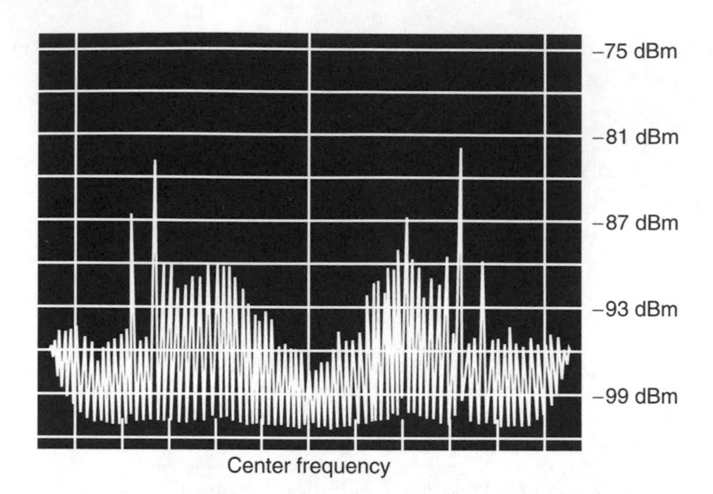

Figure 8-6 Illustration of adjacent channel spurious suppression. Center frequency: 300.1 MHz. Carrier level before suppression: 10 dB. Carrier level suppression: 90 dB. Scan width: 200 kHz/division. Reference line cable length: 160 ft.

when the delay line is several wavelengths long. Thus, for example, it is seen from Eq. (8.37) that when $\omega_0 T_1 = (2\pi/\lambda)(N\lambda)$, $(N\lambda)$ being the length of the delay line, the magnitude of e_R at an angular frequency ω_1, slightly removed from ω_0, becomes

$$|e_{R,N\lambda}| = 2B \sin\left[2(\omega_1 - \omega_0)\frac{N\pi}{\omega_0}\right]$$
$$= 2B \sin\left(\frac{2\Delta\omega N\pi}{\omega_0}\right) \tag{8.39}$$

where
$\Delta\omega = (\omega_1 - \omega_0)$.

Thus, the slope of cancellation degradation, as $\Delta\omega \to 0$, becomes proportional to N, which could be very large when N is very large.

8.3 HARMONICS FILTER

In many applications, one is required to suppress not only the interference, but several of its harmonics. In some cases, even the harmonics of the desired signal are interferences and need to be suppressed from the receive line. Adaptive cancellers can be very effectively used to perform such filter functions.

A simple approach to obtain harmonics suppression with an adaptive canceller is to use the same concept as employed for suppressing the carrier as shown in Figure 8-5. Thus, for example, if the delay introduced in branch B in Figure 8-5 corresponds to a half wavelength at the frequency whose harmonics need suppression, then from Eq. (8-39), the absolute value of the residual signal at a frequency f_1 is

$$|e_{R,\lambda/2}| = 2B \sin\left(\frac{\Delta\omega\pi}{\omega_0}\right) \tag{8.40}$$

where
$\Delta\omega = \omega_1 - \omega_0 = 2\pi f_1 - \omega_0$.

Now, if $\omega_1 = K\omega_0$, K being an integer but not equal to 1, then ω_1 will correspond to a harmonic, and from Eq. (8.40), such a harmonic will be suppressed since e_R for such a frequency will be zero. Since the residual signal will be zero for any harmonic of $\omega_0/2\pi$, the arrangement in Figure 8-5 with a delay $T_1 = \lambda/2\omega_0$ will be an effective harmonic filter.

The suppression of harmonics is also feasible by an adaptive filter, conceptually shown in Figure 2-9. We may recall from the discussion in Section 2.3 of Chapter 2 that an adaptive filter consists of a series of unit delay lines with a weight associated with each line τ. The transfer function of such a filter as given in Eq. (2.27) with N number of delay lines is

$$G(j\omega) = \sum_{n=0}^{N-1} W_n e^{-jn\tau}$$

where
W_n = the weight for the nth element.

Now, if all W_ns are equal to W_0, then

$$G(j\omega) = \sum_{n=0}^{n-1} e^{-jn\tau} \cdot W_0$$

$$= e^{-j(N-1)\omega\tau/2} \frac{\sin(N\omega\tau/2)}{\sin(\omega\tau/2)} W_0. \tag{8.41}$$

Further, if for an angular frequency ω_0, $(\omega_0\tau/2)$ is equal to π, then it is seen from Eq. (8.41) that the transfer function of the filter will be zero for all harmonics $(N > 1)$ of ω_0, making the filter a harmonic suppression filter.

A common application of the harmonic filter is one to eliminate the power line interferences in certain sensitive receive lines, where the fundamental power line frequency of 50 or 60 Hz and many of their harmonics couple into such receive lines. In some cases, a particular harmonic, being periodic in nature, is avoidable by simply adding a fixed delay line at the reference port of the adaptive canceller, as shown in Figure 8-7. This will particularly be the case when the desired signal is a very broadband signal. The fixed delay in such a case must be of sufficient length to cause the broadband signal components at the reference port of the canceller to be decorrelated from those at the receive line. The interferences, being periodic in nature, will remain correlated with each other, and hence will be cancelled at the receive line. The situation in such a case is not very different from that illustrated in Figure 8-5, where a periodic carrier signal is cancelled regardless of the length of the delay line, but the cancellation is spoiled at frequencies away from the carrier. A truly broadband desired signal, having most or all frequency components away from the carrier, will not be affected by the cancellation process. A typical example of a broadband desired signal and a periodic interference is found during the playback of speech or music in the presence of tape hum or turntable rumble, which is often periodic.

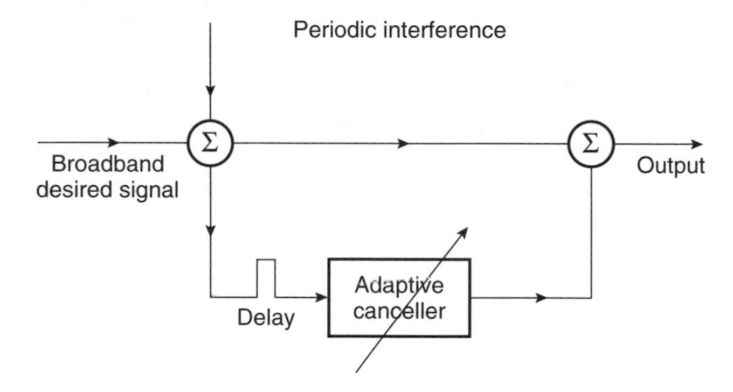

Figure 8-7 Cancellation of a periodic interference without materially affecting broadband desired signal.

There are a number of circumstances where a high-power amplifier is used at the final stage of a transmitter and the nonlinearity at the amplifier generates unwanted harmonics, the second and third harmonic being particularly objectionable. If adaptive cancellers are used to rid such objectionable interferences, one needs appropriate references at the reference ports of such cancellers. Such references are, however, not readily available and cannot, in general, be obtained from a signal generator, particularly when the harmonics of a modulated signal are of concern. To illustrate the basic nature of the problem, let the input–output relation of a high-power amplifier of concern be given by

$$e_0 = a_1 e_i + a_2 e_i^2 + a_3 e_i^3 + \cdots \tag{8.42}$$

where

e_0 and e_i = the output and input, respectively, of the amplifier
a_1, a_2, a_3, etc. = the coefficients of nonlinearity, being, in general, complex quantities, each having an amplitude and a phase.

The unwanted second and third harmonics are caused by the terms e_i^2 and e_i^3.

One way to generate the reference needed for the cancellation of the second harmonic is to multiply a sample of the input of the amplifier with a similar signal so as to obtain

$$(K_1 e_i)(K_2 e_i) = K_1 K_2 e_i^2 \tag{8.43}$$

where
K_1 and K_2 = arbitrary complex quantities.

When the output of the multiplier as given in Eq. (8.43) is used as a reference for an adaptive canceller, one may cancel the e^2 term from the output of the amplifier as given in Eq. (8.42). Similarly, when one generates a reference interference

$$K_1 K_2 K_3 e_i^3 = (K_1 e_i)(K_2 e_i)(K_3 e_i), \tag{8.44}$$

K_3 being another complex quantity, a second adaptive canceller may be used to cancel the term corresponding to $a_3 e_i^3$, as shown in Eq. (8.42), with the reference interference given in Eq. (8.44). It should be noted that when the reference interferences $K_1 K_2 e_i^2$ or $K_1 K_2 K_3 e_i^3$ are used for the adaptive cancellers, one may obtain a good cancellation of the harmonics, even when e_i is a modulated signal. Additionally, one may optimize the cancellation process so as to leave a minimum residual interference due to harmonics, even when only two cancellers with reference interferences as given in Eqs. (8.43) and (8.44) are used. For example, let the output interference of a slightly nonlinear amplifier be

$$e_0 = e_i + 0.02 e_i^2 - 0.002 e_i^3 + 0.05 e_i^4 - 0.05 e_i^5. \tag{8.45}$$

A cancelling interference e_c, given by

$$e_c = 1.0156 e_i + 0.07 e_i^2 - 0.0644 e_i^2, \tag{8.46}$$

will minimize the residual interference, although the coefficients of e_i^2 of e_i^3 in Eqs. (8.45) and (8.46) are not the same. A plot of the residual interferences $|e_0 - e_c|$ is shown in Figure 8-8 where $e_i = \sin(\omega t) = \sin(2\pi t/50)$, and three adaptive cancellers are used for suppressions of the terms associated with e_i, e_i^2, and e_i^3.

Another application of the adaptive canceller is found when a transmitter consists of a high-power amplifier at its final stage as being discussed, and the

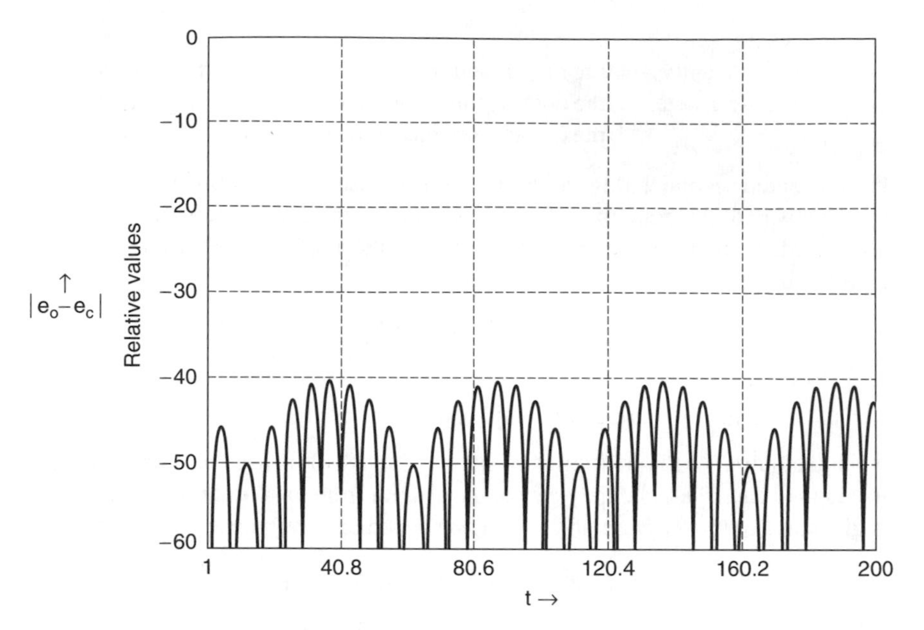

Figure 8-8 A plot of the magnitude of residual interferences when cancellations are effected for e_i, e_i^2, and e_i^3. Here, $e_i = \sin(\omega t)$, $e_0 = e_i + 0.02e_i^2 - 0.002e_i^3 - 0.05e_i^4 - 0.02875 - 0.05e_i^5$, $e_c = 1.0156e_i + 0.07e_i^2 - 0.0644e_i^3 - 0.035$.

transmitter output comprises relatively high-level spurious signals in addition to harmonics, the suppression of spurious radiation from the transmitting antenna being of more concern than the interferences.

A conventional frequency domain filter is not feasible here, even when one could design the required filter at low power level. A selective interference reduction approach [3], as shown in Figure 8-9, could be a remedy for such a case to prevent the transmitter from unnecessarily "polluting" the electromagnetic environment. As shown in the figure, a sample of the transmitting signal is obtained from the transmitting antenna line and the desired transmitting signal, that is, the transmitting signal without the spurious transmissions is removed from this sample by an adaptive canceller where the reference signal is an uncontaminated transmitting signal. The residual signals at the sample line following the cancellation are the ensemble of undesired spurious transmissions which need to be removed from the antenna line. The power levels of the spurious transmissions will be lower than those at the transmitting antenna line because of the low-level sampling. The residual signals are then led to another adaptive canceller as its reference signals, and the output of this second canceller is coupled back to the transmitting antenna line as shown in the figure. The spurious signals that remain at the transmitting antenna line following the cancellation by Adaptive Canceller 2 are monitored as error signals for Adaptive Canceller 2.

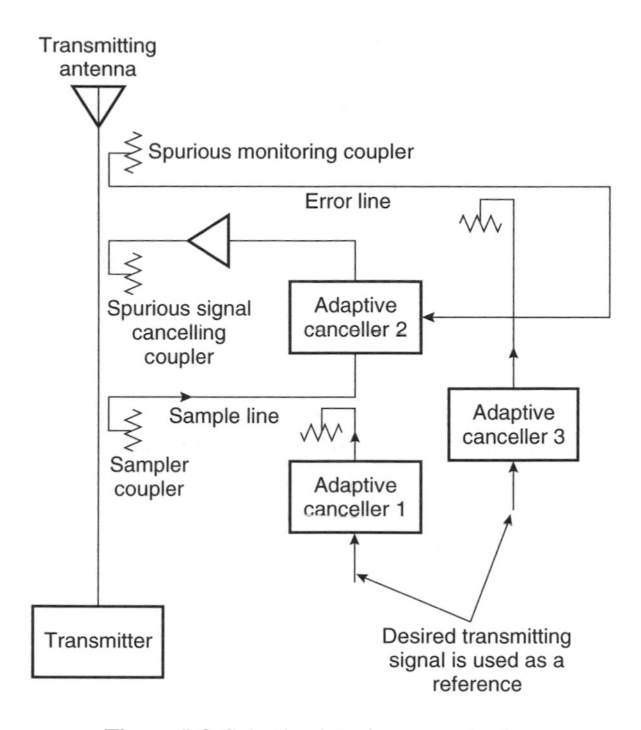

Figure 8-9 Selective interference reduction.

The output of the monitoring coupler, however, will contain the predominant desired transmitting signal, which must be removed before the error signal can be used for Adaptive Canceller 2. This enables Adaptive Canceller 2 to operate exclusively on the spurious transmissions. Adaptive Canceller 3, as shown in the figure, is used to selectively remove the desired transmitting signal from the error line. The arrangement shown in Figure 8-9, then, constitutes a remedy for suppressing undesired signals at their source, instead of at the receiver, and may be desired when there are a number of receivers in the vicinity of the transmitter emitting undesired spurious radiations. For example, in many modern radars, the modulated transmitting signal is often amplified to its full power level before it is fed to the antenna through the transmit–receive switch or its equivalent for subsequent radiation. For such cases, even a slight nonlinearity in the amplifier causes intermodulation signal products having frequencies different from those of the desired transmitting signal, and not unlike the spurious signals considered above. These intermodulation products can be sources of interferences in the neighboring communication and radar receivers, particularly when such receivers are intended to receive very weak signals. When the radar waveform is complex, the intermodulation frequency spectrum extends far beyond that of the transmitting signal spectrum for the radar, and even when the power level of the interference is much below that of the transmitting signal, the interference effect is not tolerable in many circumstances. The remedy of interference suppression at the transmitter, then, becomes an attractive solution from the engineering viewpoint.

8.4 ACOUSTIC INTERFERENCE SUPPRESSION

So far, we have discussed the use of an adaptive canceller mostly for the suppression of radiated electromagnetic interferences. It turns out that the adaptive canceller is a powerful device to rid other types of undesired signals or interferences, including acoustic interferences such as those generated from an aircraft engine or caused by automobile exhausts.

For example, noise mufflers are useful at present at the exhaust pipe to reduce the intensity of the acoustic noise generated in auto and diesel engines. While reducing the acoustic noise, however, a muffler creates a back-pressure at the exhaust pipe which, in turn, leads to a reduction of engine efficiency, and more importantly, to an increased amount of unburned hydrocarbons. Many believe that an alternative means of noise reduction that can reduce unburned hydrocarbons in exhaust and increase engine efficiency, thus requiring even less fuel, may be highly desirable. An adaptive canceller may provide such an alternative means. In this case, the acoustic noise generated at the exhaust pipe is sampled, and a "counternoise" is synthesized by the signal controller of the adaptive canceller such that when this "counternoise" is added to the noise in the exhaust pipe, the noise and "counternoise" cancel each other.

Figure 8-10 shows the schematic arrangement [3] for reducing acoustic noise at the exhaust pipe of an auto or a diesel engine. Usually, the exhaust pipe of a gasoline or diesel engine carries an intense acoustic noise covering multioctave noise components of nearly random frequencies, random phases, and random amplitudes. In Figure 8-10, this noise is sampled by a microphone-like sensor that converts acoustic noise to corresponding electric signals. The electrical signals also contain multioctave band frequency components ranging from less than 100 Hz to several kilohertz. When these signals are mixed with a CW signal at a frequency of several megahertz provided by a local oscillator as shown in the figure, the percentage bandwidth of the mixed signals, with respect to the center frequency, which is due to the local oscillator, will be relatively small, being only a fraction of an octave in total bandwidth. Such signals are easily accommodated by an adaptive canceller with an interference controller (IC) capable of controlling the amplitude and phase of the up-converted noise sample. The output of the interference controller is then down-converted to the original noise frequencies. Following the down-conversion, the counternoise is amplified, and the amplified counternoise is converted back to acoustic noise by a transducer, similar to a loudspeaker in function. For broadband operation for the canceller, the length L at the main exhaust pipe, as measured from the point of acoustic-to-electric signal conversion to the point where the counternoise is added to the noise, is made the same as L' at the branch line, as measured from the loudspeaker to the point where the noise and counternoise are added, plus the equivalent length due to the time delay introduced by the electric circuit comprised of the microphone, two mixers, filter, interference controller, IC, the amplifier, and the loudspeaker. The synchronous detector for the adaptive canceller is referenced to the interference at its original frequencies as obtained through the coupler. The error signal in this case is the residual acoustic noise left after the cancellation inside the exhaust pipe. This residual acoustic noise is sensed by a microphone-like sensor which converts the residual acoustic noise to the corresponding electric signals. When the electric signals resulting from the residual acoustic noise are not zero, direct current control signals will be generated at the output of the synchronous detector, as is common at an adaptive canceller. The control signals, in turn, will change the amplitude and phase introduced by the interference controller, IC, until the amplitude and phase are adjusted so that the error signal at the synchronous detector, and hence the residual acoustic noise at the exhaust pipe following the cancellation, approach zero.

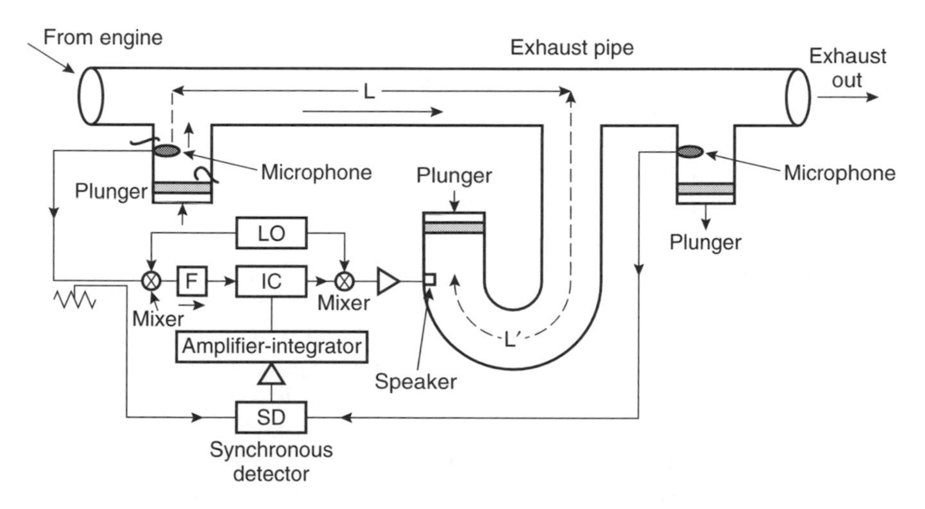

Figure 8-10 Electronic muffler for reducing exhaust noise.

The plungers shown in Figure 8-10 are used for matching purposes so that any reflections due to the sensors, such as the loudspeaker and the microphones at various parts of the exhaust pipe and its extension, are minimized. Such minimization improves the cancellation effectiveness. The functions of the synchronous detector, amplifier–integrator, etc., in Figure 8-10 are the same as those discussed in connection with an adaptive canceller.

It should be noted that the arrangement shown in Figure 8-10 does not require a minicomputer to synthesize the noise, and hence the resulting acoustic noise cancellation effectiveness is not impaired due to any sudden changes in engine performance when the minicomputer is unable to accurately synthesize the acoustic noise spectrum in real time. Also, the ability to track any changes in engine operation, no matter how sudden, along with the inherent capacity of an adaptive canceller to provide a high degree of cancellation may lead to an acoustic noise suppression which is orders of magnitude more than what is realized by a common muffler.

Another application of an adaptive canceller to minimize intense acoustic noise in certain directions caused by a moving vehicle or platform has been suggested by Ghose [4]. The acoustic noise from a moving platform, such as a large commercial aircraft or a helicopter, for example, can far exceed the reasonable, tolerable limit of noise, particularly at regions close to many airports. Often, the sources of such noise are unavoidable, and means available to combat them are limited, if any. In the case of noise due to an aircraft, the severity of the problem has compelled regulatory agencies in many areas to set upper bounds of maximum tolerable noise levels permitted by an aircraft, and to enforce procedural means to reduce noise in communities close to major airports. Although a significant reduction of aircraft noise has resulted from development efforts aimed toward the design and building of jet aircraft engines that cause less noise; the noise level still remains unacceptable and the corresponding nuisance remains unabated in many areas of the world. When the aircraft noise reduction by better design of engines, exhaust fans, etc., approaches a technological limit such that any additional noise reduction can only be made at the expense of performance degradation, complementary technologies and devices that can aid the noise reduction problem become subjects of interest. An adaptive noise canceller and the concept of creating a null in the antenna pattern of a phased array may provide a remedy for many such cases.

Figure 8-11 illustrates the approach for an adaptive noise abatement system where a cooperative sensor located at a region which is being victimized by an intense acoustic noise radiated from a distant, moving platform may help to abate the noise at that region. As shown in the figure, the acoustic noise from the platform radiates in all directions, including that of the victim region. The sensor at the victim region receives the noise. Now, if one obtains a sample of the noise generated at the platform and, having the same waveform and spectrum of the radiated noise, adjusts the amplitude and phase of the sampled noise and then reradiates the adjusted noise or "counternoise" toward the victim region with a radiator which has a directivity gain toward the victim region, the sensor at the victim region will receive both the noise and counternoise radiated from the same platform, although not from the same equivalent phase center, and not at the same amplitude. The objective of the noise abatement system is to make the victim-region-sensor output approach zero or a minimum by controlling the amplitude and phase of the "counternoise" by a signal controller.

To further illustrate the concept of adaptive, acoustic noise abatement, let us suppose that the sum of the noise and counternoise as sensed by the victim-region sensor is not zero, and the sensor output is used to modulate a radio-frequency signal from a local oscillator, as shown in Figure 8-11. The modulated signal, in turn, is then transmitted toward the platform through an antenna that has a directivity

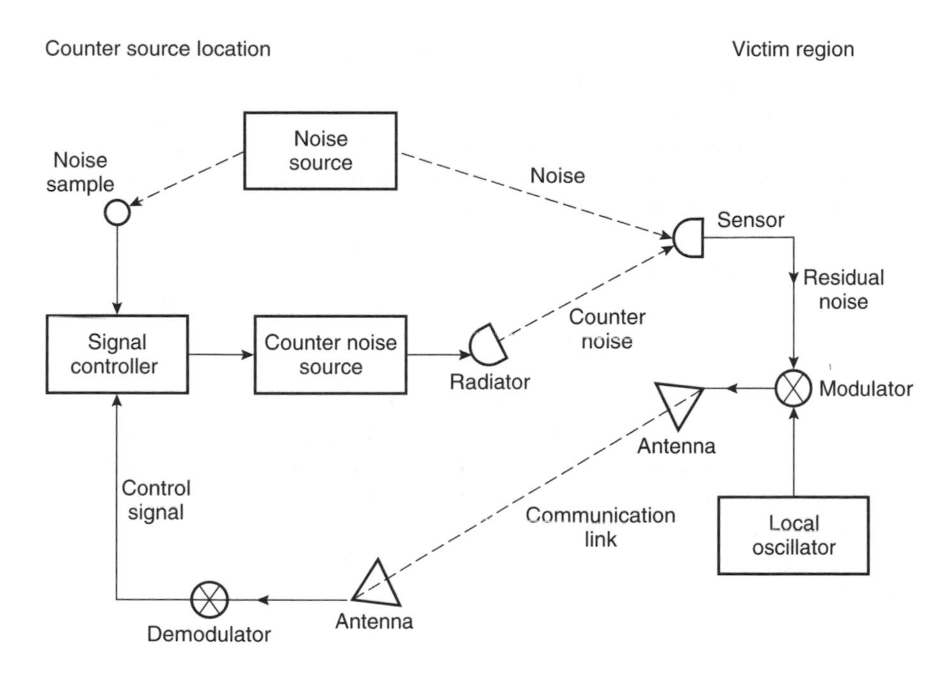

Figure 8-11 Adaptive noise abatement system.

gain along that direction. At the platform, this radio-frequency modulated signal is received and demodulated to obtain an electric signal which is a replica of the acoustic noise sensed by the victim-region sensor. The electric signal is then used as an error signal for an adaptive canceller, the primary function of the closed-loop operation of which is to adjust the signal controller settings, that is, the amplitude and phase introduced by the signal controller until the error signal approaches zero. A zero or near-zero error signal implies that the noise and counternoise cancel each other at the victim-region sensor. The radio-frequency, counternoise communication line including the two antennas, the demodulator, and the signal controller, as shown in the figure, constitute a closed-loop control system, the objective of which is to drive the error signal, and hence the residual noise at the victim-region sensor, to zero or near zero automatically.

It should be noted that the error signal which is a measure of the residual acoustic noise at the victim region is communicated to the platform by electromagnetic waves which travel at a much faster speed than the velocity of propagation of acoustic waves in air. This rapid communication to effect changes in the signal controllers is almost necessary for the feasibility of the noise abatement system since if, instead, the error signal were communicated to the platform in the form of acoustic waves, the travel time for the acoustic waves, and hence the time to effect changes in signal controllers, would be too long. In fact, the source condition could change significantly during this relatively long period to make the noise cancellation less effective, if at all.

Key considerations for the design of a noise abatement system, as illustrated in Figure 8-11, are the appropriate sampling of the acoustic noise at the platform, and the directivity of acoustic radiatiors and of the antennas for the radiation and reception of electromagnetic waves. Very often, an array [4] of acoustic sensors with appropriate weights is needed to obtain the most desirable sample of the acoustic noise waveform which, when radiated as the counternoise, will cause a maximum reduction of the victim region noise. A highly directive acoustic radiating system, comprising an array of appropriately phased loudspeakers creating a main beam along the victim region, is also needed to maximize the total acoustic power requirement for the noise cancellation. Retrodirective antennas at the victim region could also be useful for the electromagnetic-waves communication links between this region and the noise-causing platform.

It should also be remarked here that, although the location of the sensor at the victim region is essentially a point, the abatement of noise is effected over a region, which is the victim region in this case. The large distance between the victim region and the platform, which carries both noise and counternoise sources, causes the radiated acoustic waves to appear as plane waves at the victim region. The equivalent source of noise and the radiator for the counternoise, as shown schematically in Figure 8-12, create an equivalent two-element phased array setting a null along the direction of the victim region. The closed-loop operation maintains

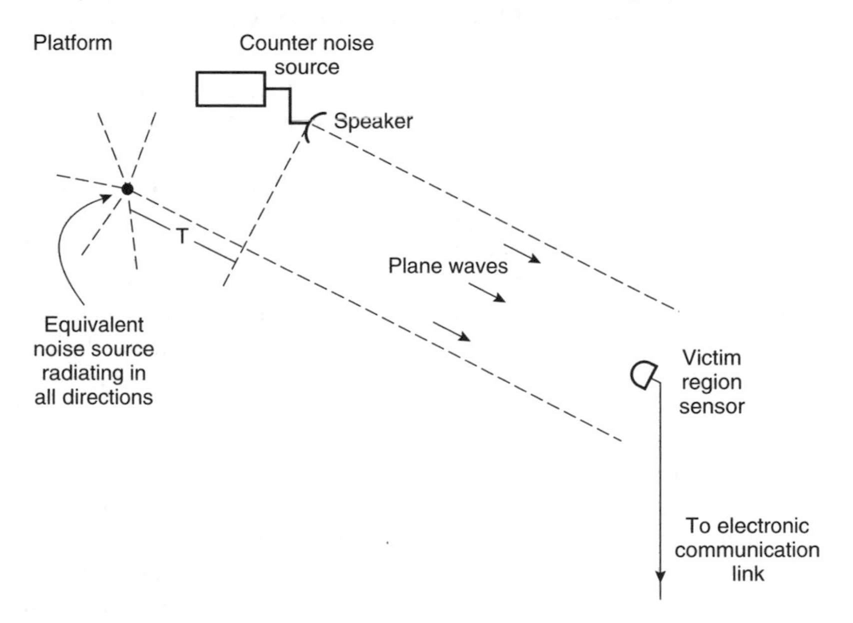

Figure 8-12 Radiation of noise and counternoise from platform. Equivalent two-element phased array with a null toward the victim region is obtained by adjusting the amplitude and phase and time delay T.

the noise null along the victim region, even when the distance between the noise-making platform and the victim region and the apparent angle of arrival of the acoustic waves from the platform to the victim region change with time due to the moving platform.

Another useful application of an adaptive canceller is found in pilot communications by radio from a cockpit [1] often overwhelmed by a high level of engine noise. The noise contains, among other things, strong periodic components, rich in harmonics, that occupy the same frequency band as speech. By placing a second microphone at a suitable location in the cockpit, one may obtain a sample of the ambient noise field free of the pilot's voice. The sample, then, can be used as a reference signal of an adaptive canceller to adaptively synthesize a countersignal which, when subtracted from the pilot-microphone line, substantially reduces the interference.

8.5 ADAPTIVE DIRECTION-FINDING SYSTEM

The technique for "angle-of-arrival" measurement for electromagnetic or acoustic waves by an interferometer, made of two sensors separated by a baseline, is well known. Since, in the case of electromagnetic waves, the electrical phases of a signal arriving at the two sensors are not the same unless the signal arrives from the boresight direction, that is, from the direction normal to the baseline at its midpoint, the phase difference between the signals received at the two sensors yields a direct measure of the "angle of arrival" of the signal. Unfortunately, several factors deter this angle measurement ability, important among them being:

- **(a)** resolution and accuracy limitation of phase difference measuring approach and available tools for this purpose;
- **(b)** presence of multipath;
- **(c)** presence of noise within the signal bandwidth;
- **(d)** lack of knowledge of the exact frequency of the signal which is needed to relate the signal phase difference between the sensors to the angle of arrival of the signal;
- **(e)** a large bandwidth of the signal, such as that of a spread-spectrum communication signal, where the concept of a single phase of the signal is not meaningful; and
- **(f)** presence of multiple signals within the reception bandwidth of the interferometer.

Often, these factors are so overwhelming that the angle-of-arrival measurement, by an interferometer requiring a fraction of degree resolution and accuracy, become impossible.

Many approaches have been used in the past to overcome some of the problems relating to the use of an interferometer. For example, a long baseline has been used to reduce the effect of noise in signal bandwidth without affecting the signal. But a long baseline introduces an ambiguity in determining the physical angle of arrival from the measurement of electrical phases. More importantly, a long baseline tends to decorrelate the signal received at the two sensors, particularly when there are local scatters in the vicinity of the sensors which affect the reception of the signal by two sensors differently. In addition, a long baseline requires a large space which may not be available, particularly on a moving platform such as an aircraft. Multiple baselines may avoid the ambiguity problem at the expense of additional system complexity.

The bias error associated with most phase detectors affects the resolution and accuracy of any phase-difference measurement, and hence the resolution and accuracy of the angle-of-arrival measurement. This problem is further aggravated when the signal level or the signal-to-noise ratio of the signal, or both, are low.

A significant improvement on the angle-of-arrival measurement is effected when a signal nulling, typical of an adaptive canceller, is used where the amplitude and phase of the signal received by one sensor are adjusted automatically until this signal becomes equal to the signal received by the other sensor in amplitude and 180° out of phase [5]. The adjustment in phase needed to create a null is a direct measure of the phase difference between the signals received by the two sensors, and hence that of the angle of arrival of the signal. This approach avoids the bias error associated with the phase detector. Also, if the reduction of signal of the second sensor by the nulling technique is 60 dB, for example, which is not atypical for an adaptive canceller, the accuracy potential for the phase difference measurement becomes 10^{-3} rad. This corresponds to an accuracy of approximately 1/6 mrad in the measurement of the angle of arrival for a baseline separation of one wavelength.

Although the nulling approach to measure phase substantially improves the interferometric measurement, it is not adequate in many cases where the frequency of the signal is not known, since to compute the physical angle of arrival from the electrical phase difference as obtained through a signal nulling approach, one needs to know the frequency of the signal. The basic concept of the angle-of-arrival measurement employing a nulling technique and associated problems and remedies may be illustrated with Figure 8-13. Let the signals received by Sensors 1 and 2 from the same signal source, the direction of which needs to be measured, be

$$\begin{aligned} e_1 &= A \sin \omega t \\ e_2 &= A' \sin.\omega t + \phi/ \end{aligned} \tag{8.47}$$

where

A and A' = the amplitudes of the two signals
ω = the angular frequency of the signal
ϕ = a phase delay.

From Figure 8-13, $\phi = \omega L/C$, L and C being, respectively, the length as shown in the figure and the velocity of propagation in free space.

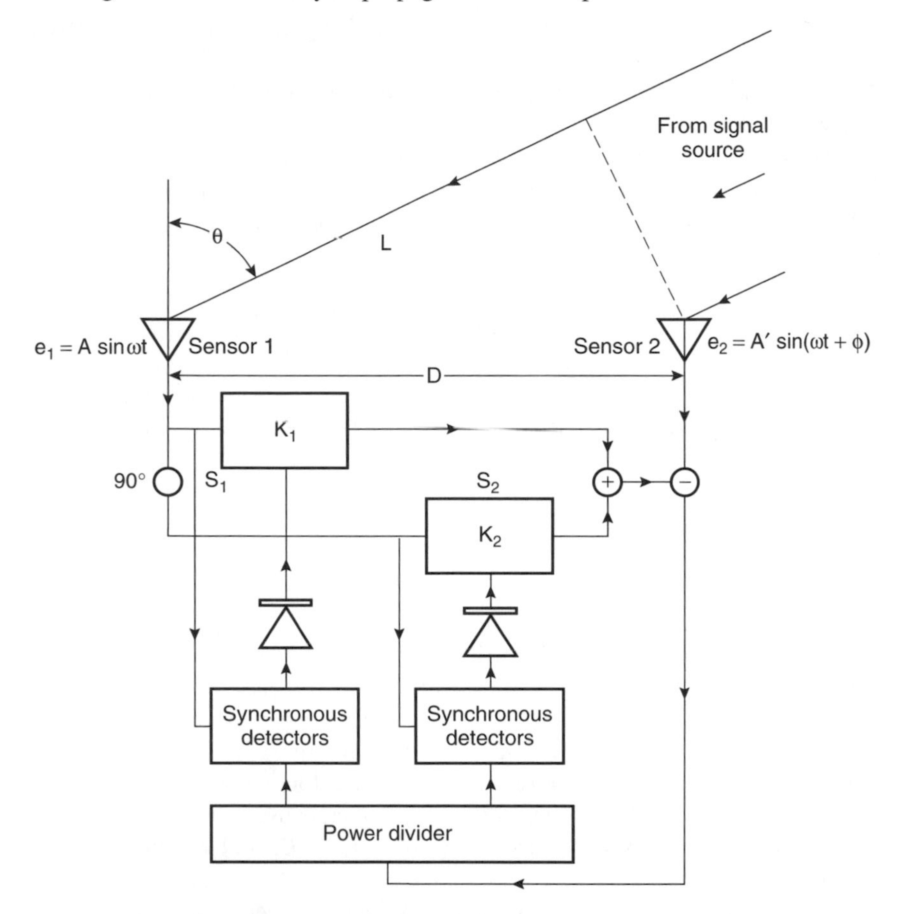

Figure 8-13 Illustration of the use of adaptive canceller for direction finding.

As shown in the figure, the signal received by Sensor 1 is adjusted in amplitude and phase by an in-phase signal controller S_1 and a quadrature-phase signal controller S_2 such that

$$K_1 A \sin \omega t + K_2 A \sin(\omega t + \pi/2) = A\sqrt{K_1^2 + K_2^2}\sin(\omega t + \tan^{-1}(K_2/K_1)) \qquad (8.48)$$

where

K_1 and K_2 = signal attenuations introduced by the signal controllers S_1 and S_2.

When the attenuations K_1 and K_2 are adjusted so that

$$\sqrt{K_1^2 + K_2^2} = A'/A \quad \text{and} \quad \tan^{-1}(K_2/K_1) = \phi, \tag{8.49}$$

the sum of the ouptut of the signal controllers creates a signal that is equal to that received by Sensor 2.

The difference in phase ϕ between the signals e_1 and e_2 as received by the Sensors S_1 and S_2, respectively, then can be determined from Eq. (8.49) when the values of K_1 and K_2 are known. This phase is also related to the angle of arrival θ since

$$\phi = \frac{\omega L}{C} = \frac{\omega}{C} D \sin\theta = \tan^{-1}\left(\frac{K_2}{K_1}\right) \tag{8.50}$$

where
D = the baseline or separation distance between the sensors.

Thus

$$\theta = \sin^{-1}\left[\frac{C}{\omega D}\tan^{-1}\left(\frac{K_2}{K_1}\right)\right]. \tag{8.51}$$

The arrangement shown in Figure 8-13 from which the required values of K_1 and K_2 may be obtained is nothing different from an adaptive canceller which, unlike cancelling an interference, cancels the desired signal reccived by Sensor 2 with a corresponding reference obtained from Sensor 1. As in the case of an adaptive canceller, when K_1 and K_2 are different from what is needed for the cancellation, there will be a nonzero residual signal following the subtraction of the cancelling signal. This nonzero difference signal is used as an error signal in a closed-loop control to make further adjustments of K_1 and K_2 until the error signal becomes zero.

When the error signal is zero, the phase difference between the received signals e_1 and e_2, that is, ϕ, becomes equal to $\tan^{-1}(K_2/K_1)$, as shown in Eq. (8.50). But K_1 and K_2 can be known from the dc control signals controlling the values of the attenuations since each control voltage bears a definite, linear relationship with respect to the attenuation introduced by the signal controller as discussed in Chapter 3. Thus, from the knowledge of the two dc control signals for the two signal controllers, one can compute the angle of arrival θ from Eq. (8.51), particularly when the baseline separation distance D and the signal angular frequency ω are known. Since the accuracy potential for the determination of θ depends on how accurately one can determine K_1 and K_2, and since an adaptive canceller with an ability to suppress a signal to 60 dB is not uncommon, the effective error in $\delta K_1/K_1$ or $\delta K_2/K_2$ will be on the order of 10^{-3}, δK_1 and δK_2 being the errors in determining K_1 and K_2, respectively.

From Eq. (8.50)

$$\tan\phi = K_2/K_1 \quad \text{and}$$

$$\frac{\partial}{\partial\phi}(\tan\phi)\cdot\Delta\phi = \frac{\partial}{\partial K_1}(K_2/K_1)\delta K_1 + \frac{\partial}{\partial K_2}(K_2/K_1)\delta K_2 \quad \text{or}$$

$$\Delta\phi = \frac{\sin(2\phi)}{2}\left[\frac{\delta K_2}{K_2} - \frac{\delta K_1}{K_1}\right] \tag{8.52}$$

where

$\Delta\phi$ = the error in determining ϕ due to errors δK_1 and δK_2 in determining K_1 and K_2, respectively.

For small values of ϕ for which $\sin(2\phi) \simeq 2\phi$

$$\frac{\Delta\phi}{\phi} = \left[\frac{\delta K_2}{K_2} - \frac{\delta K_1}{K_1}\right]. \tag{8.53}$$

Thus, when $(\delta K_1)/K_1$ and $(\delta K_2)/K_2$ are on the order of 10^{-3}, $(\Delta\phi)/\phi$ also becomes on the order of 10^{-3}. The closed-loop control for signal nulling as obtainable by an adaptive canceller and deriving the phase difference information therefrom not only eliminate the bias errors associated with the phase detectors, as referred to earlier, but also increase the accuracy potential by orders of magnitude.

In Eq. (8.51), it is seen that for known values of ω and D, θ can be determined when K_1 and K_2 are known or measured. The values of K_1 and K_2, however, cannot be measured easily. If the characteristics of K_1 and K_2 as a function of their respective dc command signals are known, then from the dc command signals following the cancellation, one may compute θ. Figure 8-14 shows the computation process from the command signals for the signal controllers. More specifically, the dc command signals are converted from the analog to corresponding digital signals by an A-to-D converter as shown in the figure. Next, the digital signals are used to compute θ in accordance with Eq. (8.51). The computer in this case also provides input data for the display of the angle θ. It should be noted here that because of the continuous tracking capability of the closed-loop nulling system, the approach shown in Figures 8-13 or 8-14 can measure the angle of a moving source.

In some situations, the angular frequency ω of the signal of interest is not known. Since θ cannot be measured without the knowledge of ω, one needs to measure both ω and θ to determine the angle of arrival of the signal of interest. A second adaptive canceller can be used to achieve this objective. Figure 8-15 shows the schematic arrangement for measuring both ω and ϕ for a stationary or a moving signal source. As in Figures 8-13 or 8-14, two sensors are used. The signal received by Sensor 1, as denoted by $A\sin\omega t$, is split into two parts by power divider 1. One part is used as a reference to create a cancelling signal for the signal received by Sensor 2. If the power dividers 1 and 2 are identical, then the signal

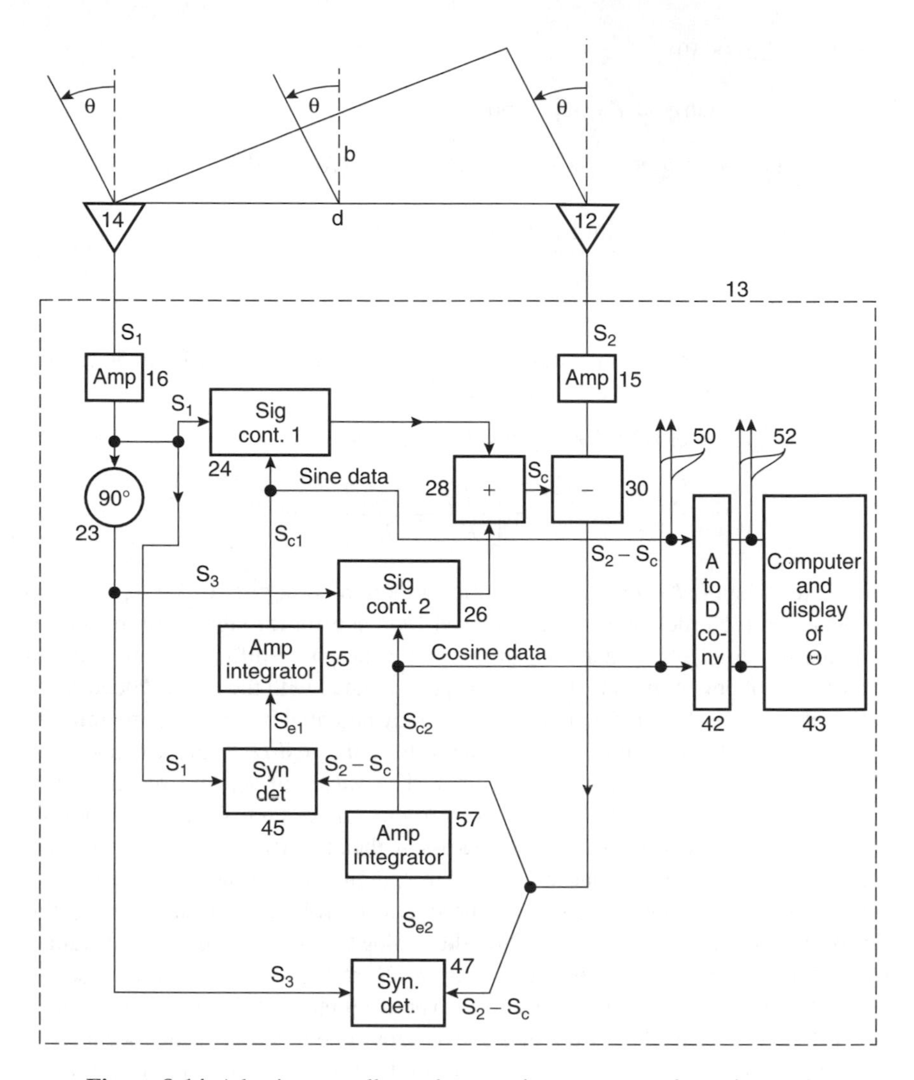

Figure 8-14 Adaptive canceller and computing process to determine angle.

controllers of Adaptive Canceller 1 must introduce the amplitude factor (A'/A) and a phase delay ϕ for the purpose of cancellation. The second part of the Sensor 1 signal is delayed by a known time delay T, and is then fed to Adaptive Canceller 2, which also tends to cancel the signal at the second branch of the Sensor 2 line. Because the input to Adaptive Canceller 2 is delayed by ωT, the signal controllers of this canceller must introduce an amplitude factor (A'/A) and a phase delay $(\phi - \omega T)$. If K_1 and K_2 are the attenuations introduced by the 0° and 90° signal controllers, respectively, of Adaptive Canceller 1, and if K_3 and

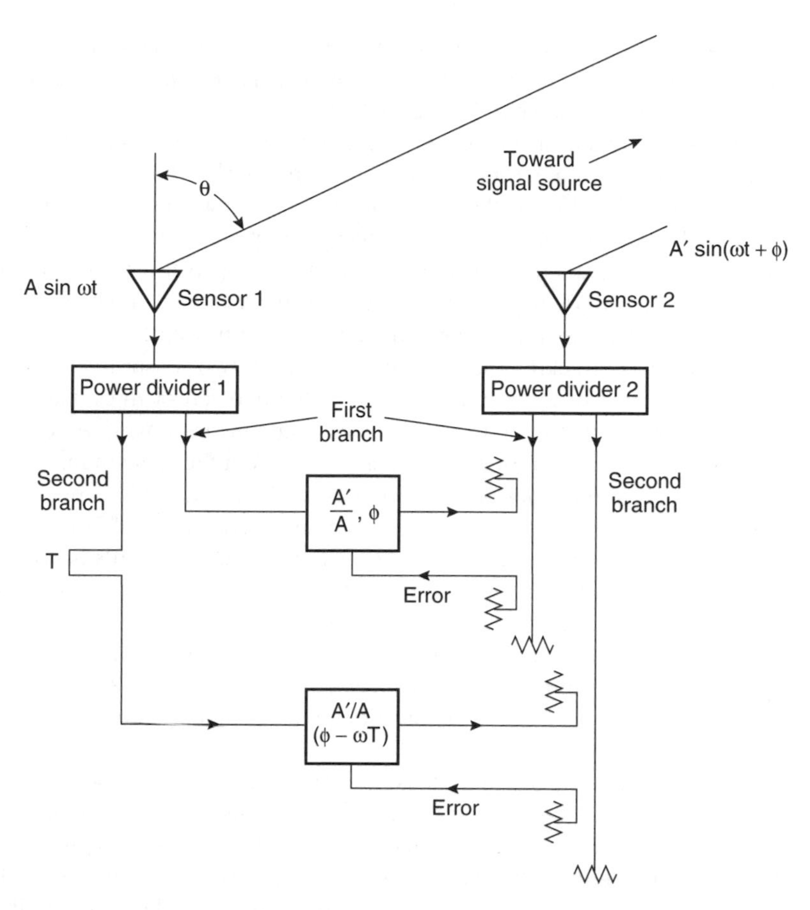

Figure 8-15 Determination of frequency and angle of arrival of a signal of interest.

K_4 are corresponding attenuations introduced by Adaptive Canceller 2, then from Eq. (8.50)

$$\phi = \tan^{-1}\left(\frac{K_2}{K_1}\right)$$

$$\omega T + \phi = \tan^{-1}\left(\frac{K_4}{K_3}\right). \tag{8.54}$$

Thus, the unknown angular frequency ω of the signal of interest becomes

$$\omega = \frac{1}{T}[\tan^{-1}(K_4/K_3) - \tan^{-1}(K_2/K_1)]. \tag{8.55}$$

From Eqs. (8.51), (8.54), and (8.55), then, one can determine ω and θ simultaneously for the signal of interest when T is known, and K_1, K_2, K_3, and K_4 are measured by the means shown in Figures 8-13 and 8-14.

Unfortunately, much of the accuracy potential for the above-described approach may not be realizable in some situations, particularly when the signal level or the signal-to-noise ratio for the signal of interest is low, and when there are multiple signals present at the same or at adjacent frequency bands. Any attempt to avoid unwanted signals interfering with the angle-of-arrival measurement for the signal of interest by frequency-domain filters usually fail, even where there is adequate frequency separation between the desired and unwanted signals, since highly accurate phase-tracking filters are needed for the two sensor lines. Almost any nontracking phase difference in the filters at the two sensor lines may result in an unacceptable reduction of accuracy for the angle measurement. Also, since the levels of signals received by the two sensors are about the same in most cases of interest, and the signal controller assemblies, including the 90° hybrid, summing coupler, and the attenuators, often introduce an insertion loss of 7–10 dB, the signal received by Sensor B has to be attenuated by at least 7–10 dB before one may expect a signal null. This requirement further deteriorates the performance capability of the angle measurement system, particularly when the input signal level or the input signal-to-noise ratio is small. Any attempt to reduce or minimize the problem by amplification of the received signal at Sensor B also reduces the accuracy potential because of the uncertainty and fluctuation of phase delay introduced by the amplifier.

An approach [6] that avoids the undesirable signal attenuation in Sensor B and aids in selecting the signal of interest from a multiple-signal environment involves creating a common reference signal from Figure 8-16. This reference signal is then used to cause signal nulls at the two sensors by two adaptive cancellers instead of one, as shown in Figure 8-13. The reference signal thus created can be amplified and filtered or tuned to selectively address the signal of interest in a multiple-signal environment without degrading the accuracy of direction finding.

As shown in Figure 8-16, the output of each Sensor A or B is divided into two branches by a balanced power divider. One branch from each sensor line leads to a summer, where the signals received by the sensors are summed and amplified as shown in the figure. The amplified signal is then filtered or selectively tuned for the frequency of the signal of interest. This tuning enables the direction-finding system to selectively address the signal of interest, particularly in a multiple-signal environment. The tuned signal of interest is then used as a reference for the two adaptive cancellers, each of which adaptively cancels the signal received by each sensor.

To illustrate how the arrangement shown in Figure 8-16 avoids the undesirable attenuation of the desired signal received by Sensor B and provides some selectivity in a multiple-signal environment, let us assume as before that the signals received by Sensors A and B are, respectively, $A\sin(\omega t)$ and $A'\sin(\omega t + \phi)$, where A and

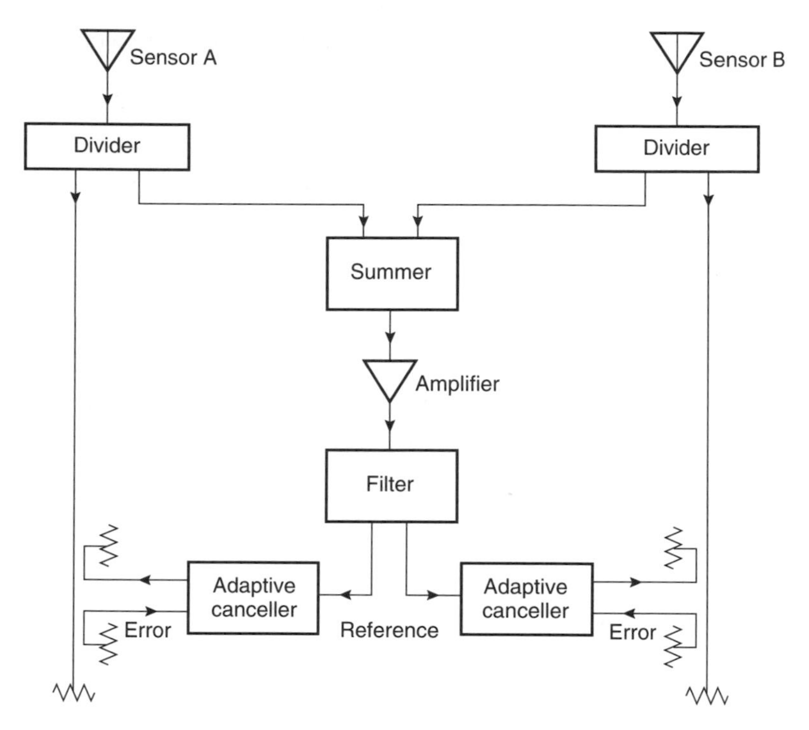

Figure 8-16 Adaptive direction-finding system employing two adaptive cancellers.

A' are the amplitudes of the received signals at the two sensors, ω is the angular frequency of the desired signal, and ϕ is the electrical phase difference between the signals received by the two sensors. Further assuming that each divider shown in Figure 8-16 causes a signal amplitude reduction by a factor $(1/\sqrt{2})$, the summed signal may be written as

$$\begin{aligned} S_s &= \frac{A}{\sqrt{2}} \sin(\omega t) + \frac{A'}{\sqrt{2}} \sin(\omega t + \phi) \\ &= \bar{A} \sin(\omega t + \theta) \end{aligned} \tag{8.56}$$

where

$$\begin{aligned} \bar{A}^2 &= (A^2 + A'^2 + 2AA' \cos\phi)/2 \\ \tan\bar{\theta} &= \frac{A' \sin\phi}{A + A' \cos\phi}. \end{aligned}$$

Now, let the gain and phase introduced by the amplifier be G and $-\gamma_1$, respectively. Similarly, let the amplitude factor and phase introduced by the filter be F and $-\psi_f$, respectively. Then the output of each branch following the filter, again assuming an amplitude reduction by a factor $(1/\sqrt{2})$ for each branch, can be written as

$$S_0 = \frac{GF}{\sqrt{2}}\bar{A}\sin(\omega t + \bar{\theta} - \gamma_1 - \psi). \tag{8.57}$$

When S_0 is used as a reference for Adaptive Canceller 1 to cancel the signal in the Sensor A line, the canceller must introduce an amplitude factor W_1 and phase β_1 such that

$$W_1\frac{GF}{\sqrt{2}}\bar{A} = A \quad \text{and} \quad \bar{\theta} - \gamma_1 - \psi + \beta_1 = 0. \tag{8.58}$$

Similarly, when S_0 is used as a reference for Adaptive Canceller 2 to cancel the signal in the Sensor B line, this canceller must introduce an amplitude factor W_2 and phase β_2 such that

$$W_2\frac{GF\bar{A}}{\sqrt{2}} = A' \quad \text{and} \quad \bar{\theta} - \gamma_1 - \psi + \beta_2 = \phi. \tag{8.59}$$

From Eqs. (8.58) and (8.59)

$$\beta_2 - \beta_1 = \phi. \tag{8.60}$$

But

$$\beta_1 = \tan^{-1}\frac{K_2}{K_1} \quad \text{and} \quad \beta_2 = \tan^{-1}\frac{K_4}{K_3} \tag{8.61}$$

where

K_1 and K_2 = the attenuations at the 0° and 90° signal controllers, respectively, for Adaptive Canceller 1

K_3 and K_4 = the attenuations at the 0° and 90° signal controllers, respectively, for Adaptive Canceller 2.

Thus, the true electrical phase difference, ϕ, between the received signal of interest by the two sensors can be obtained from the signal-controller settings of the two cancellers, and any gain and phase variations introduced by the amplifier and filter cannot affect the measured phase ϕ. Now, since the gain provided by the amplifier would be much larger than the insertion loss at the signal-controller path, there will not be any need to attenuate the desired signal received by any sensor. Additionally, the filter or tuned circuit will provide the selectivity of the signal of interest in a multiple-signal environment without affecting the measurement of ϕ, and hence without affecting the determination of the angle of arrival θ from measured ϕ in accordance with Eq. (8.51).

So far, we have considered direction finding for a source of signal that has a single or predominant angular frequency ω. If ω is known, the direction of arrival

θ can be determined from the phase angle ϕ measurable by an adaptive canceller. When ω is not known, one may determine θ and ω simultaneously by two adaptive cancellers, as illustrated in Figure 8-15. In some situations, however, there is no single or predominant ω, and the signal spectrum is broad, as it is in the case of a spread-spectrum signal, for example. The determination of the angle of arrival of such a signal by an adaptive canceller is still possible if one employs a signal controller with a variable time-delay control as discussed in Section 3.3 of Chapter 3. The concept of adaptive direction finding for a broadband signal is shown in Figure 8-17. Here, a broadband signal is shown to arrive at both Sensors A and B arriving at A, T period earlier than at B. The adaptive canceller cancels the signal received at Sensor B by controlling the amplitude K and the time delay T of the signal received at Sensor A. For the cancellation of the signal received at Sensor B, the time delay T introduced by the signal controller must equal the time interval between the arrivals of signals at the two sensors. Once this time delay is known, one can determine the angle of arrival from the relation

$$\theta = \sin^{-1}[CT/D] \tag{8.62}$$

where
D = the baseline separation distance
C = the velocity of propagation of electromagnetic waves in the medium of interest.

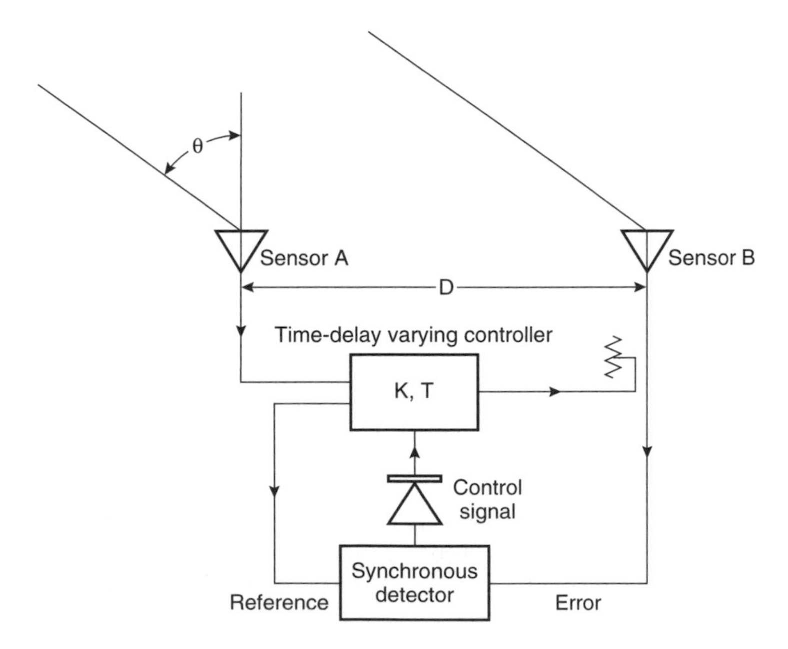

Figure 8-17 Adaptive direction-finding system with a time-delay varying signal controller.

The approach shown in Figure 8-17 then can be used to determine the angle of arrival of a broadband signal, even when the spectral characteristics of the signal are not known. It should be noted, however, that sometimes it is difficult to implement a signal controller with a variable time-delay control, particularly when the baseline is large. In the example of the time-delay-varying signal controller as discussed in Section 3.3 of Chapter 3, for example, the range of time-delay control[1] is $\pm T/2$ where

$$\Delta\omega T \simeq \pi/8 \tag{8.63}$$

and $\Delta\omega = 2\pi$ times the bandwidth of the signal of interest, the direction of which needs to be determined.

Very often, the direction finding for an unknown source of signal is a three-dimensional problem requiring angular determinations in two orthogonal planes. To examine whether adaptive cancellers may be useful for such cases, let us consider an angle-measuring system at the center of a rectangular coordinate system, having three sensors, 1, 2, and 3 located at a fixed distance L from the center, along the x, y, and z coordinates, as shown in Figure 8-18.

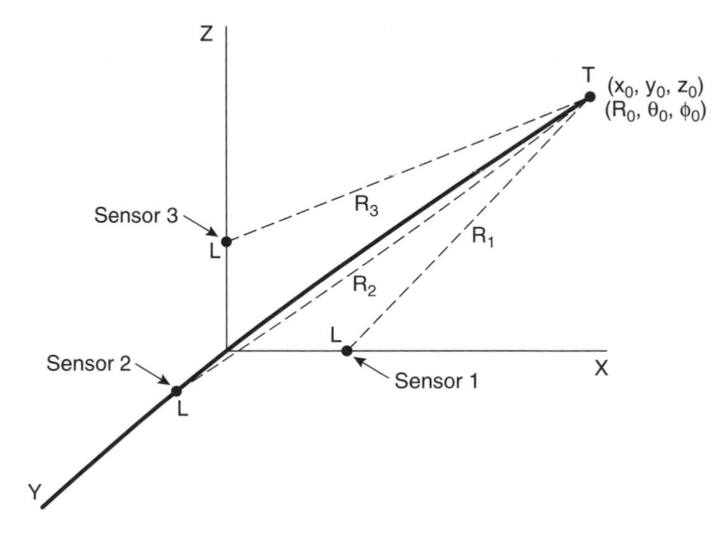

Figure 8-18 Rectangular coordinate system showing locations of three orthogonal sensors.

Let the coordinates of a source of signal T be (x_0, y_0, z_0) in rectangular coordinates and (R_0, θ_0, ϕ_0) in a spherical coordinate system, the center of the coordinate systems being the same. If $R_i, i = 1, 2, 3$ is the distance between the ith sensor

[1]This relation is given in Eq. (3.34).

and the source T, then

$$R_1^2 = (x_0 - L)^2 + y_0^2 + Z_0$$
$$R_1 \simeq R_0 - (x_0 L/R_0) \simeq R_0 - L \sin\theta_0 \cos\phi_0. \quad (8.64)$$

Similarly

$$R_2 \simeq R_0 - L \sin\theta_0 \sin\phi_0 \quad (8.65)$$

and

$$R_3 \simeq R_0 - L \cos\theta_0. \quad (8.66)$$

The electrical phases of the signal emitted from T at Sensors 1, 2, and 3 can be written, respectively, as

$$-\phi_1 = -[\beta R_0 - \beta L \sin\theta_0 \cos\phi_0]$$
$$-\phi_2 = -[\beta R_0 - \beta L \sin\theta_0 \sin\phi_0]$$
$$-\phi_3 = -[\beta R_0 - \beta L \cos\theta_0] \quad (8.67)$$

where
$\beta = 2\pi/\lambda$, λ being the wavelength of the signal emitted from the source T.

Although the phases ϕ_1, ϕ_2, and ϕ_3 cannot be measured without defining a reference phase, the phase differences $(\phi_1 - \phi_2)$ or $(\phi_2 - \phi_3)$ or $(\phi_1 - \phi_3)$ can be measured by the interferometric technique, and hence by adaptive cancellers as discussed above. Thus, one may obtain from Eq. (8.67)

$$A = \phi_1 - \phi_2 = -\beta L[\sin\theta_0 \cos\phi_0 - \sin\theta_0 \sin\phi_0]$$
$$B = \phi_2 - \phi_3 = -\beta L[\sin\theta_0 \sin\phi_0 - \cos\theta_0]$$
$$C = \phi_1 - \phi_3 = -\beta L[\cos\theta_0 \sin\phi_0 - \cos\theta_0]. \quad (8.68)$$

Since L is fixed and known, one can determine θ_0 and ϕ_0 that define the angle of arrival from the phase differences A, B, and C.

A simplified block diagram employing multiple adaptive cancellers to determine the direction of arrival of a signal in a three-dimensional geometry is shown in Figure 8-19. As shown in the figure, the signals received by three orthogonal sensors are summed to form a composite reference signal. The composite signal thus created may be written as

$$E_r = A_1 e^{-j\phi_1'} + A_2 e^{-j\phi_2'} + A_3 e^{-j\phi_3'} \quad (8.69)$$

where
A_1, A_2, A_3 = the amplitudes of the signals received by the sensors
$\phi_1', \phi_2', \phi_3'$ = the corresponding phases with respect to a common reference.

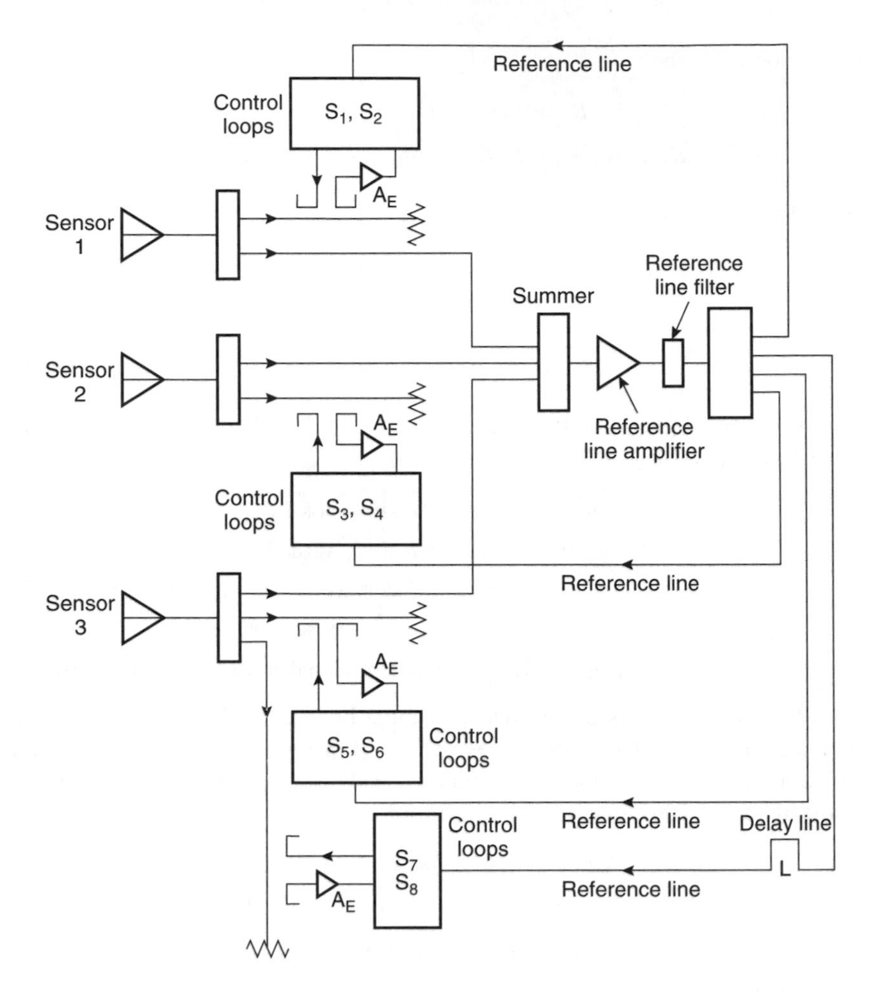

Figure 8-19 Simplified block diagram of a three-sensor direction-finding system.

One may then write

$$E_r = e^{-j\beta R_0} \bar{E} e^{-j\psi'}$$

in which

$$\bar{E}^2 = [A_1^2 + A_2^2 + A_3^3 + 2A_1 A_2 \cos(\phi_1' - \phi_2') \\ + 2A_1 A_3 \cos(\phi_1' - \phi_3') + 2A_2 A_3 \cos(\phi_2' - \phi_3')] \tag{8.70}$$

and

$$\psi' = \tan^{-1}\left[\frac{A_1 \sin\bar{\phi}_1 + A_2 \sin\bar{\phi}_2 + A_3 \sin\bar{\phi}_3}{A_1 \cos\bar{\phi}_1 + A_2 \cos\bar{\phi}_2 + A_3 \cos\bar{\phi}_3}\right] \tag{8.71}$$

where
$\bar{\phi}_i = \phi'_i - \beta R_0, i = 1, 2, 3.$

The composite reference signal thus created is amplified before feeding it to three sets of signal controller pairs. Following amplification and filtering, the reference signal becomes

$$\bar{E}_R = G\bar{E}e^{-j\beta R_0}e^{-j\psi'}e^{-i\Delta\psi} \tag{8.72}$$

where
G and $\Delta\psi$ = the gain and phase, respectively, introduced by the combined amplifier and filter, assuming that both G and $\Delta\psi$ may be functions of frequency.

The composite reference signal $\bar{E}_R$ is first split into a number of parts. Each part is adjusted in amplitude and phase to create a signal equal in amplitude and phase of the signal received by a sensor. Thus, as shown in Figure 8-19, signal controllers S_1 and S_2, yielding an in-phase attenuation K_1 and a quadrature-phase attenuation K_2, respectively, are used to create a signal equal in amplitude and phase to the signal received by Sensor 1. Similarly, the signal controllers S_3 and S_4, providing an in-phase attenuation K_3 and quadrature-phase attenuation K_4, are used to create a signal equal to that received by Sensor 2. The signal controllers S_5 and S_6, providing an in-phase attenuation K_5 and quadrature-phase attenuation K_6, are used to create a signal equal to that received by Sensor 3. The phase equations that result from the controls of the attenuations K_1–K_6, while creating a null at each sensor line, may now be written as

$$\begin{aligned}
-\beta R_0 - \psi' - \Delta\psi + \tan^{-1}(K_2/K_1) &= -\beta R_0 + \beta L \sin\theta_0 \cos\phi_0 \\
-\beta R_0 - \psi' - \Delta\psi + \tan^{-1}(K_4/K_3) &= -\beta R_0 + \beta L \sin\theta_0 \sin\phi_0 \\
-\beta R_0 - \psi' - \Delta\psi + \tan^{-1}(K_6/K_5) &= -\beta R_0 + \beta L \cos\theta_0.
\end{aligned} \tag{8.73}$$

Thus

$$\begin{aligned}
\tan^{-1}(K_2/K_1) - \tan^{-1}(K_4/K_3) &= -A \\
\tan^{-1}(K_4/K_3) - \tan^{-1}(K_6/K_5) &= -B \\
-\tan^{-1}(K_6/K_5) + \tan^{-1}(K_2/K_1) &= -C
\end{aligned} \tag{8.74}$$

where A, B, and C are the same as defined in Eq. (8.68).

A fourth adaptive canceller with signal controllers S_7 and S_8, providing an in-phase attenuation K_7 and a quadrature-phase attenuation K_8, are used to determine λ or the wavelength of the signal by utilizing the method shown in Figure 8-15. The phase equation for this adaptive canceller may be written as

$$-\beta R_0 - \psi' - \Delta\psi - \beta L + \tan^{-1}(K_8/K_7) = -\beta R_0 + \beta L \cos\theta_0. \tag{8.75}$$

Comparing Eq. (8.75) with Eq. (8.73), one obtains

$$\tan^{-1}(K_8/K_7) - \tan^{-1}(K_6/K_5) = \beta L \tag{8.76}$$

From the knowledge of K_3, K_4, K_7, K_8, and L, one determines β, and hence $\lambda = 2\pi/\beta$.

It should be noted that no attenuation of signal received by any sensor is needed with the approach under consideration, notwithstanding the provision of amplification and selective filtering. The amplification and filtering of the reference signal and the associated frequency-dependent phase shift $\Delta\psi$ do not affect the performance of the angle-measuring system. Thus, the arrangement for adaptive direction finding as schematically shown in Figure 8-19 provides a high degree of accuracy and resolution potential, even in situations where the signal level or the signal-to-noise ratio of the signal are low and there are comparable signals at the adjacent frequency bands of the signal of interest.

8.6 SAME-FREQUENCY AMPLIFICATION

Another important class of application or use of the adaptive canceller is found in problems relating to same-frequency amplification [7]. For example, the same-frequency amplification of a weak received signal, prior to its retransmission, is often needed in intracontinental microwave relay links and in airborne or satellite-borne transponders to facilitate communications well beyond the horizon. Often, the degree of such an amplification becomes limited because of the external coupling of the amplifier output to its input. Thus, in a microwave relay line, when a weak signal is received by one antenna at the left, as shown in Figure 8-20, and is retransmitted following a power amplification by another antenna at the right, the amplifier gain has to be restricted to avoid self-oscillation because of the finite decoupling between the right and left antennas. The same problem arises in a microwave transponder when a signal is received by an antenna, is amplified at the "receive" side of a "transmit–receive" hybrid as shown in Figure 8-21, and is then transmitted through the transmit side of the hybrid and the same antenna. Any reflection of the retransmitted signal from the antenna due to a possible mismatch appears at the amplifier input, and depending on the degree of mismatch, there is usually a high probability of self-oscillation, particularly for a high-gain amplifier. For a terrestrial microwave line, a high-gain amplification at each intermediate relay station permits a reduction of the total number of intermediate stations, thereby reducing both the initial investment for the link and the recurring maintenance cost for the intermediate stations saved due to high-gain amplification at the others. For a transponder, a high-gain amplification permits a longer communication range or an increased signal-to-noise ratio or signal-to-interference ratio for a given communication range. Thus, there is a strong economic motivation to design and implement a high-gain microwave amplification system that ensures the avoidance of a self-oscillation risk.

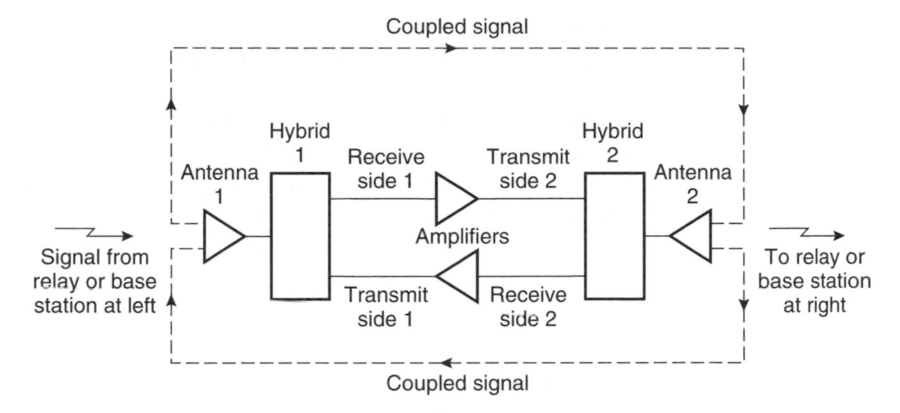

Figure 8-20 A microwave relay link providing amplifications in both directions.

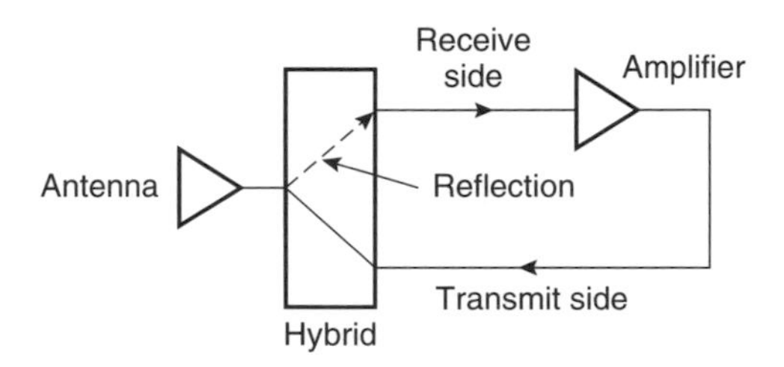

Figure 8-21 Illustration of a problem of high-gain amplification in a transponder.

If the self-oscillation that limits the high-gain amplification is due to the finite decoupling between the amplifier output to its input, as is usually the case, its avoidance may be effected by further reducing the coupling. Since passive means available, if any, in most situations to enhance decoupling are usually restricted, one seeks active approaches by adaptive cancelling the "leakage" signal at the amplifier input, resulting from the finite input–output decoupling. An adaptive canceller appears to be an ideal remedy for this purpose. For a positive feedback situation, similar to the case under consideration, a self-oscillation may be initiated by noise and may occur at any frequency, and since the amplitude and phase of the leakage signal are different for different frequencies, and they are seldom *a priori* known, any cancellation to effect a large output–input decoupling has to be adaptive. Also, to be of value, any cancellation means must be operable over a wide frequency band.

The concept [8] of amplifier gain enhancement without risking a self-oscillation where the adaptive cancellation plays a key role is illustrated in Figure 8-22. Here, the coupling of the amplifier output to its input, shown by the dotted line in the figure, may be regarded as the leakage signal path. As the amplifier gain exceeds the path loss for the leakage path, a self-oscillation occurs, making the amplifier ineffective. If, now, a countersignal which is equal in amplitude and 180° out of phase with respect to the leakage signal is synthesized and is injected at the amplifier input, the net coupled signal from the amplifier output to its input is nulled, thereby eliminating the risk of self-oscillation. Since the coupling is linear in most cases, the leakage signal coupled to the amplifier input differs from the amplifier output signal by an amplitude factor, say K, and a time delay T. The countersignal needed to null the leakage signal at the amplifier input can then be synthesized from a sample of the amplifier output signal by appropriately controlling its amplitude and phase or time delay, until such a null occurs. As noted earlier, since K and T are seldom *a priori* known, the synthesis of the countersignal has to be adaptive. This means that an error signal, defined as the difference between the countersignal synthesized by the signal controller shown in Figure 8-22 and that required for the null of the leakage signal, has to be driven to zero by a closed-loop control. The arrangement shown in Figure 8-22, then, is nothing other than an adaptive canceller.

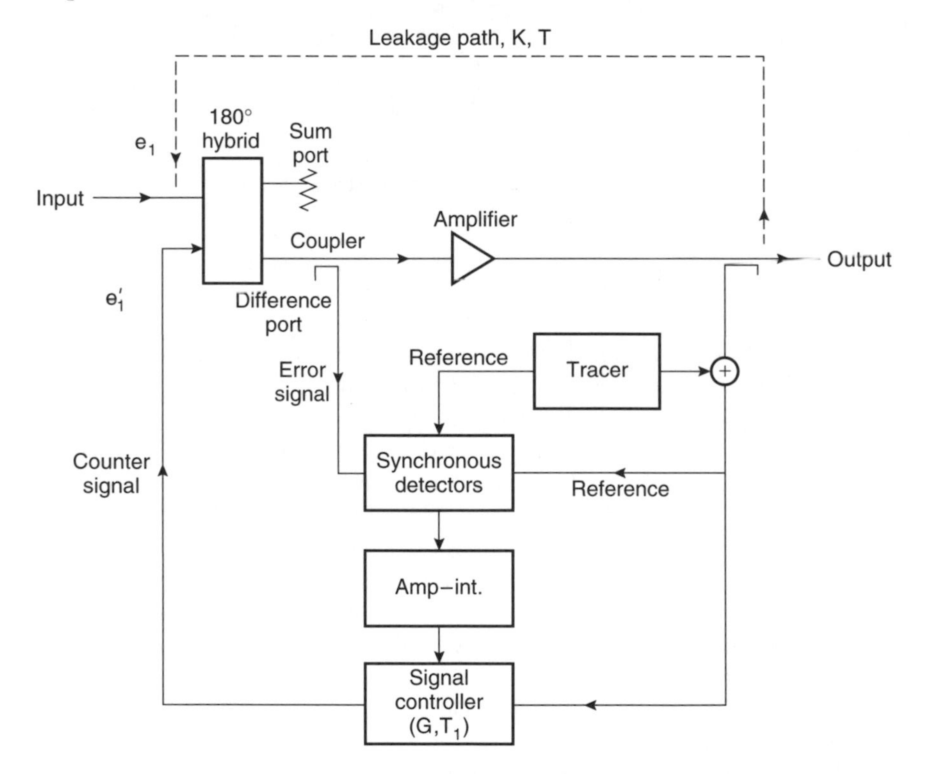

Figure 8-22 Illustration of amplifier gain enhancement by cancelling oscillation-causing feedback.

To further examine the operation of the canceller, let the leakage signal, as it appears at the amplifier input from the amplifier output, be $e_1 = KS(t-T)$ where K and T are, respectively, the amplitude reduction factor and time delay introduced by the leakage path, and $S(t)$ is the amplifier output signal as a function of time. If the signal controller shown in Figure 8-22 is capable of effecting a change in gain G and a time delay T_1 from its input to output, then the output of the signal controller will be

$$e_1' = GS(t - T_1). \tag{8.77}$$

If, now, $K = G$ and $T = T_1$, at the difference port of the 180° hybrid, and hence at the amplifier input, there will be no net leakage signal of the type $S(t)$ since e_1' will cancel e_1 at this port. If, however, the values of G and T_1, as set by the signal controller, are not exact to make $G = K$ and $T_1 = T$, as will be the case at the start, the difference port of the 180° hybrid will contain a net nonzero leakage signal, and hence a nonzero error signal. This nonzero error signal will result in dc signals at the output of the synchronous detector. The dc signals can then be amplified and integrated to create control signals for the signal controller such that G and T_1 will change until such values become what are exactly needed to drive the error signal to zero. This condition, then, constitutes simultaneously the equilibrium condition for the closed-loop control and the suppression of the oscillation causing leakage signal at the difference port output of the 180° hybrid, and hence at the amplifier input. Here, the closed loop comprises the detector, the amplifier–integrator, the signal controller, and the 180° hybrid, as shown in Figure 8-22.

It is interesting to note that the only parameters involved in the signal controller, to synthesize the countersignal for the cancellation of leakage signal, are G and T_1, both of which are independent of $S(t)$. Thus, as in a typical adaptive canceller, the operation of the adaptive control does not depend on the characteristics of $S(t)$, such as its frequency spectrum, amplitude, and modulation. Also, if the signal controller can truly synthesize the required amplitude gain and time delay, the cancellation of a wideband leakage signal is possible, since the gain and time delay required for the cancellation of one frequency component of the leakage signal are, in general, the same as needed for any other frequency components in the leakage signal spectrum. Additionally, if the characteristics of the principal amplifier, primarily its gain, or the characteristics of the leakage path change with time, the closed-loop control automatically accommodates the changes by tracking, and always suppressing the oscillation causing signal feedback.

An element of the positive feedback suppression approach as shown in Figure 8-22, significantly different from what is found in a typical adaptive canceller, is the provision of the tracer signal. Since the leakage signal, the countersignal, and the input of the amplifier have the same spectrum, it is possible, and in fact it is very likely, that the closed-loop control will make the amplifier input also zero, while eliminating the error signal. The purpose of the tracer signal is to avoid such a possibility. By design, the tracer signal is at a frequency close to that of the intended input signal for the amplifier. In this case, the closed-loop control actually operates on the tracer signal, the control needed to eliminate the tracer signal at the amplifier input being almost the same as that needed to eliminate the leakage signal, but not the amplifier input.

The approach shown in Figure 8-22 can be used to increase the amplifier gain for the problems posed in Figures 8-20 and 8-21, the increase in gain being approximately equal to the cancellation capability of the adaptive canceller, which could be 60 dB or more. Another application [8] of the same-frequency amplification is to facilitate communications among widely separated stations through a common repeater, as shown in Figure 8-23. A wideband amplifier is used at the "repeater," and each communication station has a "random" access to the repeater. Communications well beyond the horizon or line of sight are feasible in this case, particularly when the repeater is at an airborne platform. Additionally, the repeater station shown in Figure 8-23 can accommodate a large and varying number of communication channels at any time. No logic circuitry is needed to hold the repeater signal within its specified channel, and there is no restriction on the type of modulations of the signals utilizing the repeater. Yet another application of the same frequency amplification is found in the design of a "decoy," where one can increase the radar cross section (RCS) of an object or target by receiving the signal from a distant radar at the target, amplifying the signal with a high gain, and then retransmitting the signal toward the radar. The arrangement in this case is not different from the transponder shown in Figure 8-21. The high gain of the amplifier aided by an adaptive canceller permits the radar return signal to be very large, as if it is a scattering from a very large target. A retrodirective antenna can further increase the effectiveness of the RCS enhancement approach.

As seen from the above discussions, the closed-loop operation of the adaptive canceller is a very effective tool, and it is not, therefore, surprising that the adaptive canceller has been or can be used for a variety of diverse purposes. It can be used, for example, to cancel power-line interferences (50–60 Hz) from the output or recording of electrocardiograms [1]. It can also be used to provide an enhanced "look-through" [11] capability for an electronic countermeasure system, where an uninterrupted countermeasure can be maintained while continuously monitoring the effectiveness of such a countermeasure. The use of an adaptive canceller has also been suggested for adaptive network modeling [9], where a physical system with an unknown impulse response can be modeled such that the error signal, which is the difference between the unknown network response and the adaptive model response, is driven to zero by a closed-loop control that adjusts the adaptive model parameters. Other suggested uses of the adaptive canceller include an adaptive equalizer that remedies intersymbol interference in communications due to a channel having multipath or dispersion, an adaptive statistical predictor, comprising a filter that can estimate a signal at a time Δt in the future for a stationary signal, and many others.

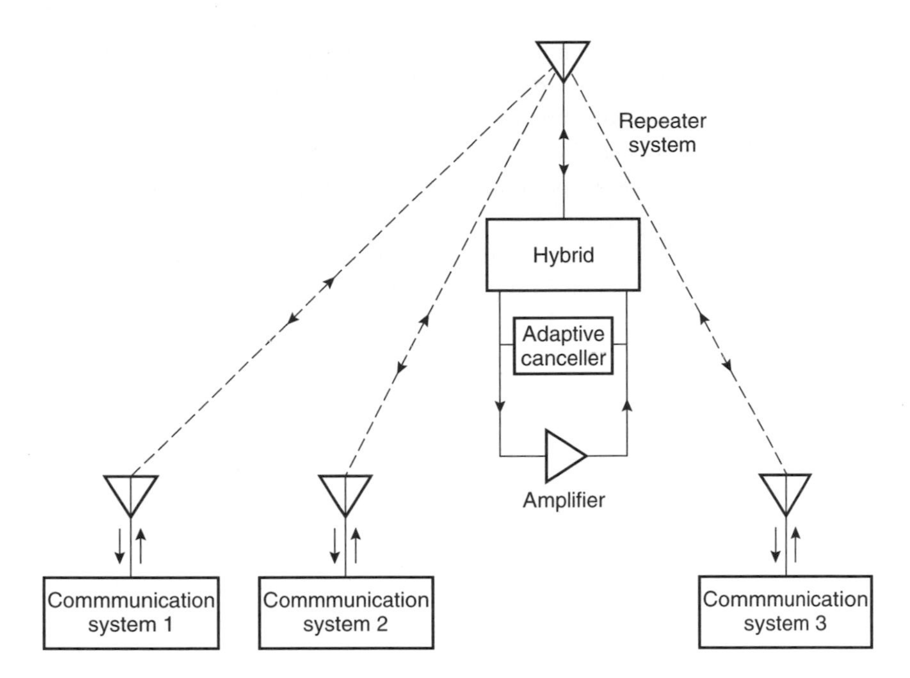

Figure 8-23 A common repeater facilitating communications among widely separated stations.

REFERENCES

[1] B. Widrow, J. Glover, Jr., J. McCool, J. Kaunitz, C. Williams, R. Hearn, J. Zeidler, E. Dong, Jr., and R. Goodlin, "Adaptive noise cancelling: Principles and applications," *Proc. IEEE*, vol. 63, pp. 1692–1716, Dec. 1975.

[2] W. Sauter and D. Martin, "HF/VHF/UHF adjacent channel measurements," Final Report under Contract F30602-69-C-0137, Rome Air Development Center, NY, Dec. 1969.

[3] R. Ghose, "Selective interference reduction in transmission lines," U.S. Patent 5,047,736, Sept. 10, 1991.

[4] R. Ghose, "Adaptive noise sbatement system," U.S. Patent 4,829,590, May 9, 1989.

[5] R. Ghose *et al.*, "Automatic direction finder," U.S. Patent 4,486,757, Dec. 4, 1984.

[6] R. Ghose, "Adaptive direction finding antenna systems," in *Proc. 1994 Journées Internationales de Nice Sur les Antennas*, Centre B/LA Turbie, Nov. 1994.

[7] R. Ghose, "Same frequency repeater amplification," *J. Inst. Electron. Telecommun. Eng.*, vol. 37, no. 4, pp. 357–362, 1991.

[8] R. Ghose, "Microwave amplification with adaptive input-output decoupling," in *Conf. Proc., Int. Microwave Conf.*, Brazil, 1991, pp. 226–231.

[9] J. McCool, "The basic principles of adaptive systems with various applications," Naval Undersea Center, San Diego, CA, Sept. 1972.

[10] B. Widrow and S. Stearns, *Adaptive Signal Processing*, Englewood Cliffs, NJ:Prentice-Hall, 1985.

[11] "Adaptive interference cancellation for enhanced ECM look through," WP-223, American Nucleonics Corp., May 1981.

ABOUT THE AUTHOR

Rabindra Ghose earned his BEE degree from Jadavpur University. He also received a DIIS degree from the Indian Institute of Science, an MSEE from the University of Washington, an MA (Math) Ph.D., and EE degrees from the University of Illinois, and an MBA degree from the Golden Gate University.

Dr. Ghose has been elected a Fellow of the Institute of Electrical and Electronics Engineers; the Institute of Electrical Engineers (London); the Institute of Engineers (India); Institute of Physics (London); American Physical Society: American Association for the Advancement of Science, and the Institute of Electronics and Telecommunication Engineers. He is a registered Professional Engineer in Electrical and Nuclear Engineering.

He served the Radio Corporation of America and Ramo Wooldridge Corporation as a member of the technical staff, graduate faculty of the University of Southern California, Space General Corporation as Director of Research and Advanced Development, and American Nucleonics Corporation as its President and Chairman of the Board. He also served as a member of the Advisory Board of the U.S. Defense Intelligence Agency; Scientific Advisory Board of the U.S. Air Force, (Divisional Advisory Group); and Research and Technology Advisory Council of the NASA, (Guidance, Control and Information). Currently, he is the President of Technology Research International, Inc.

Dr. Ghose holds more than a score of U.S. patents as an inventor and is author of scores of technical and scientific papers, book reviews and text books, including Microwave Circuit Theory and Analysis and EMP Environment and System Hardness Design. He is a member of the Bar of the California and U.S. Supreme Courts and is a registered patent attorney.

INDEX

A

D

G

H

I

J

K

L

M

N

O

P

Q

R

S